T0313976

Microwave and Millimeter-Wave Antenna Design for 5G Smartphone Applications

IEEE Press
445 Hoes Lane
Piscataway, NJ 08854

Microwave and Millimeter-Wave Antenna Design for 5G Smartphone Applications

Wonbin Hong
Pohang University of Science and Technology
Pohang, South Korea

Chow-Yen-Desmond Sim
Feng Chia University
Taichung, Taiwan

Published by John Wiley & Sons, Inc., Hoboken, New Jersey.
Published simultaneously in Canada.

For general information on our other products and services or for technical support, please contact our Customer Care Department within the United States at (800) 762-2974, outside the United States at (317) 572-3993 or fax (317) 572-4002.

Wiley also publishes its books in a variety of electronic formats. Some content that appears in print may not be available in electronic formats. For more information about Wiley products, visit our web site at www.wiley.com.

Library of Congress Cataloging-in-Publication Data Applied for
Hardback ISBN: 9781394182428

Cover Design: Wiley
Cover Image: © Yuichiro Chino/Getty Images

Set in 9.5/12.5pt STIXTwoText by Straive, Pondicherry, India

Contents

About the Authors

Wonbin Hong received his B.S. in electrical engineering from Purdue University, West Lafayette in 2004 and Masters and Ph.D. in electrical engineering from the University of Michigan, Ann Arbor in 2005 and 2009, respectively. As of 2016 February, Dr. Hong is with the Department of Electrical Engineering at Pohang University of Science and Technology (POSTECH) as an associate professor. He currently holds the Mueunjae Chaired Professorship. From 2009 to 2016, he was with Samsung Electronics as a principal and senior engineer. Since 2021, he is also the CEO of Kreemo Inc., a University startup specializing in 360° coverage mm-wave antennas and measurement solutions.

He has authored and co-authored more than 150 peer-reviewed journals, conference papers, and multiple book chapters and is the inventor of more than 50 granted and 180 pending patent inventions.

He has received numerous awards during his tenure at Samsung, and as a faculty and CEO including multiple recognitions for his contributions in the field of 5G antennas by the Government of Republic of Korea. His students were recipients of multiple paper awards including 1st Best Paper Award from numerous conferences including the 2020 IEEE AP-S/URSI, 2020 IEEE EuCAP, 2018 IEEE ISAP, and 2nd Best Paper Award in 2021 IEEE AP-S/URSI.

Email: whong@postech.ac.kr

Chow-Yen-Desmond Sim was born in Singapore in 1971. He received the B.Sc. degree from the Engineering Department, University of Leicester, the United Kingdom, in 1998, and the Ph.D. degree from the Radio System Group, Engineering Department, University of Leicester, in 2003. In 2007, he joined the Department of Electrical Engineering, Feng Chia University (FCU), Taichung, Taiwan, and he was promoted to Distinguished Professor in 2017. He co-founded the Antennas and Microwave Circuits Innovation Research Center in FCU and served as the Director between 2016 and 2019. He has served as the Head of Department of Electrical Engineering in FCU between 08/2018 and 07/2021. He has authored or coauthored over 190 SCI papers. His current research interests include 5G (sub-6 and mmWave) antenna for smartphone/base-station/AiP, WiFi-6E laptop antenna, and RFID applications. He is a Fellow of the Institute of Engineering and Technology (FIET), a Senior Member of the IEEE Antennas and Propagation Society, and a Life Member of the IAET (Taiwan). He has served as the Associate Editor of IEEE Access between 08/2016 and 01/2021. He is now serving as the Associate Editor of IEEE AWPL, IEEE Journal of RFID, and (Wiley) International Journal of RF and Microwave Computer-Aided Engineering. Since October 2016, he has been serving as the technical consultant of SAG (Securitag Assembly Group), which is one of the largest RFID tag manufacturers in Taiwan. He is also serving as the consultant of ZDT (Zhen Ding Technology, Taiwan) since August 2018, which is ranked no. 1 global PCB enterprise in 2021. He was the recipient of the IEEE Antennas and Propagation Society Outstanding Reviewer Award (IEEE TAP) for eight consecutive years between 2014 and 2021. He also received the Outstanding Associate Editor Award from the IEEE AWPL in July 2018.

Email: cysim@fcu.edu.tw

Preface

With the development of the fifth generation (5G) of mobile communications and artificial intelligence of things (AIoT), it has driven the industry to confront and develop new diversified innovative application services, which will allow the industry to move from the automation era of the third industrial revolution into a new era of digital and intelligent Industry 4.0. According to the Global mobile Suppliers Association (GSA) report, as of January 2021, 144 operators worldwide have provided 5G services in 61 countries and regions, ringing the bell of the arriving 5G commercial era. Although killer applications for 5G mobile services are still in the exploratory stage, the release of the 3GPP standard Rel-16/17 will more fully support industry applications such as autonomous driving, industrial networks, smart logistics and telemedicine, and vertical markets. Therefore, new business opportunities based on the new 5G mobile services and applications are expected to bring about a whole new situation.

Observing the world's leading 5G development countries such as South Korea, the United States, China, Japan, and Germany, global governments and operators regard the expansion of 5G applications as an important development goal and actively plan relevant regulatory measures and development strategies to accelerate the landing of 5G applications. According to the Japan Electronics and Information Technology Industries Association (JEITA), the global 5G private network market will reach US$99.1 billion in 2030, and the global manufacturing industry's use of 5G for digitization is more active than other industries. Besides participating in industrial applications and the formulation of the 5G standard, 5G private networks have been established in multiple factories for smart factory application verification. Even though the development prospects for the vertical application of the 5G private network system seem to be bright, there are still many challenges to be overcome.

The cellular service providers released the first 5G-enabled smartphones in the middle of 2020, and it was slowed down due to the COVID-19 pandemic; however, countries such as South Korea and China have been making very good progress in

2021, especially in the development and deployment of wireless infrastructure technologies such as the massive multiple-input multiple-output (MIMO) and millimeter-wave (mmWave), both of which contribute to the realization of 5G characteristics such as very high data transmission rate and extremely low latency rate. Besides the challenges of building expensive 5G infrastructure, another challenge that has raised the eyebrows of the consumer end-user is the development of new 5G smartphones with sub-6 GHz and mmWave operational bands. Since 2015, many researchers have begun developing antennas designed for 5G smartphone applications, but most are working in the narrow C-band (3.5 GHz). However, in the past five years, many mmWave antenna designs have been studied as well, especially after the announcement of 5G New Radio (NR) bands in the frequency range 1 (FR1) and frequency range 2 (FR2).

As early as January 2019, we have begun corresponding with each other regarding the writing of a book specifically for 5G smartphone antenna design, as we have found no such book available in the market with a clear 5G antenna design guideline for smartphone applications. But at that time, there wasn't enough published open literature to support this cause. Even though in early 2020, when we have finalized the topics and chapters of the book, in which Chapters 1–5 (for sub-6 GHz antenna) and Chapters 6–10 (for mmWave antenna) will be written separately by us (Desmond and Wonbin, respectively), due to the outbreak of the COVID-19 pandemic, all preparation and writing up of the book have come to a halt, until early 2021 when we resumed the writing of this book.

This book is a complete reference book on 5G antennas for smartphone applications. It provided the 5G smartphone antenna engineer with a clear design guide and knowledge on how to build a multiple or wideband antenna from sub-6 GHz to mmWave bands. It can also be used in an advanced antenna design course. This book has presented many works that were also reported by the authors of this book, and detailed illustrations with many examples from the open literature are also given. One point to note, this book has covered almost all reported works in the area of 5G smartphone antenna design (including the new mmWave 5G Antenna-in Package design) up to date and the latest development are explicitly presented.

Acknowledgments

Chow-Yen-Desmond Sim wishes to thank his family (in Taiwan and Singapore, especially his beloved wife Shu-Fen) for all the support that he has received during the process of writing this book, especially when the family had to go through a difficult period amid the COVID-19 pandemic. He is also indebted to many of his postgraduate students from Feng Chia University (Taiwan), who had provided him with all the necessary aid in producing many of the figures and tables depicted in Chapters 1–5. Lastly, he is grateful to the co-author (Prof. Wonbin Hong) of this book for his invitation to co-write this book. Even at one time when he has ceased writing the book and pondering if he should give up the hope of completing the book chapters during the pandemic, the patience and assistance given by Wonbin have very much encouraged him to complete his very first international book in 5G antenna design for smartphone applications.

Wonbin Hong wishes to first thank his past and current students at Pohang University of Science and Technology (POSTECH). In particular, this book would not have been possible without the participation of Dr. Junho Park, who was deeply involved in Chapters 6, 7, and 10. He is extremely grateful to Dr. Jaehyun Choi and Dr. Jae-Yeong Lee for their participation in Chapters 8 and 9, respectively. The three aforementioned contributors are recognized by Wonbin Hong as co-authors of the mentioned Chapters. Ms. Youngmi Kim and Ms, Yemin Park of POSTECH were extraordinarily helpful during the editing and the final preparation would not have been possible without them. He thanks his beloved wife InKyung and daughter Yonwoo for their support and encouragement. It has been an amazing experience to work with Desmond, a very close friend and incredibly talented colleague. And last but certainly not least, the authors thank the editorial board at Wiley for their faith, friendship, and support during this exciting journey.

1

Introduction

Since the appearance of the first cellular network technology (mobile telecommunications) in the 1980s, a new generation (G) will emerge approximately every decade. 1G refers to the first wireless cellular technology that used the analog cellular network to allow voice transmission only, and the maximum data rate was 2.4 Kbps. 2G utilized digital signals for voice transmission in the early 1990s, and the analog world was discarded. At this point in time, peak data rates with General Packet Radio Service (GPRS) of up to 50 Mbps can be sent/transmitted by phones, including text messages and emails. 3G has brought the internet to mobile phone users in late 2001, based on Global Systems for Mobile (GSM) and Universal Mobile Telecommunications System (UMTS). The peak data rate was estimated to be 2 Mbps. 4G Long-Term Evolution (LTE) was commercialized in 2010, and the end-users can experience Downlink (DL) and Uplink (UL) data rates of up to 100 and 50 Mbps, respectively. Mobile phone users can perform online gaming, watch movies online, and conduct video conferences everywhere. With the arrival of the 4G LTE-Advanced (LTE-A) Pro (sometimes known as 4.5G) that comes with a new technology known as carrier aggregation, it allows Gbps data rates and an even faster internet connection.

The 5G cellular network infrastructure was deployed by global operators in early 2019. Even though the initial stage is the 5G Non-Standalone (NSA) architectures (see Figure 2.3), the rollout of 5G Standalone (SA) architectures is expected to be completed in 2025 with 1350 million 5G connections [1]. Interestingly, mobile phone users globally are forecasted to amount to 7.49 billion in 2025 [2]. It is also expected that future innovation devices/technologies such as wearable devices and the massive Internet of Things (IoT) will be emerged in full force, especially when the Fifth-Generation (5G) SA networks are fully established. As indicated by the International Telecommunication Union (ITU), future

Microwave and Millimeter-Wave Antenna Design for 5G Smartphone Applications,
First Edition. Wonbin Hong and Chow-Yen-Desmond Sim.

Published 2023 by John Wiley & Sons, Inc.

development of International Mobile Telecommunications (IMT) for 2020 and beyond would support different user-experienced data rates covering a variety of environments for enhanced Mobile Broadband (eMBB) that has high mobility up to 500 km/h with acceptable Quality of Service (QoS) [3]. The initial stage of 5G environment is expected to deliver ultra-low latency of below 1 ms, at least 1 Gbps peak data rate (up to 10 Gbps), and able to connect approximately 1 million devices per kilometer square. This ultra-low latency feature is expected to transform autonomous vehicles, industrial and manufacturing processes, traffic systems, etc., while higher data rates will expect 50 to 100 times faster than current 4G networks. Therefore, the 5G mobile network is predicted to revolutionize the Industrial Internet of Things (IIoT) and automotive industry and enable advanced mobile broadband by delivering Machine-to-Machine (M2M) and machine-to-person communications on a massive scale.

To enable higher data transmission rates (throughput) for 5G New Radio (NR) Over-the-Air (OTA), one of the key techniques is the advancement in spatial processing, which includes the massive Multiple-Input Multiple-Output (MIMO) network and its related hardware devices such as the antenna [4–6]. In a massive MIMO communication system, the antenna designs of the mobile device and base station are the key elements of the air interface, as higher throughput (channel capacity) can be made possible by increasing the number of antenna array elements on both the transmitting and receiving ends (see Figure 5.1). However, as the volume size of a smartphone is very limited, MIMO antenna array designs at 5G NR band n77/n78/n79 (3300–5000 MHz) for 5G smartphones can generally accommodate approximately 4 to 8 array elements, not to mention the additional factors to consider the new Unlicensed National Information Infrastructure 6 GHz band (UNII-6GHz) that has recently been included into the 5G frequency range 1 (FR1) band as the 5G NR Unlicensed (NR-U) band n46 (5150–5925 MHz) and n96 (5925–7125 MHz). As for the 5G FR2 band (5G mmWave band > 24 GHz), a compact-size antenna array of up to 16 elements can be easily implemented due to the smaller electrical wavelength required for the antenna. However, the metal casing of the smartphone may pose a serious threat to antenna performance. From the practical point of view, increasing the number of 5G FR1 antenna array elements will increase the difficulty of maintaining good isolation/decoupling between adjacent antenna elements [7]. Therefore, a full investigation and solutions to resolve the decoupling problems for different scenarios are vital to the 5G smartphone antenna engineers. As the present flagship smartphone has included up to three mmWave 5G Antenna-in-Package (AiP) for the 5G FR2 cellular network, the various design techniques and approaches to realize the integrated mmWave antenna array are essential to the antenna community. In addition, the challenges for mmWave 5G AiP to achieve frequency agility and coexist with the metallic chassis of the mobile device are vital topics for the Radio Frequency (RF) packaging industry.

In this book, we have focused specifically on the designs of antenna arrays for 5G mobile devices at two vital 5G frequency bands, 5G FR1 and 5G FR2 (mmWave). This book is organized into 10 chapters:

a) Chapter 1 gives a brief introduction to the 5G cellular network technology.
b) Chapter 2 illustrates the present 5G frequency characteristics and the frequency bands designated for 5G NR. An overview of the general channel models for 5G FR1/FR2 bands is provided, along with a brief description of 5G network architecture and mobile devices' evolution.
c) Chapter 3 explores the multi-antenna placement on an actual smartphone, followed by showing the topologies of 5G FR1 band array antenna designs for smartphone applications, especially those working at C-band 3.5 GHz (3400–3600 MHz).
d) Chapter 4 investigates the most recently reported dual-band, multi-band, and wideband antenna array designs for smartphone applications working at 5G FR1 band, including 5G NR band n77/n78/n79 and 5G NR-U band n46/n96. A complete review of the co-existence designs with 3G/4G antennas and mmWave 5G AiP is illustrated with various case studies.
e) Chapter 5 discusses in detail the MIMO antenna diversity performances, followed by a full review of the various isolation enhancement techniques applied by the antenna array designs operating in the 5G FR1 band. Practical considerations, existing problems, and new design approaches on the 5G FR1 band antenna array are briefly illustrated.
f) Chapter 6 investigates the mmWave 5G AiP for mobile applications, reviewing the design strategies, packaging considerations, and antenna-IC feeding mechanisms. The selection of appropriate materials and future challenges are illustrated.
g) Chapter 7 presents the multi-physical approach for the mmWave 5G AiP, including a brief background review and the challenges encountered by the packaging industry. As heat dissipation of AiP is a critical matter, the strategies to improve heat dissipation are illustrated with some practical examples.
h) Chapter 8 reveals the background and challenges for mmWave 5G AiP with tunable frequency. Its related matching network and design topologies are demonstrated with examples.
i) Chapter 9 discusses the design constraint and major challenges in realizing cost-effective and compact mmWave 5G antennas for mobile applications. The methods to resolve the limitations are illustrated with case studies involving mmWave 5G antenna miniaturization and utilization of conventional Printed Circuit Board (PCB) fabrication and laminations.
j) Chapter 10 gives a clear illustration and design methodology on mmWave Antenna-on-Display (AoD) design for 5G mobile devices. An overview of the antenna element/array design and packaging of the display-integrated AoD are included with related detailed examples.

References

1 The 5G guide, a reference for operators, GSMA, Apr. 2019.
2 Statista (Online) https://www.statista.com/statistics/218984/number-of-global-mobile-users-since-2010/
3 IMT Vision – Framework and overall objectives of the future development of IMT for 2020 and beyond, Recommendation ITU-R M.2083-0, Sept. 2015.
4 E. G. Larsson, O. Edfors, F. Tufvesson, and T. L. Marzetta, "Massive MIMO for next generation wireless systems," *IEEE Commun. Mag.*, vol. 52, no. 2, pp. 186–195, Feb. 2014.
5 L. Lu, G. Li, A. Swindlehurst, A. Ashikhmin, and R. Zhang, "An overview of massive MIMO: benefits and challenges," *IEEE J. Sel. Top. Sign. Process.*, vol. 8, no. 5, pp. 742–758, Oct. 2014.
6 Bleicher, "The 5G phone future," *IEEE Spectrum*, vol. 50, no. 7, pp. 15–16, 2013.
7 Y. Huo, X. Dong, and W. Xu, "5G cellular user equipment: From theory to practical hardware design," *IEEE Access*, vol. 5, pp. 13992–14010, Aug. 2017.

2

Considerations for Microwave and Millimeter-Wave 5G Mobile Antenna Design

Due to the ongoing deployment of the 5G mobile services since early 2019 and the recent exponential growth in the number of mobile phone users globally, all major cellular industries are actively commercializing and competing for their new flagship 5G mobile phones (or smartphones) with better features, such as the foldable phone from Samsung, Wing T-shaped dual-display phone from LG, and powerful processor that outperforms any chipset found in an Android phone from Apple iPhone. Even though the promises of delivering higher multi-Gigabit/s peak data rate communication, negligible latency for real-time interaction with ultra-reliable communication services, and high densities of connected devices/sensors are appealing to the end-users, to realize the potential of 5G, a different smart antenna system must be adopted. Here, one of the key smart antenna technologies known as the multiple-input multiple-output (MIMO) antenna system is applied to exploit smart beamforming for high data rate transmissions in the 5G mmWave frequency bands, as well as the 5G mid-band frequency ranges. The MIMO antenna system deploys multiple antennas at both the transmitter and receiver to increase the throughput and channel capacity of the radio link. It also applies the techniques known as spatial diversity and spatial multiplexing to transmit independent and separately encoded data signals (streams), by reusing the same time and frequency resources. More descriptions of the MIMO technology and the motivation and requirement of applying the MIMO antenna technology to a mobile phone are explicitly illustrated in Chapter 5.

In this chapter, we review the present 5G frequency characteristics and the frequency bands for 5G New Radio (NR), and summarize the general channel models for 5G FR1 bands and in the 5G FR2 mmWave bands. A brief description of the network architecture for the 5G system is presented with an overview of the 5G network structure. The evolution and history of mobile devices, and their cellular

Microwave and Millimeter-Wave Antenna Design for 5G Smartphone Applications,
First Edition. Wonbin Hong and Chow-Yen-Desmond Sim.

Published 2023 by John Wiley & Sons, Inc.

generations and featured sizes are revealed. Finally, we look at some of the advanced materials that the 4G/5G antennas have applied, and a comparison is made between these materials.

2.1 Frequency Characteristics and Channel Models

As there is no single technology or solution that can ideally be suited to all the different potential 5G applications and their spectrum availability, the third Generation Partnership Project (3GPP) has taken evolutionary steps on the network and device sides. In December 2017, the 3GPP Technical Specification Group Radio Access Network (TSG RAN) Plenary Meeting successfully approved the first 5G NR specification, known as 3GPP Release 15, meaning that the completion of this very first 5G NR standard will enable full-scale development of 5G NR [1]. According to TS 38.104, Section 5.2, the frequency bands for future 5G NR are separated into two different frequency ranges, namely Frequency Range 1 (FR1) and Frequency Range 2 (FR2). The FR1 has initially been known as the sub-6 GHz frequency bands (450–6000 MHz) but has now moved to (410–7125 MHz), and it is also known as mid-band/low-band in which some of these bands are traditionally used by previous wireless communication standards, such as the LTE band 46 (5150–5925 MHz) is now included in the 5G NR-U (NR-Unlicensed) spectrum. On the other hand, the FR2, also known as the mmWave band (24 250–52 600 MHz), has possessed a very short electrical wavelength that would yield a shorter transmission range, but it can give very wide operating bandwidth than those in the FR1. Even though the Frequency Range 3 (FR3) and Frequency Range (FR4) have recently been imposed for the upcoming 6G technology, they are outside the scope of this Book and will not be further discussed.

The 5G NR can also be further classified into three bands, namely Frequency Division Duplex (FDD) Bands, Time Division Duplex (TDD) Bands, and Supplementary Bands that include Supplementary Downlink (SDL) Bands and Supplementary Uplink (SUL) Bands. The detailed classification of each band in relation to the 5G NR frequency bands is shown in Tables 2.1–2.4 [1, 2].

Even though the above tables have shown a very wide bandwidth for each 5G NR bands category (FDD, TDD, SDL, and SUL), it is still worth observing the different frequency bands considered by different countries. The 5G bands (FR1 and FR2) outlined by country/region are shown in Figure 2.1 [3, 4]. In December 2018, China issued the permit to use the spectrum between 3300 and 3600 MHz, whereas the United States and Japan have allocated (3600–4100 GHz) and (4500–4900 MHz), respectively; thus, both the 5G NR band n78 (3300–3800 MHz) and n77 (3300–4200 MHz) are now highly supported by many major telecommunication operators in the world [4]. For European countries, the 5G NR band n78 (3300–3800 MHz), in particular, is being rolled out as their preferred

Table 2.1 FR1 FDD (frequency division duplex) frequency bands for 5G new radio.

5G NR band	Uplink frequency (MHz)	Downlink frequency (MHz)	Bandwidth (MHz)
n1	1920–1989	2110–2170	60
n2	1850–1910	1930–1990	60
n3	1710–1785	1805–1880	75
n5	824–849	869–894	25
n7	2500–2670	2620–2690	70
n8	880–915	925–960	35
n12	699–716	729–746	17
n14	788–798	758–768	10
n18	815–830	860–875	15
n20	832–862	791–821	30
n25	1850–1915	1930–1995	65
n26	814–849	859–894	35
n28	703–748	758–803	45
n30	2305–2315	2350–2360	10
n65	1920–2010	2110–2200	90
n66	1710–1780	2110–2200	90
n70	1695–1710	1995–2020	15/25
n71	663–698	617–652	35
n74	1427–1470	1475–1518	43
n91	832–862	1427–1432	30/5
n92	832–862	1432–1517	30/85
n93	880–915	1427–1432	35/5
n94	880–915	1432–1517	35/85

mid-band frequencies. As for the 5G NR band n79 (4400–5000 MHz), it is due to the two bands imposed by China (4800–5000 MHz) and Japan (4500–4900 MHz), and, therefore, the NR band n79 is only supported by the following operators: China Mobile, China Unicom, China Telecom, NTT DOCOMO, KDDI, and Softbank Mobile. Nevertheless, some other countries have considered using the 5G NR band n79 as their private 5G network (also referred to as "non-public networks" by 3GPP), as the operating band between 4400 and 5000 MHz are not shared by many mobile network operators. Furthermore, as the organization/industry that owns the wireless spectrum can have full control over the network, it can completely isolate its users from other public networks.

Table 2.2 FR1 TDD (time division duplex) frequency bands for 5G new radio.

5G NR band	Uplink frequency (MHz)	Downlink frequency (MHz)	Bandwidth (MHz)
n34	2010–2025	2010–2025	15
n38	2570–2620	2570–2620	50
n39	1880–1920	1880–1920	40
n40	2300–2400	2300–2400	100
n41	2469–2690	2496–2690	194
n46	5150–5925	5150–5925	775
n47	5855–5925	5855–5925	70
n48	3550–3700	3550–3700	150
n50	1432–1517	1432–1517	85
n51	1427–1432	1427–1432	5
n53	2483.5–2495	2483.5–2495	11.5
n77	3300–4200	3300–4200	900
n78	3300–3800	3300–3800	500
n79	4400–5000	4400–5000	600
n90	2496–2690	2496–2690	194
n96	5925–7125	5925–7125	1200

Table 2.3 FR1 supplementary downlink bands (SDL) and supplementary uplink bands (SUL) for 5G new radio.

5G NR band	Uplink frequency (MHz)	Downlink frequency (MHz)	Bandwidth (MHz)	Type
n75	–	1432–1517	85	SDL
n76	–	1427–1432	5	SDL
n80	1710–1785	–	75	SUL
n81	880–915	–	35	SUL
n82	832–862	–	30	SUL
n83	703–748	–	45	SUL
n84	1920–1980	–	60	SUL
n86	1710–1780	–	70	SUL
n89	824–849	–	25	SUL
n95	2010–2025	–	15	SUL

Table 2.4 5G NR frequency bands in FR2 above 24 GHz.

5G NR band	Band alias (GHz)	Uplink band (GHz)	Downlink band (GHz)	Bandwidth (GHz)	Type
n257	28	26.5–29.5	26.5–29.5	3	TDD
n258	26	24.25–27.5	24.25–27.5	3.25	TDD
n259	41	39.5–43.5	39.5–43.5	4	TDD
n260	39	37–40	37–40	3	TDD
n261	28	27.5–28.35	27.5–28.35	0.85	TDD

Besides the above sub-6 GHz NR Bands, the FCC has seen the potential of implementing the new unlicensed mid-band spectrum (5925–7125 MHz) for next-generation wireless broadband services [5, 6], aiming to combine the 5G NR with low-band LTE that may meet the growing needs of mobile broadband subscribers and deliver new 5G-based services. Notably, this band is sometimes referred to as the 6 GHz band, as it represents 1200 MHz of available spectrum from 5925 to 7125 MHz. Because the 802.11ax standard (Wi-Fi 6) has covered the 2.4 and 5 GHz bands, the Wi-Fi that works within this 6 GHz band was given a new name, Wi-Fi 6E. This name was chosen by the Wi-Fi Alliance to avoid confusion for 802.11ax devices that also support the 6 GHz band. The "6" represents the sixth generation of Wi-Fi, and the "E" is referred to as "Extended." Based on the concept of coexistence of the unlicensed spectrum with the 5G networks, as seen in Table 2.2, besides incorporating the 5 GHz unlicensed band (5150–5925 MHz) as the 5G NR-U band n46, the 3GPP Release 16 has also included the new 5G NR-U band n96. It is believed that the Standalone (SA) 5G NR-U bands can extend the potential of high-performance 5G to private networks, and with the unlocking of this 5G NR-U in the 6 GHz band, it can also be expected to play a major role in the Industrial Internet of Things (IIoT) due to its wide bandwidth characteristics [7]. Once a fully reliable 5G wireless connectivity is developed, the industry can stand a chance to unlock the distinctive capabilities of Industry 4.0, such as large-scale Autonomous Mobile Robots (AMRs) and Machine-to-Machine (M2M) communication.

The 5G NR bands in the FR2 (above 24 GHz) are a cornerstone of the future 5G networks, enabling a faster data rate and much wider operational bandwidth. Presently, three popular 5G NR bands, namely n257, n258, and n260, are included in the FR2 mmWave band, in which both bands n257 and n258 cover from 24 250 to 29 500 MHz, while bands n259 and n260 cover from 37 000 to 43 500 MHz. As for band n261, it is a subset of band n257 with a narrower bandwidth of 27 500–28 350 MHz.

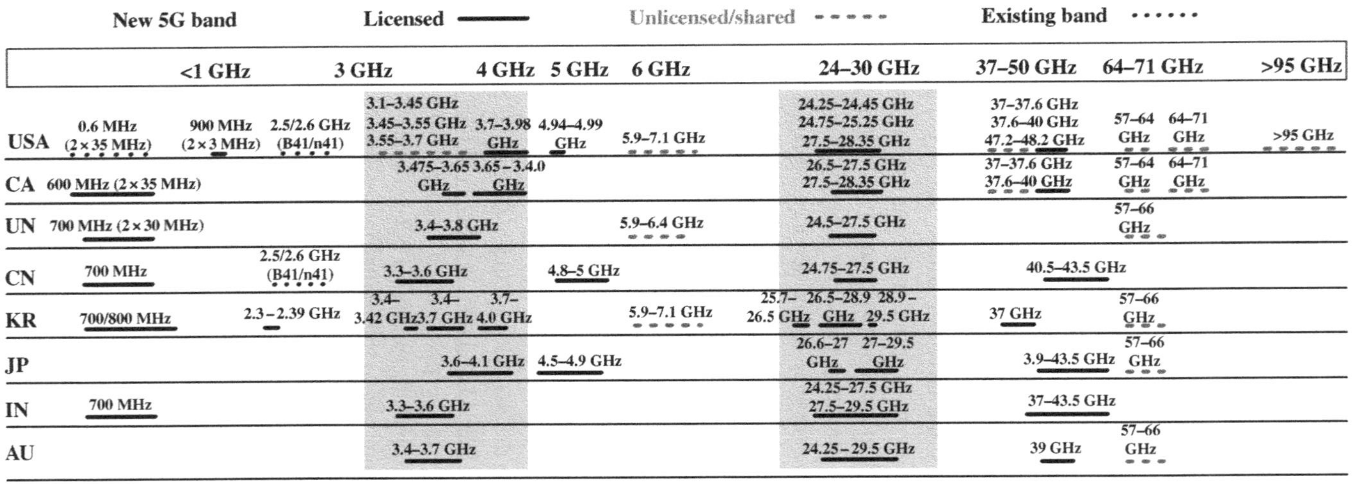

New 5G band | Licensed —— | Unlicensed/shared - - - - - | Existing band · · · · · ·

	<1 GHz	3 GHz	4 GHz	5 GHz	6 GHz	24–30 GHz	37–50 GHz	64–71 GHz	>95 GHz
USA	0.6 MHz (2×35 MHz) 900 MHz (2×3 MHz)	2.5/2.6 GHz (B41/n41) 3.1–3.45 GHz 3.45–3.55 GHz 3.55–3.7 GHz	3.7–3.98 GHz	4.94–4.99 GHz	5.9–7.1 GHz	24.25–24.45 GHz 24.75–25.25 GHz 27.5–28.35 GHz	37–37.6 GHz 37.6–40 GHz 47.2–48.2 GHz	57–64 GHz 64–71 GHz	>95 GHz
CA	600 MHz (2×35 MHz)	3.475–3.65 GHz	3.65–3.4.0 GHz			26.5–27.5 GHz 27.5–28.35 GHz	37–37.6 GHz 37.6–40 GHz	57–64 GHz 64–71 GHz	
UN	700 MHz (2×30 MHz)	3.4–3.8 GHz			5.9–6.4 GHz	24.5–27.5 GHz		57–66 GHz	
CN	700 MHz	2.5/2.6 GHz (B41/n41) 3.3–3.6 GHz		4.8–5 GHz		24.75–27.5 GHz	40.5–43.5 GHz		
KR	700/800 MHz	2.3–2.39 GHz 3.4–3.42 GHz 3.4–3.7 GHz	3.7–4.0 GHz		5.9–7.1 GHz	25.7–26.5 GHz 26.5–28.9 GHz 28.9–29.5 GHz	37 GHz	57–66 GHz	
JP			3.6–4.1 GHz	4.5–4.9 GHz		26.6–27 GHz 27–29.5 GHz	3.9–43.5 GHz	57–66 GHz	
IN	700 MHz	3.3–3.6 GHz				24.25–27.5 GHz 27.5–29.5 GHz	37–43.5 GHz		
AU		3.4–3.7 GHz				24.25–29.5 GHz	39 GHz	57–66 GHz	

Figure 2.1 5G spectrum outlined by country. *Source:* Quadcomm [3].

As aforementioned, the 5G network systems are designated with many different NR bands from low-band/mid-band frequencies to mmWave spectrums. The future 5G users will experience seamless coverage with very high-quality connectivity beyond the capability of its predecessors. At that point in time, the full potential of the IoT would be released, as well as enabling smart city innovation and Industry 4.0. As the 5G systems should be adapted to a wide range of scenarios, such as indoor, urban, suburban, and rural areas, its corresponding new requirements for 5G channel modeling should be investigated. Here, the work in [8] has summarized ten new requirements for a 5G channel model as follows:

1) Wide frequency range: It should support a wide frequency range from 410 MHz to 100 GHz. The channel models at higher frequency bands (>6 GHz) should maintain compatibility with the ones at lower frequency bands (<6 GHz).
2) Broad bandwidths: It should support large channel bandwidths from 500 to 4000 MHz.
3) Wide range of scenarios: It should support a wide range of scenarios, such as indoor, urban, suburban, and rural areas, and High-Speed Train (HST) scenarios.
4) Double-directional Three-Dimensional (3D) modeling: It should provide full 3D modeling that includes 3D antenna modeling and 3D propagation modeling.
5) Smooth time evolution: It has to evolve smoothly over time, involving parameters drifting and cluster fading in and fading out, which are important to support mobility and beam tracking for 5G communications.
6) Spatial consistency: The spatial consistency refers to two closely arranged transmitters or receivers possessing similar channel characteristics.
7) Frequency dependency and frequency consistency: The parameters and statistics of a new 5G channel model should vary smoothly with the frequency. Channel parameters and statistics at adjacent frequencies should have strong correlations.
8) Massive MIMO: It must support massive MIMO.
9) Direct D2D/V2V: In D2D/V2V scenarios, both the transmitter and receiver are fitted with antennas that exhibit lower profile (height) and may experience larger multipath signals. The D2D/V2V channel models have to consider the mobility of both ends, which significantly increases the modeling complexity.
10) High mobility: It should support high mobility scenarios, such as HST scenario train speed of over 500 km/h. The channel model should capture certain characteristics such as high mobility channels, large Doppler frequency, and non-stationarity. In addition, the HST scenarios must also include other scenarios, such as open space, viaduct, cutting, hilly terrain, tunnel, and station.

Even though, at present, many works on 5G channel measurement and channel modelling have been reported [8], which includes massive MIMO communication scenarios, vehicle-to-vehicle (V2V) communication scenarios, HST communication scenarios, and mmWave communication scenarios, this book focuses more on the antenna designs for 5G mobile devices (especially those for smartphone applications), therefore, for brevity, we would not illustrate further on the 5G channel modeling.

2.2 5G Network Architecture

The 5G network architecture was introduced by the 3GPP, a consortium with many standard organizations that develop protocols for mobile communications. It is a new mobile technology that is extended and enhanced upon its predecessor, the 4G LTE network architecture. As the design considerations for a 5G network architecture are highly complex due to the many different challenging applications, such as those that may require a very large data throughput with very long signal propagation distance. Therefore, to fully realize the 5G network systems, the 5G network architecture must support the FR1 (low-band, mid-band) and FR2 (high-band, mmWave) spectrum, from licensed, unlicensed/shared, and existing bands [8], as shown in Figure 2.1.

Figure 2.2 shows the three frequency bands that are designated at the core of 5G networks with three circles indicating the distances required by each spectrum [9]:

a) The outermost circle refers to the 4G LTE (low-band and lower mid-band) used today that operates between 698 and 2690 MHz, and it can yield a broad coverage of up to 150 km. Notably, as indicated in Tables 2.1–2.3, these spectrums are now included in the 5G NR bands. As the performance of the 5G low-band is analogous to its 4G LTE counterparts, it essentially supports the LTE/5G architecture for present 5G mobile devices.
b) The 5G mid-band operates in two different frequency ranges, namely 5G NR band n77/n78/n79 (3300–5000 MHz) and 5G NR-U band n46/n96 (5150–7125 MHz). At the moment, the range between 3300 and 5000 MHz provides a capacity layer for urban and suburban areas; with the deployment of Microcell for urban areas, the coverage area outdoors is between 500 and 2500 m, with peak data rates in the hundreds of Mbps.
c) The 5G high-band (mmWave) delivers the highest frequencies of 5G, as it operates between 24 and 100 GHz. At such high-frequency bands, the mitigation of the traveling signal strength is significantly large due to atmospheric absorption. Furthermore, it can be easily absorbed by objects such as the tree (trunk and leaves), and it cannot penetrate through buildings and walls. Therefore, as depicted in Figure 2.2, the 5G high-band is a short-range wireless

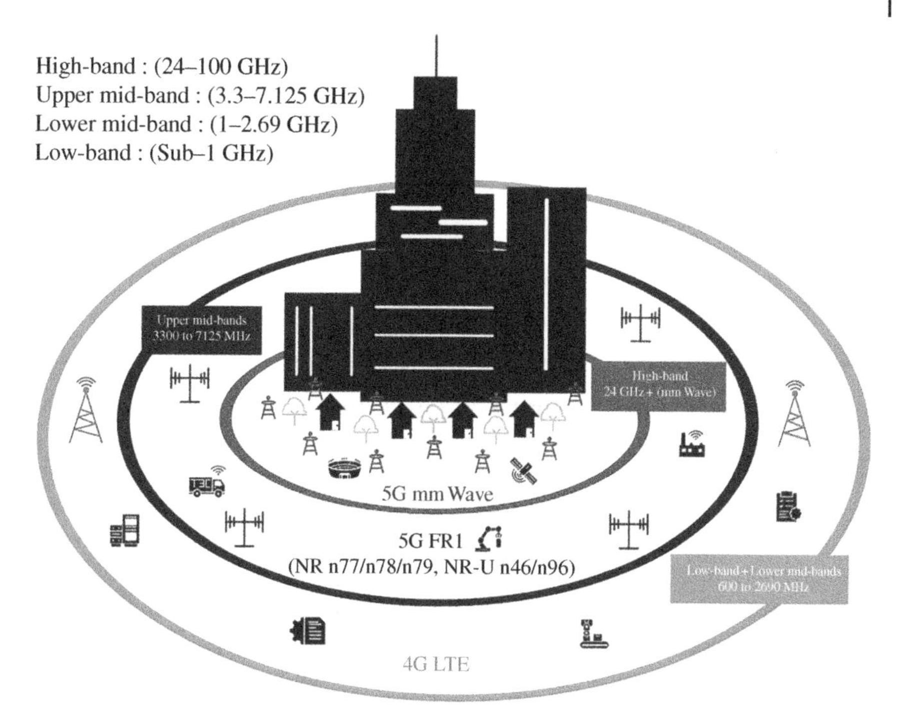

Figure 2.2 The three frequency bands at the core of 5G network. *Source:* DIGI [9].

communication, and the 5G high-band mobile deployments will require dense network topologies with Inter-Site Distances (ISD) of approximately 150 to 200 m. As a result, commercializing the 5G high-band is very challenging because it requires many additional small cells (such as the Femtocell and Picocell) that usually have a coverage radius of fewer than 500 m. Based on the aforementioned facts, wide-scale commercial deployments of the 5G high-band would be inevitably delayed due to the demand for very large investments from the 5G mobile network operators.

To build a true 5G network system, its corresponding network infrastructure is required to be evolved from its predecessor (4G LTE). Figure 2.3 illustrates the evolution from 4G LTE network infrastructure to 5G SA network infrastructure. In this figure, the initial 5G network is a Non-Standalone (NSA) type that depends on the 4G LTE network facilities. Through EN-DC dual connection technology (E-UTRAN New Radio-Dual Connectivity), besides a direct connection to the 5G base stations, the 5G users can also obtain greater capacity from the 4G LTE networks and 5G networks. During the transition period when the 5G cellular base station infrastructure has not been widely deployed, the users of the 5G NSA

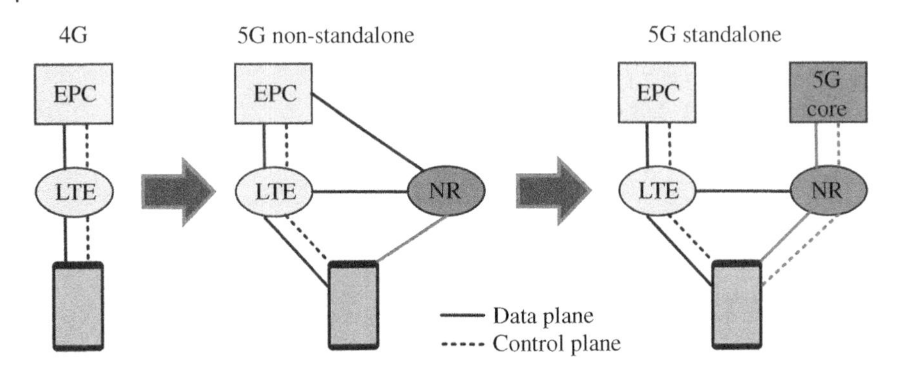

Figure 2.3 Evolution from the 4G network to 5G SA infrastructure. *Source:* OPPO [10].

network can also enjoy a faster network speed as compared to the original 4G LTE network that is only connected to the 4G base station. On the other hand, the 5G SA is the true 5G network as it involves a 5G core that can yield enormous channels throughout of up to multi-Gigabit/s and ultra-low network latency. As the 5G SA network is independent of the 4G network, when a mobile device supports both the 4G and 5G SA networks, it is known as a dual-mode 5G-supported device [10].

To further comprehend the cellular architecture and key technologies for the 5G network, [11] has proposed a new heterogeneous 5G cellular architecture with separated indoor and outdoor applications using Distributed Antenna System (DAS) and massive MIMO technology. The readers are encouraged to read [10, 11] as they have illustrated some of the potential key challenges in the 5G wireless technologies, such as cognitive radio networks, mobile Femtocell, edge computing, and 5G network slicing.

2.3 Evolution of Mobile Devices

Since the appearance of the very first mobile phone (Motorola DynaTAC 8000x) in 1983, today's mobile devices have come a long way from a heavy and bulky type to a full-screen lightweight smartphone type. In 2008, the iPhone 3G (or iPhone 2) had paved the way for present smartphones, and thereafter, smartphones had dominated the entire mobile device market. From the statistics, the number of smartphone subscriptions in 2020 was approximately 6055 million, and it is projected to be over 7500 million in 2026. Between 2017 and 2021, the number of smartphones sold to end-users worldwide each year was approximately 1530 million, except for 2020 (approximately 1378 million), due to the COVID-19 pandemic.

As many books and online websites have described the evolution of mobile devices from 1G to 5G [12–16], instead of repeating the same narrative, we would

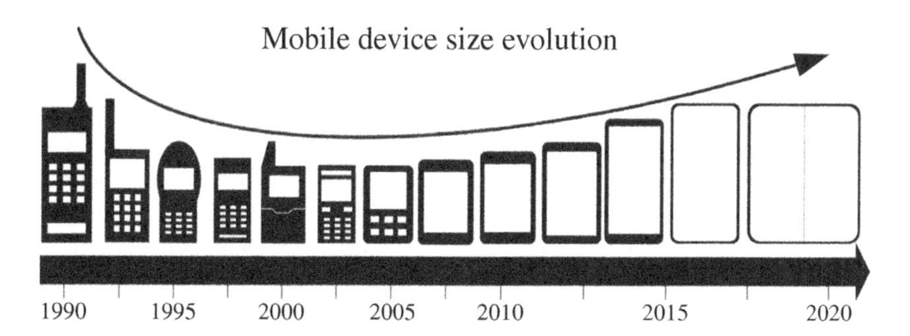

Figure 2.4 The evolution of the mobile device dimension and screen size.

prefer to illustrate the development of the mobile device dimension and screen size. As depicted in Figure 2.4, one can see that the dimension of mobile phones has gone through a U-shaped trend, from the initial portable type in the 1980s that weighs almost 1 kg with a planar size of $30 \times 9\,\text{cm}^2$ into the 1990s that celebrated the flipped phone and slider-style phone by Nokia 8810 that was among the first phones to have a built-in antenna, and it weighs only 145 g with a length under 15 cm. Between 2004 and 2006, when the term "smartphone" began to emerge in the market, a significant milestone in the evolution of mobile phones was the release of the iPhone2 that has a planar size of $138 \times 67\,\text{mm}^2$ and weighs around 148 g. Thereafter, one can see that the smartphone's screen size has been increasing linearly, and the screen-to-body ratio of a smartphone has gone up to near 92% today, whereas it was limited to 70% in 2015. In 2020, Samsung achieved the next revolutionary step by introducing a large foldable screen smartphone with a weight of 271 g and a flexible touch screen. The unfolded planar screen size is approximately $158 \times 128\,\text{mm}^2$ (or 7.6 in.), and the folded one is approximately $158 \times 67\,\text{mm}^2$ (6.2 in.). While this form has pushed the smartphone to function more like a tablet computer, on the other hand, it extends the interaction mode between end-user and device considerably.

2.4 Antenna Materials

The selection of a desirable material is vital to the performances of the 5G antenna, especially those that are working in the FR2 mmWave band. When considering which type of material should be applied for the 5G antennas, one should begin with the nature of the antenna design type, such as patch antenna, monopole/dipole/loop antenna, flexible antenna, and dielectric resonator antenna, followed by evaluating the selected 5G frequency, as frequency below 6 GHz usually will experience lower substrate loss. For example, typical material such as the low-cost FR4 is not suitable for mmWave antennas, as they will incur high dielectric losses. Therefore, when selecting a Printed Circuit Board (PCB) substrate material for your 5G antennas, the following properties should be considered: dielectric constant (Dk), loss tangent (Df, dissipation factor), thermal conductivity, manufacturability, and thickness. Higher Dk values can support smaller antenna structures for a given frequency, and lower Df can prevent further dielectric losses at higher frequencies (such as mmWave). High thermal/temperature changes usually affect the Dk value (especially for FR4) and result in frequency variation, and minor fabrication errors may incur large impedance mismatching. As thicker substrate may yield better impedance bandwidth, it may increase the manufacturing cost and difficulty.

In this section, a brief description and properties of different types of materials that have been applied for the 5G antenna designs are illustrated, and they are categorized as follows:

1) Laminate sheet
2) Ceramic
3) Organic
4) Others

2.4.1 Laminated Sheet

a) FR4:

The FR, in FR4, stands for flame retardant. As it is a composite material composed of woven fiberglass cloth with an epoxy resin binder that is flame-resistant (self-extinguishing), FR4 is a widely applied insulating material for making PCBs. The FR4 has been used in many 5G printed antenna array designs (see Chapters 3–5), especially those working below the 6 GHz band. The reason for that is that FR4 is a low-cost material and can be fabricated very easily. The FR4 usually has a Dk value of between 3.9 and 4.7 (typical value is 4.4), and a Df value of between 0.02 and 0.03 (standard value is 0.02). One specific problem for FR4 is the moisture absorption from the environment, as it will considerably affect the Dk and vary the antenna's frequency. According to [17], the amount of moisture absorbed by the FR4 (from different manufacturers) usually would not exceed 0.1–0.3% in weight of the sample. As aforementioned, the FR4 material is unsuitable for a 5G mmWave band of over 24 GHz due to its electric dispersion properties that will incur high losses [18].

b) Rogers substrate:

Rogers is a company that manufactures laminate materials for making PCBs [19]. Typical Rogers materials have the properties such as low electrical signal loss, cost-effective PCB fabrication, lower dielectric loss, better thermal management, a wide range of Dk values from 2.2 to 10.2, and improved impedance control. Therefore, it is a better material than the FR4 as it demonstrated much lesser loss at mmWave, but it is more expensive [20]. As there are many different types of Rogers material, we only feature the following Rogers substrates typically used for 5G mmWave antenna, namely RO4003, RO4350, and RT5880.

i) The R04003 caters to cost-sensitive microwave/radio-frequency (RF) design, as the cost is only a fraction of conventional microwave laminates. It has a Dk value of 3.38 (±0.05) and low-loss characteristics with Df = 0.0027 (@10 GHz). As the fabrication process is analogous to the FR4, the RO4003 is always a popular substrate for 5G mmWave antenna engineers.
ii) The R04350B is also a popular substrate for 5G mmWave antenna designs. It has very similar features compared with RO4003, and the Dk and Df values are 3.48 (±0.05) and 0.0037 (@10 GHz), respectively.
iii) The RT5880 has exhibited very low Dk and Df values of 2.2 (±0.02) and 0.0009 (@10 GHz), respectively. As it shows uniform electrical properties (such as Dk) over a wide frequency range with a very low moisture absorption of 0.02%, the RT5880 is well suited for 5G mmWave/wideband applications. Nevertheless, the cost of the material is high.

Even though other Rogers substrates such as RO3003G2 and RO4830 that are commercially available have slightly different Dk values of 3.00 (±0.04) and 3.24, respectively, with low Df values of 0.0011 and 0.0033, they are more suitable for use in the 76–81 GHz automotive radar sensor PCB applications, as they have shown very low insertion loss of approximately 2 dB/in. at 77 GHz.

2.4.2 Ceramic

Among the many types of ceramic substrates for antenna design, the low-temperature co-fired ceramic (LTCC) is the most popular one because it has a very high Dk value that can help reduce the antenna size for a given frequency, low Df (or high-quality Q factor) and good temperature stability. Besides the ability to incorporate passive components such as capacitors and resistors into the LTCC structure, an LTCC-based antenna can be easily integrated with a 5G antenna module [21]. Notably, the LTCC is also popular in the RF packaging industry because it offers low-loss dielectrics, good thermal conductivity, and a high degree of integration due to cavities and embedded passives [22].

As the LTCC process uses ceramics as the substrate material, the circuit is integrated into the ceramics by screen printing technology, and via low-temperature sintering, integrated ceramic components are formed. Presently, most of the commercial LTCC has a relative Dk value between 5 and 10 (with a few between 10 and 20) and a very low Df value of <0.002. From work reported in [23], the Ferro k6.5 – ERS3702 has the lowest Df in the market (0.00078 at 69 GHz). The features of using the LTCC are listed as follows:

a) Reduce size and weight, improve product performances, and can be designed in different shapes and sizes according to requirements.
b) It has an excellent high Q factor at mmWave.
c) The use of high-conductivity metal materials as conductor materials is conducive to the quality factor of the system.
d) It can handle high current and high temperature and has better thermal conductivity than general circuit boards.
e) Passive parts can be embedded into multilayer circuits to increase circuit density.
f) Better temperature characteristics, such as a smaller Coefficient of Thermal Expansion (CTE describes how the size of an object changes with a change in temperature), and a smaller temperature coefficient of resonance frequency (T_f). For an ideal resonator, the T_f should be near zero.
g) Very low moisture absorption, and it is used by the current mainstream mmWave antenna design solution for the Antenna-in-Package (AiP).

Table 2.5 shows the performance comparison of the typical substrate employed by many antenna engineering for the fabrication of their 5G antennas. As shown in this table, FR4 has attractive low-cost characteristics, but its other properties, such as Dk, Df, and moisture absorption, are poorer than that of the Rogers. In contrast, Rogers RT5880 has exhibited very low Dk and Df values, but the material cost is very high. Nevertheless, due to stable Dk and low Df, the Rogers substrates

Table 2.5 Performance comparison of Rogers, FR4, and LTCC substrate.

	Dk	Df	HTC (W/m K)	Moisture absorption	Remarks
RO3003G2	3.00 (±0.04)	0.0011	0.6–0.8	0.06	Stable Dk
RO4350B	3.48 (±0.05)	0.0037	0.6–0.8	0.06	Stable Dk
RO4003C	3.38 (±0.05)	0.0027	0.6–0.8	0.06	Stable Dk
RO4830	3.24	0.0033	0.6–0.8	0.15	Stable Dk
RT5880	2.2 (±0.02)	0.0009	0.6–0.8	0.02	Low-loss
FR4	3.9–4.7	0.02	0.25	0.1–0.3	Low-cost
LTCC	5–10	>0.002	2–3	0	High Dk

HTC, heat transfer coefficient.

are very suitable for 5G mmWave antenna. On the other hand, due to its high Dk, low Df, and zero moisture absorption, the LTCC is more suitable for working as the base of the antenna array designated for the 5G mmWave AiP.

2.4.3 Organic

Recently, many different types of organic low-loss materials are selected for 5G devices, namely Polytetrafluoroethylene (PTFE), Polyimide (PI), Modified Polyimide (MPI), and Liquid Crystal Polymer (LCP) [24]. The features of these organic materials are as follows:

a) The PTFE is a synthetic fluoropolymer of tetrafluoroethylene that has numerous applications. It has high-temperature resistance characteristics, and the working temperature can reach up to 250 °C. Under the high frequency of 5G communication, the Dk value is below 2.4, and the Df value is below 0.0006. In the 5G industry chain, the PTFE is used as the raw material for the three intermediate products: high-frequency Copper-Clad Laminates (CCLs), semi-flexible coaxial cables, and fine RF coaxial cables. It is mainly applied to the 5G base stations' Active Antenna Unit (AAU), the mainboard of mobile devices, smartphone dielectric materials, and the RF connection components of the 5G smartphone.
b) The PI and MPI are CCLs with lightweight, thin, and flexible properties. Furthermore, they have also exhibited the features of high thermal stability, allowing processing temperature of up to 500 °C, low Dk value, and resistance to acids and organic solvents. The Flexible CCL (FCCL) is a core component of Flexible Printed Circuits (FPCs). Therefore, the FPCs are commonly applied in many electronic products such as smartphones and laptop computers. As the FPC is suitable for manufacturing complex multi-band antennas with limited shape and low fabrication costs, it has become the mainstream of current mobile devices.
c) LCP is a new and promising thermoplastic organic material. It has shown excellent properties such as ultra-low moisture absorption, good chemical resistance, and high gas barrier. Most importantly, the LCP exhibits a stable Dk value of approximately 3.0 and a low Df value of 0.004 across a wide frequency range from entire RF up to 110 GHz [25]. As the LCP has also demonstrated good flexibility that leads to the convenient deployment of antennas in space, it is considered the ideal substrate for antennas working in the microwave and mmWave frequency. However, due to the low processability caused by the stable structure of LCP and the small number of suppliers, the cost of LCP film can be several times higher than that of general PI film.

Table 2.6 shows the performance comparison between PI, MPI, and LCP. Here, the LCP is superior to PI and MPI in terms of transmission loss and stability. However, due to the relatively high cost and process difficulty, it is rather impractical to replace MPI with LCP completely. Furthermore, as the

Table 2.6 Performance comparison between PI, MPI, and LCP.

	Transmission loss	Flexibility	Dimensional stability	Moisture absorption	Heat resistance	Cost
PI	Poor	Poor	Poor	Better	Better	×1
MPI	General	General	General	General	General	×1–2
LCP	Better	Better	Better	Poor	Poor	×2–2.5
Remarks	LCP is suitable for high frequency and high speed	LCP is suitable for miniaturization	LCP is more reliable	LCP has better performance	LCP is difficult to process	LCP is more expensive

Source: From: [26].

performances of MPI are between PI and LCP, it is more feasible for both MPI and LCP to coexist in the future 5G era in which the MPI will be used for mid-and low-frequency bands (below 7 GHz), and the LCP will be used for high-frequency bands (for mmWave above 24 GHz). Therefore, in the short term, the 5G antenna substrate materials market will be expecting a situation where multiple materials are competing at the same time.

Table 2.7 shows the trend of antenna technology, the number of required antennas, and antenna types from 1G to 5G [26]. This table shows that the number of antennas needed from 1G to 5G has changed drastically, as the 5G antenna system requires up to more than 16 antennas, whereas the 1G antenna system is a single antenna type. The 5G antenna system has included two antennas (main and diversity) for 4G LTE band, eight antennas for 5G FR1 band, two to three antenna arrays (1×4) for 5G FR2 mmWave band, and other typical antennas such as BT, WiFi, GPS, NFC, and WC. Notably, the antenna fabrication/material technology has also altered during these periods, from 3G that mainly applies the FPC, to 4G that uses mainly the LDS process to fabricate the main and diverse 4G antennas, to 5G that are now a mix of different materials, including MPI/LCP for 5G FR1 antenna array, and LTCC for 5G FR2 mmWave antenna array.

2.4.4 Others

Besides the above major materials for 5G devices, especially those that are suitable for 5G antenna fabrication, there are still other potential materials that have been reported for 5G devices [24], such as low-loss thermoset materials, hydrocarbon, poly(*p*-phenylene ether) (PPE or PPO), and glass. For brevity, the authors would

Table 2.7 Antenna technology and characteristics from 1G to 5G.

	1G	2G	3G	4G	5G
Antenna technology	External	Shrapnel	FPC, metallic chassis	LDS	MPI, LCP
No. of antenna	1	1–2	4–5	7–8	>16
Antenna type	Main	Main, BT	Main×2, BT, WiFi, GPS, radio	Main×2, BT, WiFi, GPS, radio, NFC	Main×8, BT, WiFi, GPS, radio, NFC, WC

BT, bluetooth; FPC, flexible printed circuit; GPS, global positioning system; LDS, laser direct structuring; NFC, near field communication; MPI, modified polyimide; LCP, liquid crystal polymer; WC, wireless charging. *Source:* From: [26].

recommend the work in [24] to the readers, as it illustrated many different types of low-loss materials for various 5G devices, such as 5G base station, 5G smartphone, and 5G Customer Premise Equipment (CPE).

2.5 Conclusion

In this chapter, we have illustrated a few major concerns when designing the microwave and mmWave 5G mobile antenna. The first concern is the frequency selection of the antenna in which the 5G NR band spectrum outlined by the 3GPP Release 16 must be strictly upheld. Next, due to the huge variation in the mobile device dimension from 1G to 5G, as well as the enlarging screen size of the smartphone, they should not be overlooked by the antenna engineers because such a large screen-to-body ratio for the smartphone will eventually lead to changes in the antenna structure, design, and placement. Finally, we have outlined the various typical materials that are now applied in the 5G smartphone antenna and other high-frequency/mmWave devices. The properties of these materials, as well as their suitability for the 5G FR1 band or 5G FR2 mmWave band, are illustrated and compared as well.

References

1 (Online) https://www.everythingrf.com/community/5g-nr-new-radio-frequency-bands.
2 ETSI TS 138 101-1 V16.5.0 (2020-11) (online) https://www.etsi.org/deliver/etsi_ts/138100_138199/13810101/16.05.00_60/ts_13810101v160500p.pdf.
3 (Online) https://www.qualcomm.com/media/documents/files/spectrum-for-4g-and-5g.pdf.
4 (Online) https://www.everythingrf.com/community/5g-frequency-bands.
5 5G Spectrum Public Policy Position, Huawei, 2020.
6 "Expanding Flexible Use in Mid-Band Spectrum Between 3.7 and 24 GHz," FCC, Aug. 2017.
7 G. Naik, J. M. Park, J. Ashdown, and W. Lehr, "Next generation Wi-Fi and 5G NR-U in the 6 GHz bands: opportunities and challenges," *IEEE Access*, vol. 8, pp. 153027–153056, Aug. 2020.
8 C. X. Wang, J. Bian, J. Sun, W. Zhang, and M. Zhang, "A survey of 5G channel measurements and models," *IEEE Commun. Survey Tutorial*, vol. 20, no. 4, pp. 3142–3168, Aug. 2018.
9 (Online) https://www.digi.com/blog/post/5g-network-architecture?utm_source=subscribe&utm_medium=rss.

10 5G SA versus 5G NSA: What's the Difference?, Oppo.

11 C. Wang, F. Haider, X. Gao, X. H. You, Y. Yang, D. Yuan, H. M. Aggoune, H. Haas, S. Fletcher, and E. Hepsaydir, "Cellular architecture and key technologies for 5G wireless communication networks," *IEEE Commun. Mag.*, vol. 52, no. 2, pp. 122–130, Feb. 2014.

12 M. S. Sharawi, Printed MIMO Antenna Technology, Artech House, 2014.

13 (Online) https://www.tigermobiles.com/evolution/#start.

14 (Online) https://www.netstar.co.uk/mobile-phones-years/

15 (Online) https://minutehack.com/opinions/the-evolution-of-the-mobile-phone.

16 (Online) https://www.cengn.ca/timeline-from-1g-to-5g-a-brief-history-on-cell-phones/

17 J. Bortfeldt, F. Dubinin, P. Lengo, J. Samarati, and K. Zhukov, "Study of hygroscopic expansion of anode readout boards of gaseous detectors based on FR4," *J. Instrum.*, vol. 16, P03016, Mar. 2021.

18 K. Bharath Kumar and T. Shanmuganantham, "Four dielectric substrate analysis for millimeter wave application," *Mater. Today Proc.*, vol. 5, no. 4, pp. 10771–10778, 2018.

19 Rogers Corporation, https://rogerscorp.com/

20 Selection of PCB materials for 5G, Microwave Journal E-Book, Feb. 2018.

21 S. Chen and A. Zhao, "LTCC based dual-polarized magneto-electric dipole antenna for 5G millimeter wave application," in *13th European Conf. Antennas Propag. (EuCAP)*, Krakow, Poland, Mar.–Apr. 2019.

22 D. G. Kam, D. Kiu, A. Natarajan, S. Reynolds, and B. A. Floyd, "Low-cost antenna-in-package solutions for 60-GHz phased array systems," in *19th Topical Meeting on Electrical Performance of Electronic Packaging and Systems*, Austin, TX, USA, Oct. 2010.

23 P. M. Marly, E. S. Tormey, Y. Yang, and C. Gleason, "Low-K LTCC dielectrics: novel high-Q materials for 5G applications," in *IEEE MTT-S Int. Microw. Workshop Series on Advanced Materials and Processes for RF and THz App. (IMWS-AMP)*, pp. 88–90, Bochum, Germany, Jul. 2019.

24 L. Jiang, "Low-loss materials for 5G 2021–2031. Trends, technologies, and forcast," IDTechEx, 2021.

25 Y. Ji, Y. Bai, X. Liu, and K. Jia, "Progress of liquid crystal polyester (LCP) for 5G application," *Adv. Ind. Eng. Polym. Res.*, vol. 3, no. 4, pp. 160–174, Oct. 2020.

26 (Online), Anue, https://news.cnyes.com/news/id/4411672.

3

Basic Concepts for 5G FR1 Band Mobile Antenna Design

Since the World Radiocommunication Conference (WRC-15, Nov. 2015) has identified frequency bands in the lower part of the C-band (3400–3600 MHz) allocated for Mobile Broadband Services (MBS) [1], many design structures for narrow C-band (3400–3600 MHz) antennas have been reported between 2015 and 2020. In this chapter, we start with a brief overview consideration in designing an antenna element for 5G FR1 mobile devices in the C-band and their design topologies, which also includes the placement of other antennas such as GPS, WiFi, and 4G LTE, followed by introducing the chassis, battery, and LCD screen effects that may result in antenna performance deterioration. As these C-band antenna designs are the very first batches of 5G antenna designs that appeared in the open literature, they have laid a foundation for the upcoming multi-band and wideband antenna designs for 5G NR band n77/n78/n79 (3300–5000 MHz), not to mention the new extended 5G NR band n46 (5150–5925 MHz) and n96 (5925–7125 MHz). Next, the various feeding mechanisms that have been applied by these C-band 5G antennas will be shown, followed by discussing the effects and consideration from the chassis (or frame of the mobile device), especially when the chassis is a metallic type, and how a different feeding approach can be applied for such metallic frame/chassis. Finally, the exposure of the MIMO antenna to the user's hand is studied, and how the user's hand would mitigate the antenna performances, especially the impedance matching and antenna efficiency, would be fully discussed as well.

3.1 Design Considerations

Before designing a 5G FR1 MIMO antenna array into an actual mobile phone (or smartphone) device, one needs to consider many other passive/active elements located in a smartphone besides the geometry and location of 5G

Microwave and Millimeter-Wave Antenna Design for 5G Smartphone Applications,
First Edition. Wonbin Hong and Chow-Yen-Desmond Sim.

Published 2023 by John Wiley & Sons, Inc.

FR1 MIMO antenna array, such as the location of other antennas (including WiFi, GPS, Bluetooth, and 4G LTE).

3.1.1 Antenna Placement

Figure 3.1 shows the collocation of an eight-antenna array (for C-band 3.5 GHz) in which four of the C-band array elements are distributed at the four corners of the mobile device, while the other four are integrated with the 4G LTE and WiFi antennas. Figure 3.2 shows a detailed recent development in antenna distribution consideration of the 5G NR band n77/n78 (3300–4200 MHz) that has a similar distribution as in Figure 3.1, showing four of the 5G NR band n77/n78 antenna elements allocated at the four corners (Ant. 2, Ant. 3, Ant. 5, and Ant. 6) [2]. However, the other three 5G antenna elements are placed at the top longer side edge frame (Ant. 7, Ant. 8, and Ant. 9), and the other one is allocated at the bottom longer side edge frame (Ant. 10) along with the WiFi MIMO array. Notably, the antennas located at four corners are the integration of LTE and 5G NR n77/n78 bands.

By further observing the mmWave antenna distribution from Figures 3.1 and 3.2, one can see that Figure 3.1 has indicated that the two mmWave antennas will be located along with the two side frames, whereas the locations of the two mmWave antennas in Figure 3.2 are laid on the Printed Circuit Board (PCB) (in orthogonal position). Both solutions stated in the two figures are possible, except

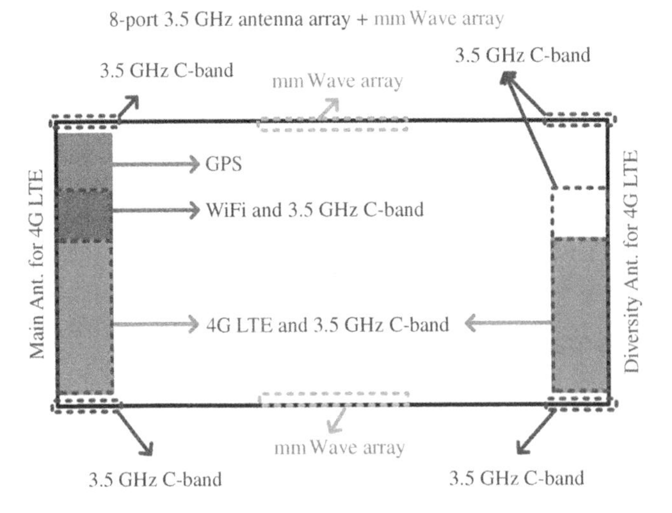

Figure 3.1 Conceptual diagram for 8-port C-band (3.5 GHz) antenna array and mmWave antenna array placement with other antennas.

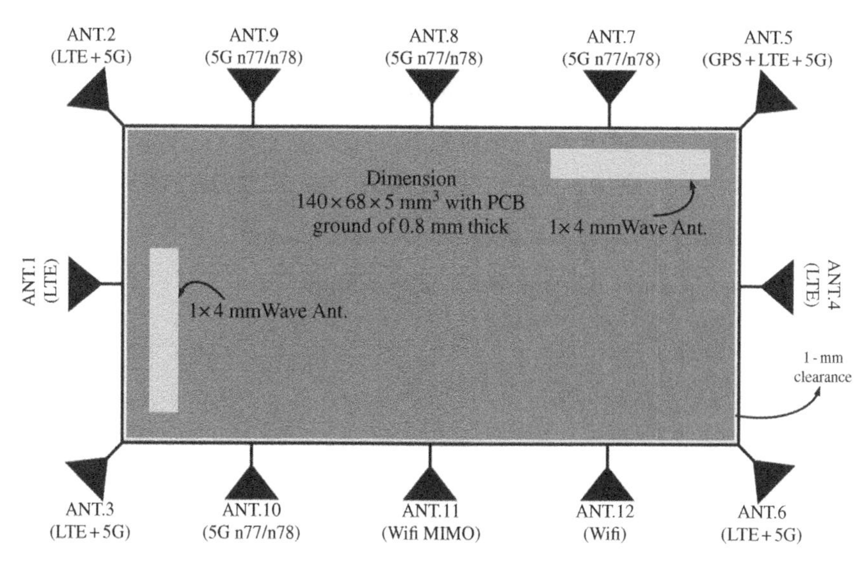

Figure 3.2 Conceptual diagram for integration of 5G NR band n77/n78 antenna array and mmWave antenna array placement with other antennas.

that at the moment, the three Antenna-in-Packages (AiPs) for the 5G mmWave antenna array are allocated on the two longer side frames while the other one is placed on the back of the LCD panel, as shown in Figure 3.3, in which the tear-down of a Samsung Galaxy S20 has uncovered the placement of the QTM AiP Module [3].

3.1.2 Smartphone Components and Their Effects

It is well known that the smartphone is a combination of several hundreds of active/passive components and hardware. Figure 3.4 shows a few of the major components that may yield serious mitigation on the 5G FR1 antenna array performances, and they are, namely, the frame, PCB components and shielding, modules (camera, speaker/receiver, etc.), battery, glass/display screen, back panel (aluminum, glass, or full metal), and connectors (Type 3 USB port, SIM slot, ear jack connector, etc.). A similar diagram has also been illustrated in Figure 3.5, showing the exact positions of these major components, as well as the locations of the 4G LTE antennas, GPS antennas, and WiFi/Bluetooth antennas [4].

A general discussion on these major components is as follows:

1) The chassis (or frame) of the smartphone must be seriously considered, especially when the back panel is a metallic type as well, meaning that the smartphone is a full-metallic type. If only the chassis is a metallic type, the 5G FR1

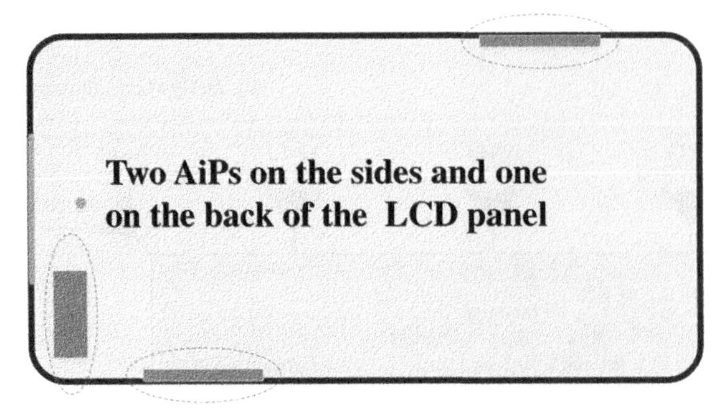

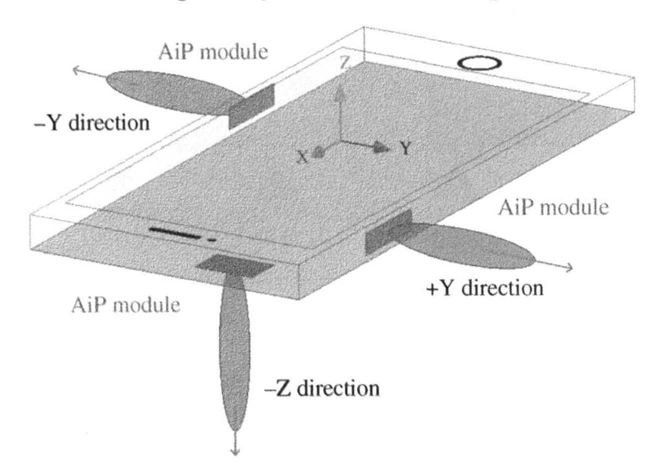

Figure 3.3 The placement of 5G mmWave AiP module (Qualcomm QTM) in Samsung Galaxy S20. *Source:* IFIXIT.

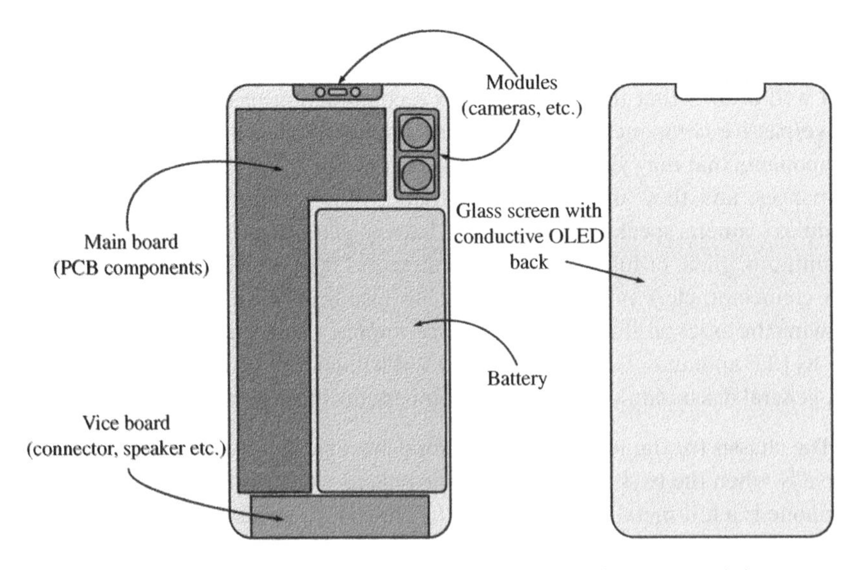

Figure 3.4 The various major components in a smartphone that may result in mitigation of 5G FR1 antenna performance.

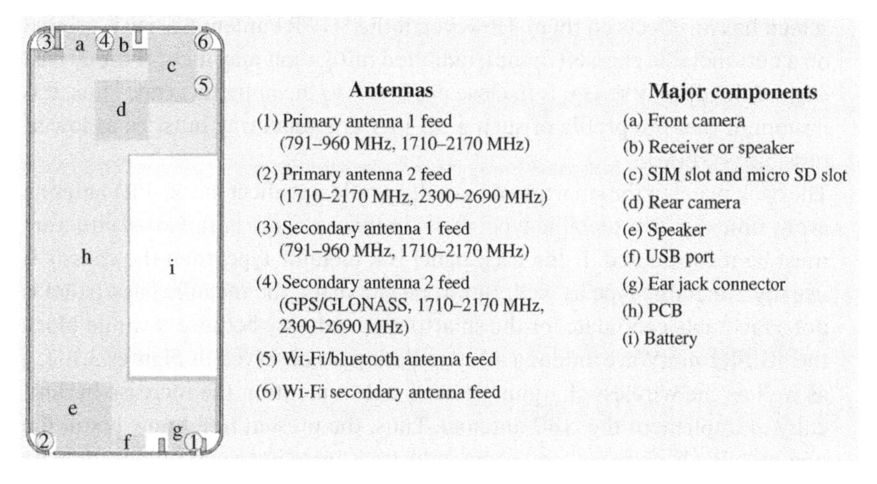

Figure 3.5 Major components' configuration and positions of antennas in a smartphone. *Source:* From [4]. ©2017 IEEE. Reproduced with permission.

antenna array design will be very different because the antenna structure cannot be designed as a PIFA, monopole, or loop structure type. Section 3.4 will fully discuss the antenna design considerations when metallic or non-metallic chassis type is applied.

2) The PCB components and shielding can be considered as a Perfect Electric Conductor (PEC) during the process of simulating the 5G FR1 antenna array by simply constructing a thin metal block above the system ground. By observing most of the reported papers, the authors can see that these reported works have ignored the PCB components and shielding if necessary.
3) The modules can seriously affect the antenna performance especially when the antennas are in close proximity to it. From the open literature, very few papers have discussed their effects on 5G FR1 antenna array, because these modules (camera, etc.) are usually placed below the back panel or placed at one corner (or very top section) of the smartphone, which are usually much closer to the 4G LTE, GPS, and WiFi antennas.
4) The position of the battery is vital to the performance of the 5G FR1 antenna array, especially when the antenna array is in close proximity to the battery, the radiation and impedance matching of the antenna array will be affected. Thus, one must first consider the positions of the battery before placing the 5G FR1 antenna array. During the process of simulating the 5G FR1 antenna array, the battery can also be considered as a thin metal block, similar to the PCB components.
5) The glass/display screen can hardly affect the 5G FR1 antenna array if the antenna structures are printed on the PCB (shared with the system ground), meaning that they may be shielded by the PCB ground, and thus the display

screen has no effects on them. However, if the 5G FR1 antenna array is printed on a non-metallic chassis (frame), radiation mitigation may incur if the curved edge of the display screen is in close proximity to the antenna array. Thus, it is imminent that the profile of such a 5G FR1 antenna array must be as low as possible (<5 mm).

6) The back panel of the smartphone usually would not affect the 5G FR1 antenna array, unless it is of metallic type, while in this case, the entire array structure must be reconsidered. If the back panel is a metallic type, then the chassis is usually a metallic type as well. But in recent years, the metallic back panel is not a favorable candidate for the smartphone industry because it would block the 5G FR2 mmWave antenna array radiation, as observed in Figures 3.1–3.3, as well as the wireless charging devices, not to mention the increase in difficulty to implement the NFC antenna. Thus, the present trend now is still the non-metallic back panel, and some industries are using a certain type of glass to replace the plastic type.
7) The connectors, such as the USB port, SIM slot, and ear jack connector, are usually small in size, and thus it is always a good practice to avoid placing the 5G FR1 antenna array in their positions. As long as these connectors are located approximately 1 cm away from the 5G FR1 antenna array structures, they should not be considered a problem.

Figure 3.6 shows the conceptual illustration of a 5G smartphone (incorporated with the aforementioned major components) that has integrated the 5G FR1

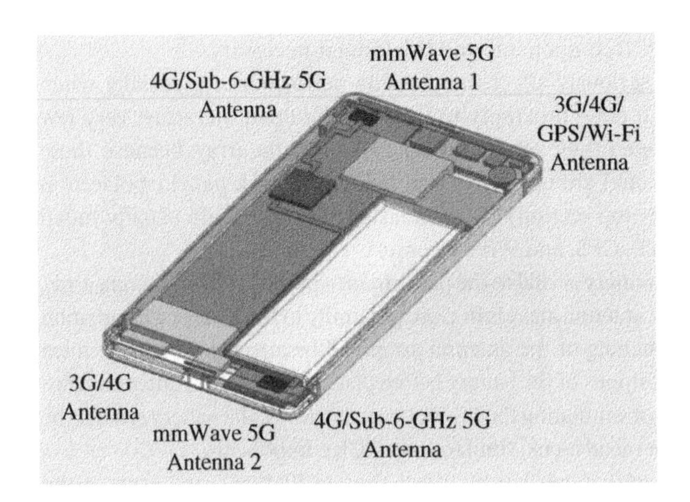

Figure 3.6 Conceptual illustration of a 5G mobile smartphone with full incorporation of 5G FR1 and 5G FR1 antennas. *Source:* From [4]. ©2017 IEEE. Reproduced with permission.

(sub-6 GHz) and FR2 antennas (mmWave) [4]. Notably, the configuration of the entire mobile device structure can be used as a reference during the design and simulation of the 5G FR1 antenna array. Here, one can see that the battery and PCB components are located on the right and left sections, respectively, and those smaller components (square-, rectangular-, and circular-shaped structures) are the modules and connectors.

One good example of showing an actual detailed structure of the components in a practical mobile phone as aforementioned can be seen in Figure 3.7. In this reported work in [5], a 4G LTE antenna was devised to integrate with a full-metallic chassis (frame and back panel are metallic) and all of the module components aforementioned. A zoomed diagram is also included in Figure 3.7 with different layers of the mobile phone. The 4G LTE antenna is configured using the metallic chassis as the main radiator and the metallic back panel acts as the antenna ground. During the design and simulation process, major components such as the camera, receiver, and other metallic components/modules or structures that occupy the finite space of the smartphone have been considered and drawn in detail, in which the proposed 4G LTE antenna-feeding mechanism is a coupling feed type and a tuner is required for tuning the low-band frequency. The effects of applying a metallic chassis to a mobile phone and the advantage of applying that will be further discussed in Section 3.4.

Even though the work in [5] has considered many practical components in the design of a 4G LTE antenna for metallic chassis, however, the simulation of larger components such as the LCD panel and battery has not been considered. As aforementioned, during the process of designing the 5G FR1 and FR2 antennas, there is a need to consider the effects of these larger components. Thus, the work reported in [6] has studied the impact of loading the LCD module and the battery. As depicted in Figure 3.8, the simulated LCD module is mainly composed of two parts, namely the LCD panel and the LCD shield. For modern smartphones, the screen-to-body ratio (updated in June 2021) has already raised to above 90%, and therefore, during the simulation process, the LCD shield and the LCD panel must have the same planar dimension with a similar thickness of approximately 1 mm. As indicated in [6], the LCD shield is typically considered as a metallic object that adheres to the system ground plane, and the LCD panel can be considered as a glass (with relative permittivity of 7, and loss tangent of 0.02) attached to the LCD shield. From the simulated results in [6], it concludes that the LCD module that is extended to the edges of the slot radiators has relatively large impacts on the antenna array because it is in close proximity to the slot radiators. Furthermore, slight resonance frequency shifts to the lower frequency spectrum are expected with minor impedance mismatching as well. These phenomena are also verified by the work reported in [7] in which a dual-loop antenna for WWAN/LTE band full-metallic chassis was loaded by major

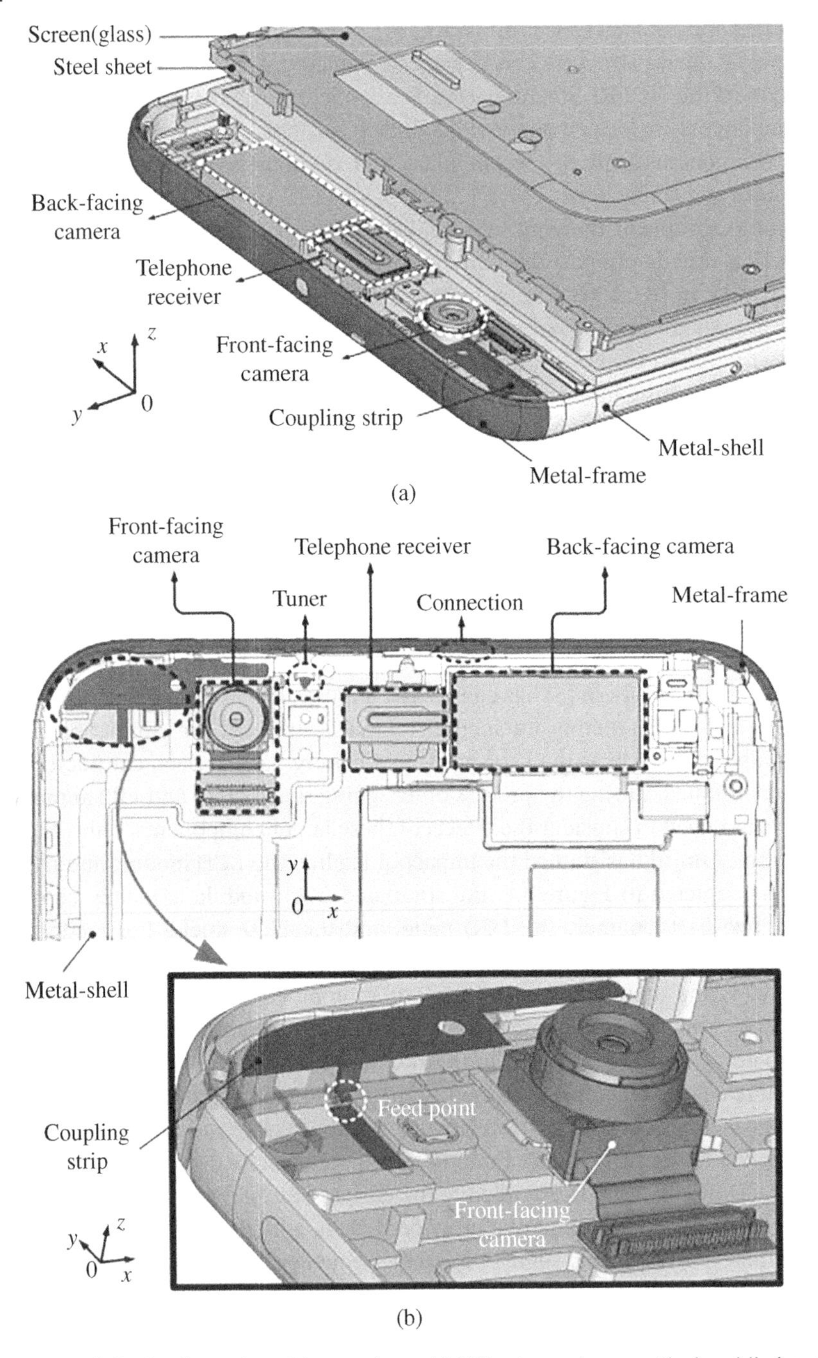

Figure 3.7 Configuration of integrating a 4G LTE antenna in a practical mobile handset/phone environment: (a) explosive view, and (b) zoomed view. *Source:* From [5]. ©2018 IEEE. Reproduced with permission.

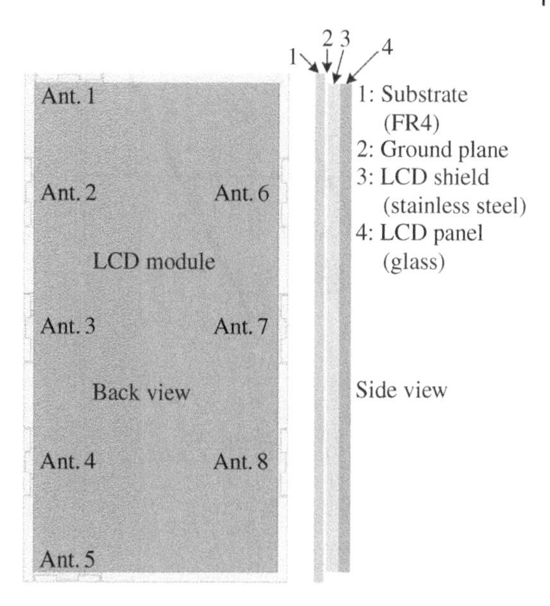

Figure 3.8 Simulation model of the 5G FR1 slot antenna array type loaded by the LCD module. *Source:* From [6]. ©2019 IEEE. Reproduced with permission.

components such as display panel, speaker, and USB connector, as depicted in Figure 3.9. In this figure, one can see that they have minor impacts as well on the reflection coefficient and total efficiency, as long as the major components are not in close proximity to the radiator.

As indicated earlier, the simulation model of the phone battery in a smartphone can be considered as a thin metal block. As shown in Figure 3.10, the antenna array (with slot antenna radiators) reported in [6] was loaded by a battery (assumed as a thin metal block) that has a planar size of $118 \times 40\,\text{mm}^2$ with a battery thickness of 4 mm. As the simulated battery model was placed on the front surface of the PCB, it must be electrically shorted to the bottom ground plane via several shorting pins that are distributed evenly. Notably, the position of the battery is not in the positions shown in Figures 3.4–3.6, and the reason for that is that if it is to be allocated to one edge of the system ground, the battery model will stack over the $50\,\Omega$ feeding line of the antenna array, not to mention it will also stack over the slot radiator, which eventually results in poor impedance matching and radiation performances. Therefore, one way to overcome this situation is to relocate the radiators, which in this case would be Ant. 3 to Ant. 5, so that the battery is not in close proximity to the radiators. Nevertheless, [6] has concluded that as long as the battery model is not near the feeding line or the radiator, negligible impacts on the antenna array are expected.

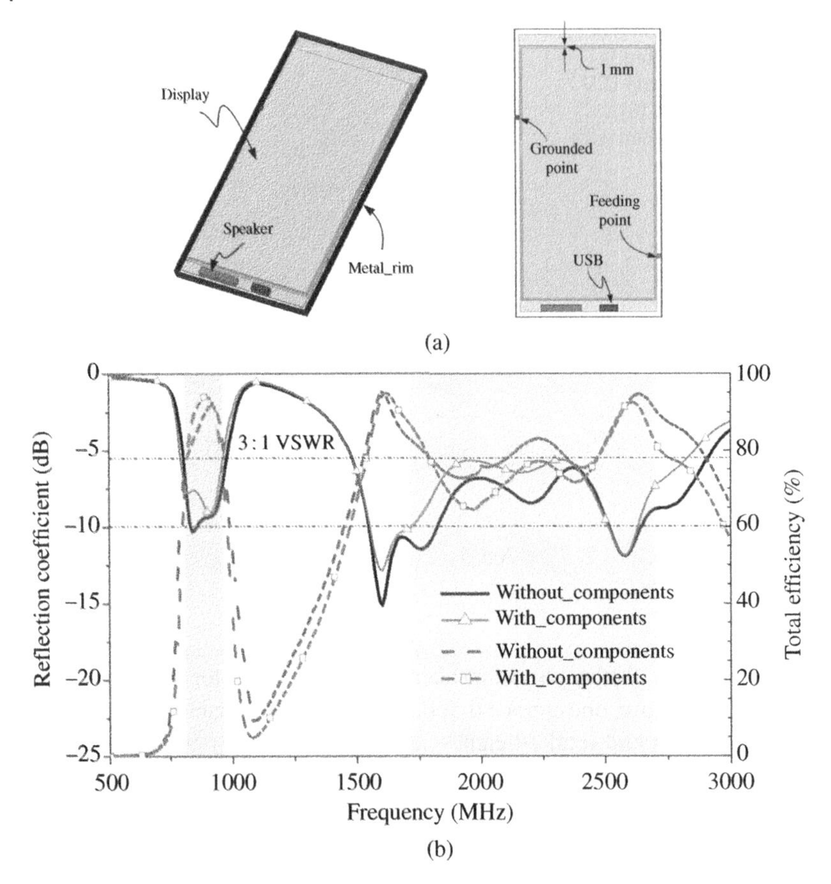

Figure 3.9 (a) Simulation model of the dual-loop array for WWAN/LTE full-metallic chassis smartphone loaded by various major components and, (b) its effects on reflection coefficient and total efficiency. *Source:* From [7]. ©2016 Wiley. Reprinted with permission.

3.2 Antenna Element Design and Topologies

Since the announcement of C-band (3400–3600 MHz) for mobile broadband applications in the WRC-2015, the antenna element design methods and techniques that have been applied for this band or working within the 5G NR band 78 (3300–3800 MHz) can be segregated by the following categories: slot antenna, monopole antenna, loop antenna, and inverted-F antenna. One important point to take note of is that the implemented antenna design must be fitted into the smartphone casing or its corresponding system ground.

3.2.1 Slot Antenna Design

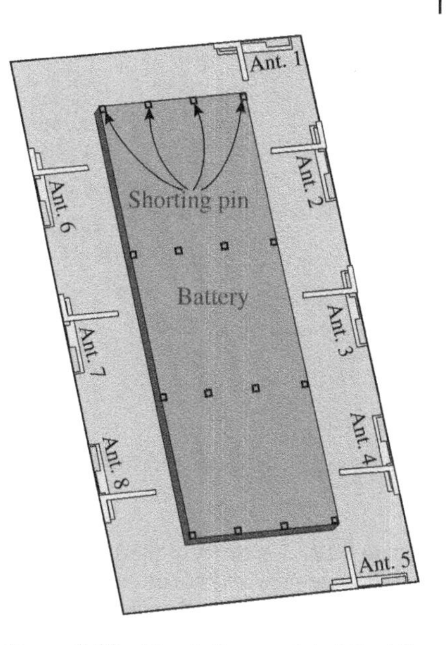

Figure 3.10 Simulation model of the 5G FR1 slot antenna array type loaded by the battery. *Source:* From [6]. ©2019 IEEE. Reproduced with permission.

One of the earliest antenna design types that are selected for realizing the 5G smartphone applications is the slot antenna design. The reason for that is due to its simplicity in exciting an open slot resonance mode (which is quarter-wavelength long), and thus the physical slot size of the slot antenna can be greatly reduced [8–11]. Even though such a design can easily achieve a broad operating bandwidth, the downside of applying the slot design is the need to embed/integrate the slot into the smartphone system ground, meaning that much of the two longer side edges of the system ground must be set aside for the slot antenna array. Figure 3.11 shows the design of the single open-end slot element design reported in [11], which was fed by a simple L-shaped feeding strip, and the simulated (surface current and electric field) distributions in this figure have clearly demonstrated that it is exciting an open slot resonance mode. Besides applying a single open-end slot method, the work reported in [6] has introduced a dual open-ended slot design (which is also known as a balanced slot) for 10-antenna MIMO array smartphone applications, as depicted in Figure 3.12, and its corresponding figures with loaded LCD and battery are shown in Figures 3.8 and 3.10, respectively. By observing

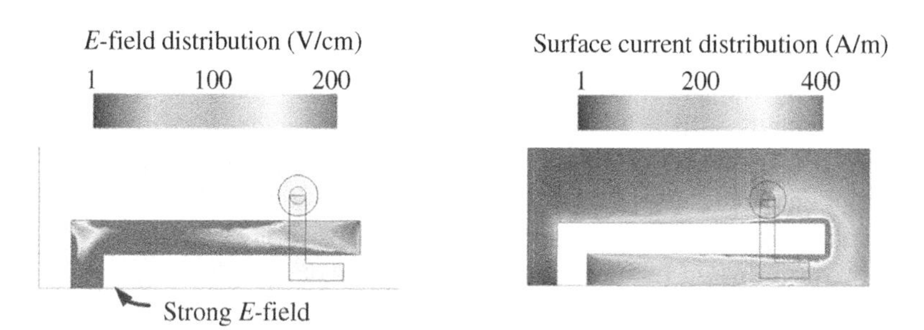

Figure 3.11 Simulated surface current and electric field distributions of the L-shaped open slot antenna. *Source:* From [11]. ©2016 IEEE. Reproduced with permission.

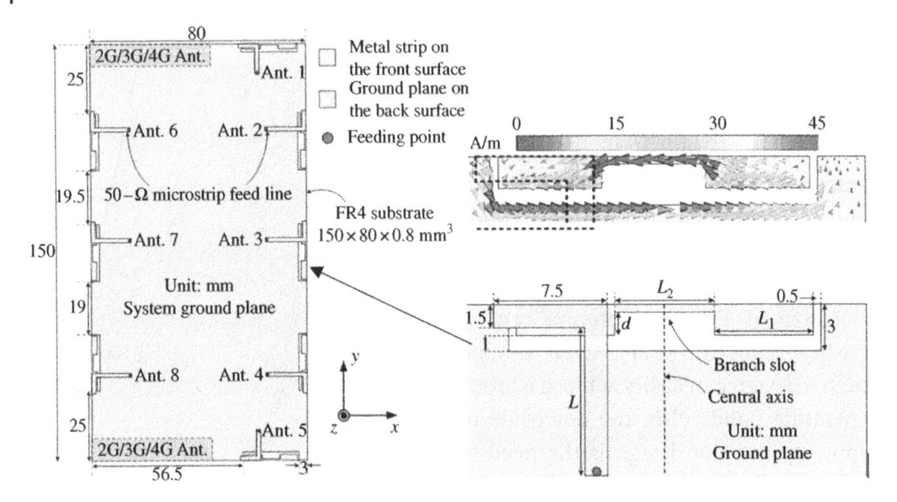

Figure 3.12 Geometry of the 10-antenna MIMO array with dual open-ended slot antenna element and its corresponding current distributions. *Source:* From [6]. ©2019 IEEE. Reproduced with permission.

Figure 3.12, one can see that the peak currents are located at the middle position of the slot, while two nulls are demonstrated at both open ends. Thus, the slot design is a half-wavelength balanced slot mode excited at only 3.5 GHz.

To allow the slot designs depicted in Figures 3.11 and 3.12 to excite two resonance modes at 3.5 and 5.5 GHz (instead of only one mode), a modified open slot antenna design was implemented in [12] for 10-antenna array sub-6 GHz MIMO applications in 5G smartphones. As shown in Figure 3.13, the open slot section is very narrow and the feeding element is placed near the edge and open slot positions. By doing so, two distinct half-wavelength E-field distributions are excited separately on the left and right sections of the slot.

3.2.2 Monopole Antenna Design

Besides applying the slot antenna design as introduced in sub-section 3.2.1, another easy method to achieving a single resonance mode (with quarter-wavelength) and compact size for 5G FR1 smartphone applications is the implementation of monopole antenna design. A hybrid antenna type that combines the slot antenna and monopole antenna has been reported in [13]. As depicted in Figure 3.14, the proposed 12-antenna array is composed of 10 typical single open-ended slot antenna types and two dual-mode monopole slot antenna types. By further observing the current distribution and S-parameters of this figure, it is clear that the two monopole strips, strip 1 and strip 2, are responsible for the excitation of the two resonance modes at 3.5 and 5.5 GHz, respectively.

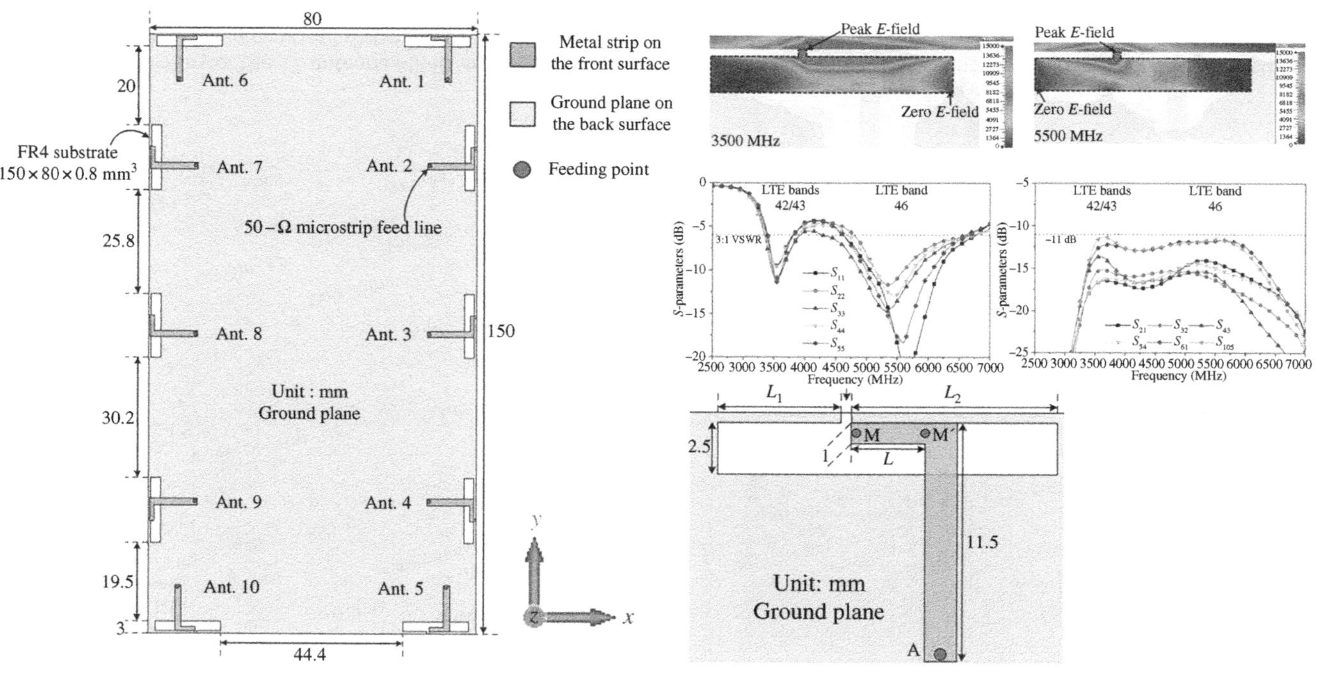

Figure 3.13 Geometry of the 10-antenna MIMO array with modified single open-ended slot antenna element. Simulated peak *E*-field distributions and measured *S*-parameters are included. *Source:* From [12]. ©2018 IEEE. Reproduced with permission.

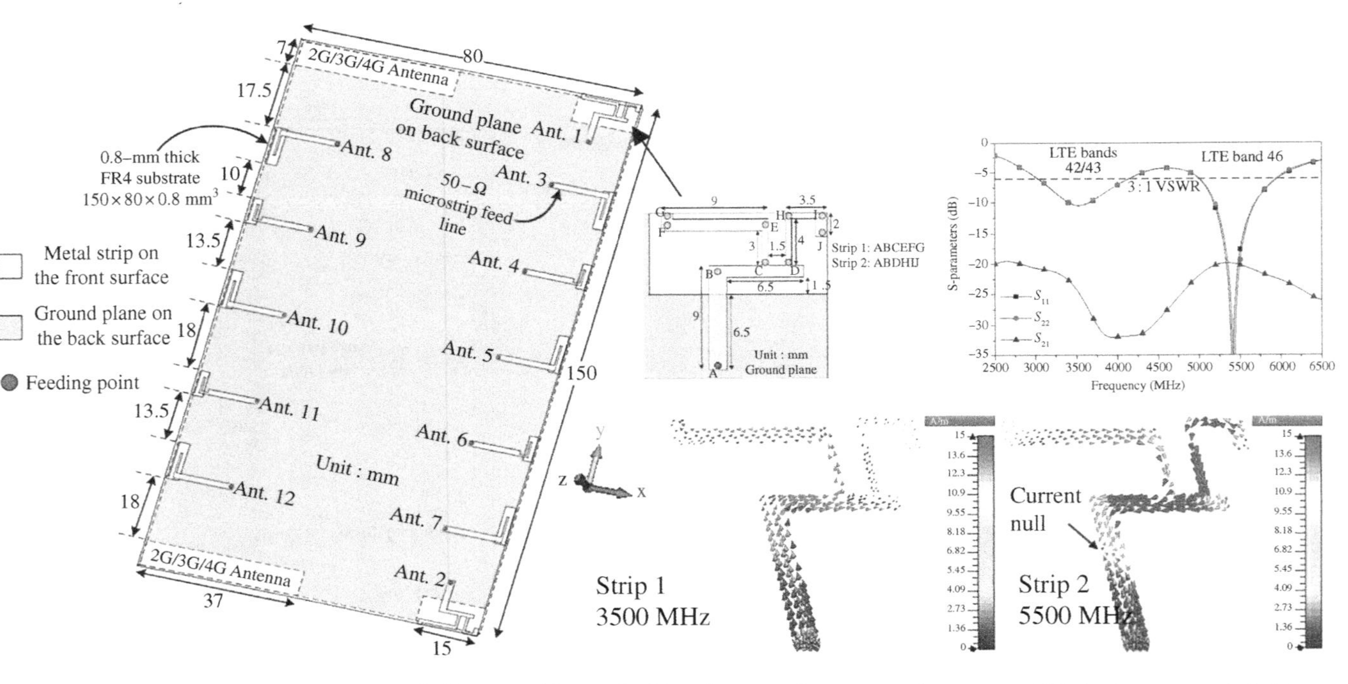

Figure 3.14 Geometry of the 12-antenna MIMO array with two dual-band monopole antenna elements and 10 single open-ended slot antenna elements. Simulated current distributions on the two monopole strips and measured S-parameters are included. *Source:* From [13]. ©2017 IEEE. Reproduced with permission.

Besides applying the above standard/typical quarter-wavelength resonance mode monopole design type, two unique designs have also been introduced for 5G smartphone applications. The first unique design is the coupled-fed strip, which is a modification of a direct-fed monopole, as depicted in Figure 3.15 [11]. Here, if a direct-fed mechanism is applied to the C-shaped strip, it will yield a quarter-wavelength resonance mode, however, in this case, because the feed mechanism applied is a coupled-fed type, the C-shaped strip has yielded a half-wavelength resonance mode instead (which can be considered as a coupled dipole). The second unique technique is to allow the monopole structure design to excite at higher-order resonance mode, as shown in Figure 3.16 [14]. In this figure, the fundamental mode is at 1.75 GHz, and thus the higher-order resonance mode with a current null shown in the current distribution diagram is 3.5 GHz. Other monopole structures for the excitation of this 3.5 GHz resonance for 5G smartphone applications are also discussed in [15, 16], and the coupled dipole was also introduced in [17].

3.2.3 Loop Antenna

A conventional loop antenna (of any shape such as circular loop, square loop, or folded-dipole loop) can excite a full wavelength resonance mode with its maximum current (I_{max}) location typically distributed at the upper section and bottom section (feeding point), while its corresponding current null (I_{null}) or voltage maximum point is located on the two sides, as depicted in Figure 3.17.

To allow this loop antenna design to be realized in a finite limited spacing smartphone, a good practice is to cut the loop antenna into halve across the two I_{null} points and shorted them into the system ground of the smartphone, as shown in Figure 3.17. By doing so, the previous two I_{null} points will be reversed and become I_{max}, whereas the previous I_{max} point at the top section will become I_{null}, which results in exciting a resonant mode analogous to a half-wavelength loop mode. This easy technique was first applied to the 5G smartphone in [18], in which a modified folded dipole loop design was fed by a T-shaped coupled-fed element, and it has successfully excited a 3.5 GHz resonance mode (half-wavelength loop mode). In this case, the I_{null} point section of the coupled-fed modified folded dipole loop can be removed because it is a null point and will not have any effects on the antenna performance. Nevertheless, to reduce the length of the coupled-fed modified folded dipole loop from $L = 24.8$ mm to $L = 17.4$ mm, a further modification was applied to [18], in which two stubs protruded from the antenna structure, as depicted in Figure 3.18 [19]. The antenna structure of [18] was also included in this figure (with current distribution), and one can see that the stubs can successfully reduce the total length L without disturbing the current null (I_{null}) and current maximum (I_{max}) positions. Figure 3.19 shows the configuration of the eight-antenna MIMO array of [19] positioned along the two longer side edges of the system ground of a 5G smartphone. Thereafter, many other coupled-fed loop antenna design structures have been reported as well in [20–25].

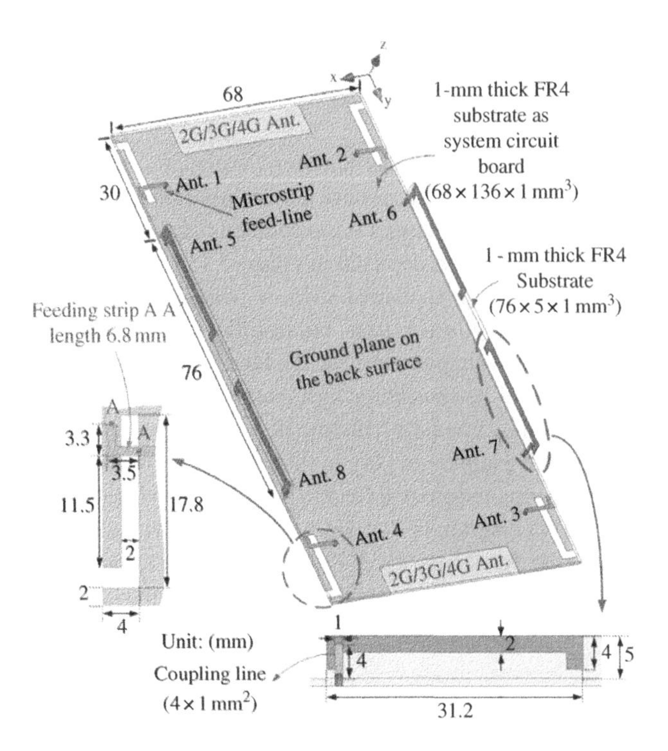

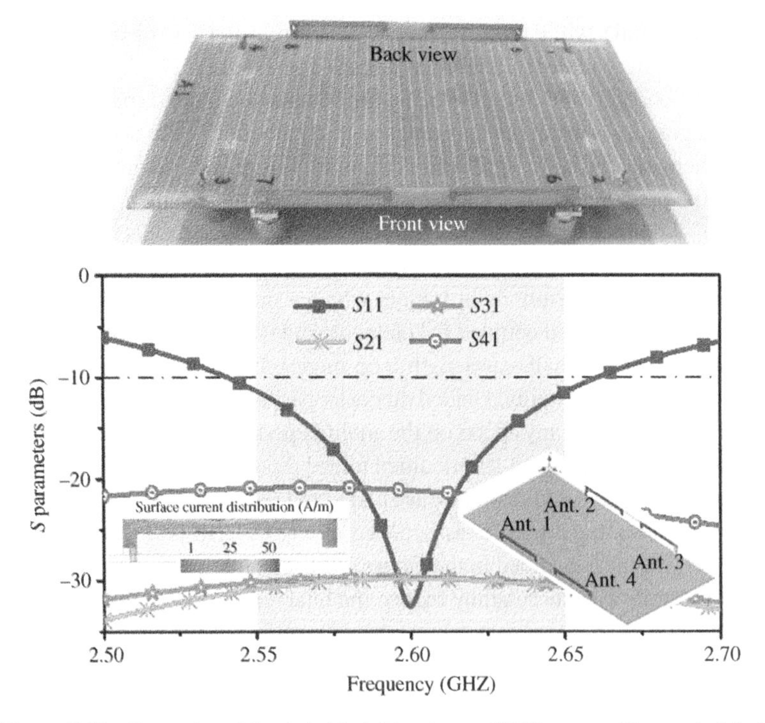

Figure 3.15 Geometry of the hybrid eight-antenna MIMO array with coupled-fed strip antenna elements and open-ended slot antenna elements. Simulated current distributions of the coupled-fed strip and S-parameters are included. *Source:* From [11]. ©2016 IEEE. Reproduced with permission.

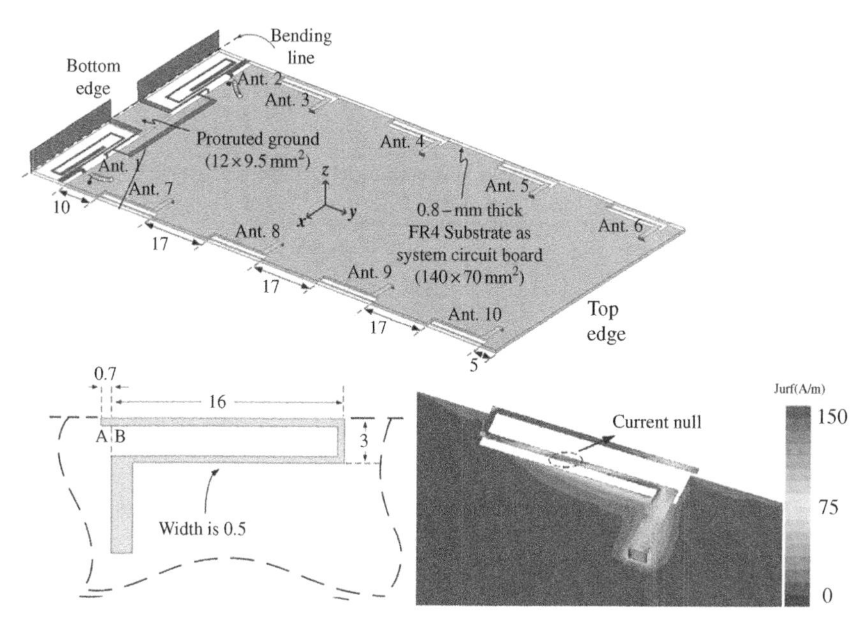

Figure 3.16 The layout of the eight-antenna MIMO array, including the geometry of the monopole antenna element and its simulated current distributions at higher-order resonances. *Source:* From [14]. ©2016 IEEE. Reproduced with permission.

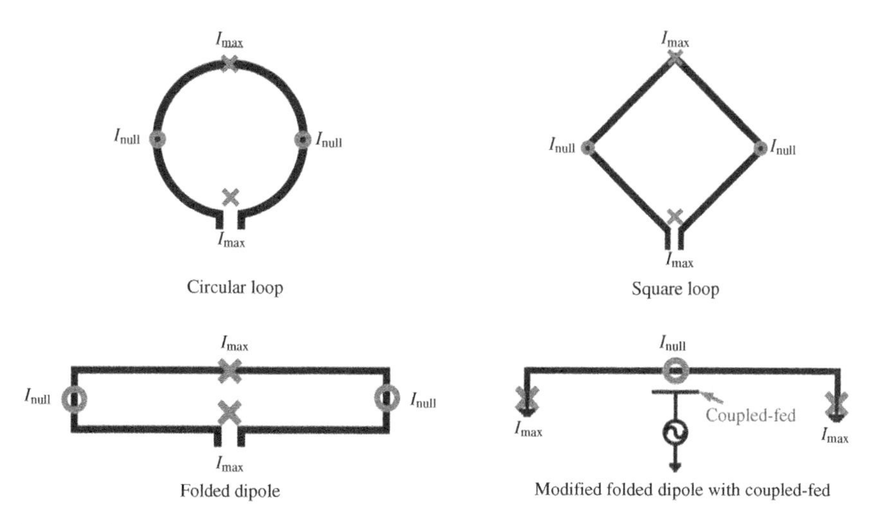

Figure 3.17 Configuration of the three typical types of loop antennas, and the modified folded dipole loop antenna with coupled-fed structure.

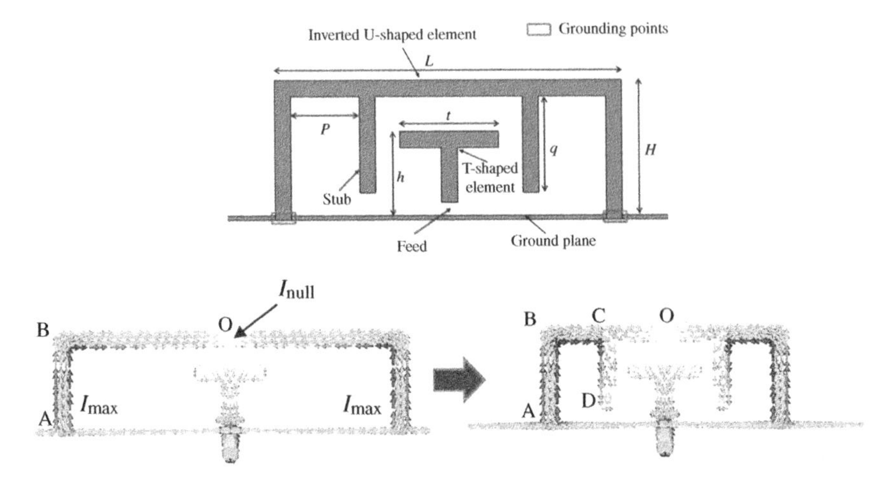

Figure 3.18 Configuration of the modified folded dipole loop antenna, and its corresponding current distribution diagrams from [18, 19]. *Source:* From [19]. ©2019 IEEE. Reproduced with permission.

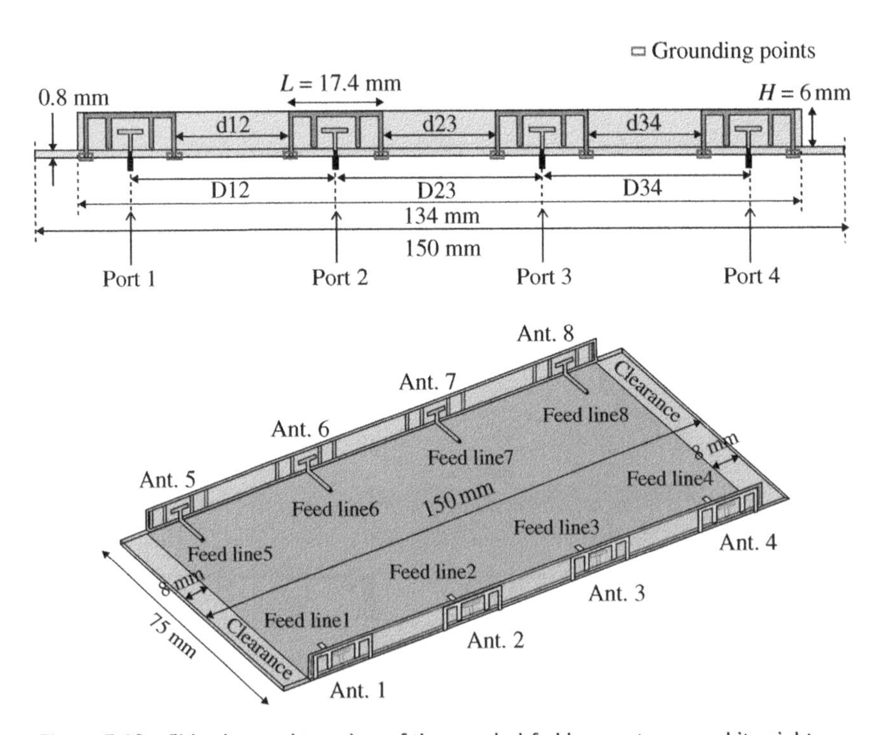

Figure 3.19 Side view and top view of the coupled-fed loop antenna and its eight-antenna MIMO array configuration on the frame of 5G smartphone. *Source:* From [19]. ©2019 IEEE. Reproduced with permission.

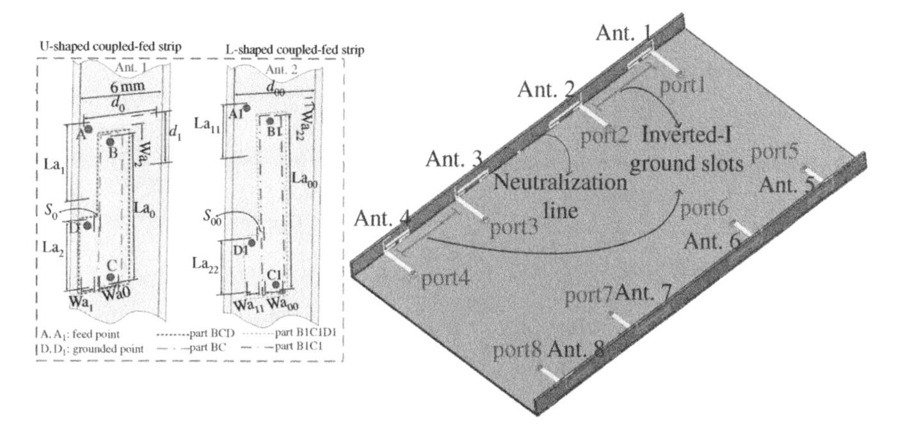

Figure 3.20 Configuration of two coupled-fed loop antenna designs with dissimilar coupled-fed strips (Ant1: U-shaped type, Ant2: L-shaped type), and the eight-antenna MIMO array placement on the fame of 5G smartphone. *Source:* From [25]. ©2019 IEEE. Reproduced with permission.

Amid these 5G FR1 antenna designs with single-band operation in the 3.5 GHz that is related to a loop structure, one needs to take notice that [25] have proposed two types of four-antenna arrays with dissimilar coupled-fed loop elements, namely U-shaped and L-shaped, which are symmetrically distributed in the inner surface of the smartphone frame, as depicted in Figure 3.20. Interestingly, instead of using a different coupled feeding structure for the excitation of the loop antenna as introduced in [19, 25] (T-shaped coupled-fed, U-shaped coupled-fed, and L-shaped coupled-fed), the work reported in [16] has introduced a parasitic gap-coupled open-loop branch structure for the excitation of the 3.5 GHz mode, which is coupled-fed by the main radiator (folded monopole), as shown in Figure 3.21. From the current distribution and reflection coefficient diagrams in Figure 3.21, one can see that the main radiator has excited the 4.9 GHz band, and the parasitic gap-coupled loop branch has excited the 3.5 GHz band, which can cover the assigned 3.3–3.6 and 4.8–5.0 GHz bands for China's 5G sub-6 GHz band.

3.2.4 Inverted-F Antenna

The inverted-F antenna (IFA) design is a very simple quarter-wavelength antenna that has been applied in the 2G, 3G, and 4G mobile phones. The IFA structure printed on a planar substrate is depicted in Figure 3.22, and in this case, it is known as the Planar Inverted-F antenna (or PIFA). In this figure, as long as one end of the structure is "open" and the other end is "shorted" to the ground, and it realizes a quarter-wavelength current distribution, a typical PIFA is formed. As this antenna design topology is very simple and usually excite a single resonance

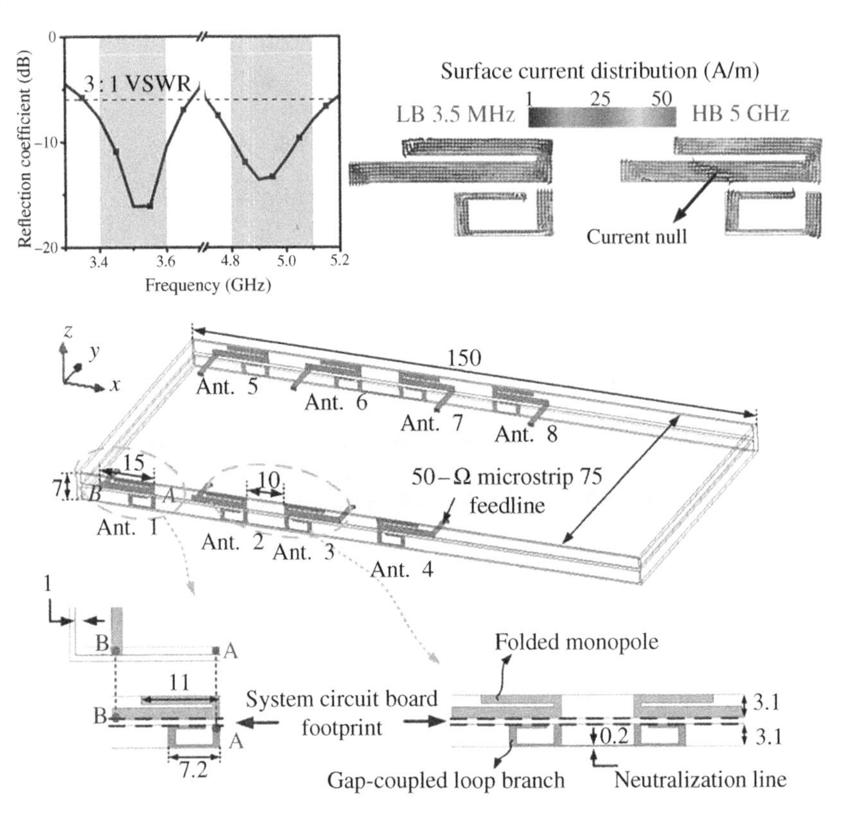

Figure 3.21 Configuration of parasitic gap-coupled loop branch antenna coupled-fed by the main radiator and its corresponding eight-antenna MIMO array placement on the frame of 5G smartphone. The current distribution of the antenna at the low-band (3.5 GHz) and high-band (5.0 GHz), and its corresponding simulated reflection for Ant1. *Source:* From [16]. ©2018 IEEE. Reproduced with permission.

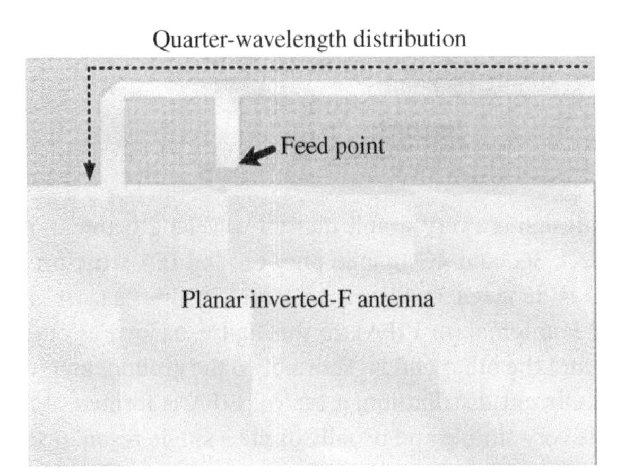

Figure 3.22 Configuration of a planar inverted-F antenna printed on a substrate with conventional quarter-wavelength current distribution.

mode, this sub-section will not further introduce the PIFA or IFA design for 5G smartphone, but one can see the works reported in [26–28].

3.3 Antenna-Feeding Mechanism and Impedance Matching

This section is to summarize the various feeding mechanisms that have been applied in the open literatures and those reported in [2–28]. So far, if one needs to design a 5G FR1 antenna, a couple of antenna design structures can be applied, and each design structure may require a different feeding mechanism or tuning method for achieving optimum impedance matching. Figure 3.23 summarizes most of the feeding mechanisms that we have observed so far for the 5G FR1 antenna element. The reference number is indicated for each feeding method as well. Here, the conventional method is the coupled-fed [11] and direct-fed method [14], as shown in Figure 3.23a,b, respectively, and the coupled-fed method is mostly implemented for 5G FR1 antenna

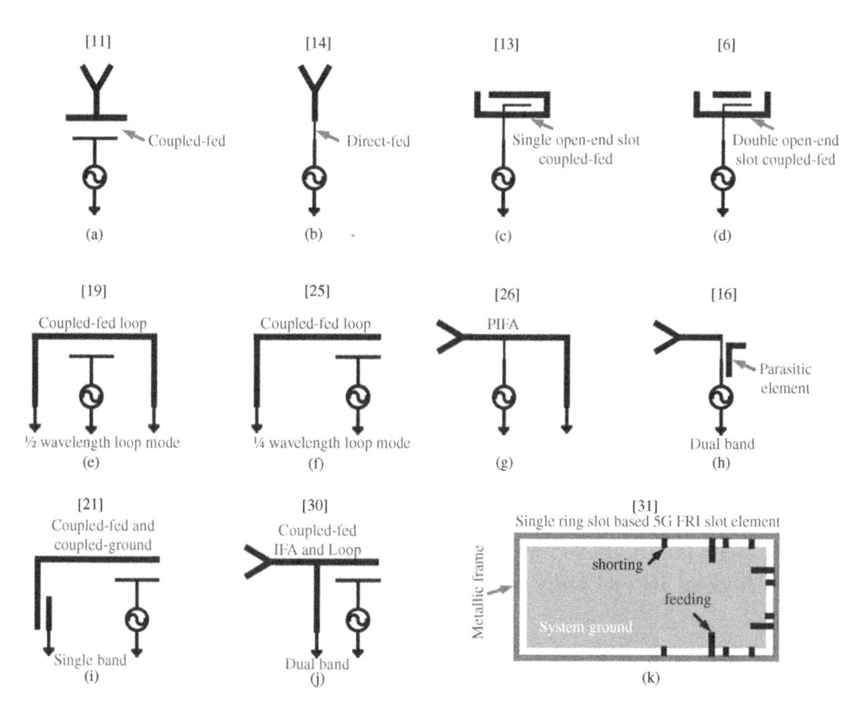

Figure 3.23 Different feeding mechanisms for 5G FR1 antenna elements summarized from the open literature. Reference numbers are also included for each feeding design. (a) Coupled-fed, (b) Direct-fed, (c) Single open-end slot coupled-fed, (d) Double open-end slot coupled-fed, (e) Coupled-fed loop (1/2 wavelength loop mode), (f) Coupled-fed loop (1/4 wavelength loop mode), (g) PIFA or IFA, (h) Parasitic element (for dual band), (i) Coupled-fed and coupled-ground (for single band), (j) Coupled-fed IFA and loop (for dual band), and (k) Single ring slot based 5G FR1 slot element (for metallic frame).

element as it can yield larger impedance bandwidth. For the slot antenna design method, as depicted in Figure 3.23c,d, the single open-end slot coupled-fed method [13] is the most highly applied feeding mechanism because it can yield compact size (due to quarter-wavelength slot mode), whereas the double open-end slot coupled-fed method [6] is least preferred (due to half-wavelength slot mode), but it can achieve higher isolation between adjacent slot antenna elements. Notably, such double open-end slot design [6] can also be extended to a metallic frame (with open slot), and it is further introduced in Section 3.4. Next, there are two types of coupled-fed feeding techniques applied to loop antenna design (sometimes known as a gap-coupled loop), as shown in Figures 3.23e,f. The first feeding type is to position the coupled feed at exactly the middle section of the loop, in which the advantage is to excite a self-decoupled balance mode that can yield better isolation between adjacent elements [18, 19]. The second feeding type is to position the coupled feed at exactly one end of the loop, and it is worth noting that the coupled-fed structure (usually a C-shaped or L-shaped type) must be carefully selected for better impedance matching [25, 29]. Figure 3.23g is a typical IFA/PIFA type [26–28], and Figure 3.23h is showing the method of loading a parasitic element for exciting an additional resonance for exciting the 3.5 GHz band [16]. Next, Figure 3.23i shows an unusual method of implementing coupled-fed and coupled-ground method, in which this method reported in [21] has shared the same coupled ground with the adjacent antenna, which results in good isolation at the 3.5 GHz band. The coupled-fed IFA and loop antenna feeding method as shown in Figure 3.23j was reported in [30], and it is for exciting two different resonance modes, namely IFA and loop mode so that dual operating bands can be achieved. Such a design method can also be modified as a two-antenna type decoupling building block that will be discussed in the later chapter. Last but not least is the slot-coupled feeding technique applied for a full-metallic frame (or rimmed) 5G smartphone [31], as shown in Figure 3.23k. Here, multiple shorting points between the metallic frame and system ground are loaded for creating the four slots that can excite the 3.5 GHz band with desirable isolation between adjacent slot elements. Such a design method will be discussed further in Section 4.3.1.

The following points are vital to the antenna engineer when simulating or fabricating the devised 5G FR1 antenna for smartphone applications:

1) It is always a good practice to include the SMA connector during the simulation process, as it may work as an extended ground. Figure 3.18 is a good example, as the simulated current distribution diagram has exhibited that part of the ground current may have flowed toward the SMA ground.
2) The soldering of the 50 Ω SMA connector ground to the system ground must be firm and smooth without any gaps between them, as shown in Figure 3.24.

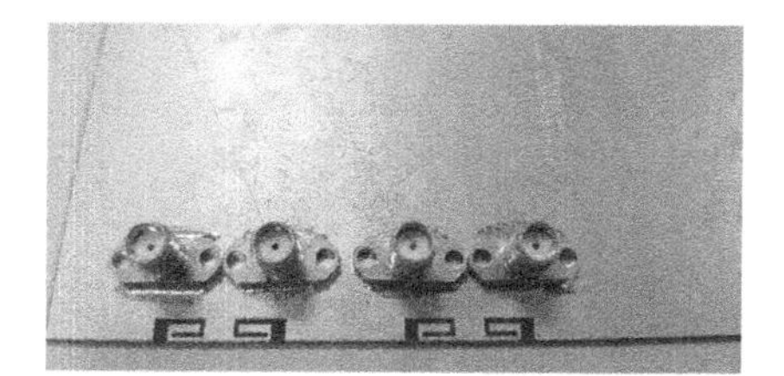

Figure 3.24 50 Ω SMA and coaxial cable feeding for 5G FR1 antenna elements.

3) If a coaxial cable must be used for the measurement of the 5G FR1 antenna element in the actual fabrication process, it is crucial to keep the coaxial cable as short as possible, as shown in Figure 3.24. If a longer coaxial cable is required, soldering the coaxial cable ground to the system ground is always a good practice.
4) For single or double open-end slot coupled-fed design, the use of an L-shaped strip line is crucial, as the length of the L-shaped strip (especially the horizontal section of the strip) depicted in Figures 3.23c,d are usually tuned for achieving good impedance matching.
5) For the very wideband 5G FR1 slot antenna element applied to the metallic frame, the use of tuning stub [32, 33], or quarter-wavelength impedance transformer [34] will be a better choice.

3.4 Chassis Consideration and Effects

Before the use of a modern smartphone, the chassis (or sometimes known as the frame or rim) of an older mobile phone is only meant for holding the screen display, keypad, and other major components, and it was not even applied as an antenna radiator. It was not until the emerging of the smartphone that the metallic chassis (using aluminum or stainless steel) was designed as part of the antenna radiator for 3G and 4G bands. Basically, there are two major types of chassis for modern 4G/5G smartphones, namely metallic chassis and non-metallic chassis. If the smartphone has applied a non-metallic chassis (e.g. plastic) instead of a metallic type, the results can be observed in the extended results of [12], as shown in Figure 3.25. Here, the 5G FR1 slot antenna array was surrounded by a 1 mm-thick plastic frame (with relative permittivity of 3 and loss tangent of 0.01) that has a height of 6 mm. The simulated results have concluded that due to the dielectric loading of the plastic chassis, the resonant modes (S_{11} to S_{55}) of the slot array have shifted to the lower frequency spectrum. Notably, only minor effects were observed in other performances such as isolation, gain, and efficiency.

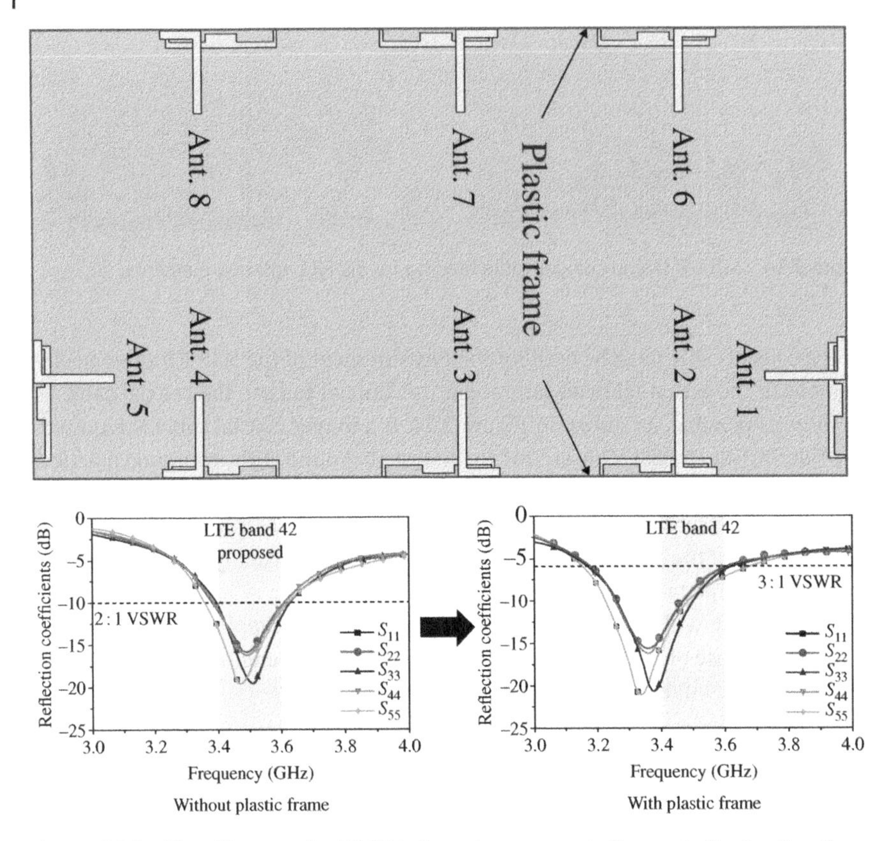

Figure 3.25 The effects on the 5G FR1 slot antenna array before and after loading the plastic chassis (or plastic frame). *Source:* Extended results of [12].

The effects of the metallic chassis on the antenna performance on the other hand need much careful attention because of the infamous "Antennagate" scandal in July 2010, when the new Apple iPhone 4's antenna encountered reception problems for the first time when the user's hand was in contact with certain locations of the chassis. Figure 3.26 shows the iPhone 4 smartphone and its conceptual antenna design within the metallic chassis. The problem with the metallic chassis that was used to excite both the Bluetooth/WiFi/GPS and 3G (UMTS/GSM) operational bands lies within the left gap, in which the touching (covering) of this left gap using the user's finger or palm will result in poor impedance matching of the two operational bandwidths, followed by poor reception [35]. Notably, this problem can be resolved by applying the full-metallic chassis (without any opening slot in the chassis) antenna design that has been reported in [7], as shown in Figure 3.9, requiring only a feeding line and a shorting point embedded into the metallic chassis (via the system ground) for exciting the bands of interest.

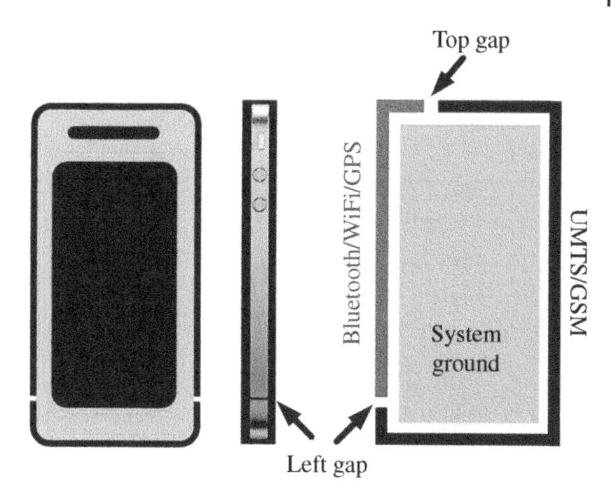

Figure 3.26 Apple iPhone 4 and its metallic chassis antenna design. *Source:* Apple Inc.

Figure 3.27 further shows the top section of the work reported in [5] with the metallic chassis and various components, and its corresponding metallic back panel (with an inverted U-shaped gap) is a modern type presently applied by the antenna industry. Notably, the inverted U-shaped gap is usually 2-mm wide (for good antenna radiation), which, in turn, separates the metallic chassis and the metallic back panel shell that works as the metal clearance for the devised antenna. One point to be taken is the connector at point B as it should be wide enough ($W_2 = 2\,\text{mm}$) to achieve a better separation between the devised antenna and the top-left corner of the handset, as the latter will generally be used to insert the Global Positioning System (GPS) antenna. During the design process in the actual mobile handset environment, it is also noteworthy that the gap distance ($G_6 = 4.0\,\text{mm}$) between the coupling strip of the devised 4G LTE antenna and the left corner metallic chassis is a crucial parameter for achieving better antenna efficiency. Although a typical gap distance of $G_1 = 1.5\,\text{mm}$ (between the coupling strip and the metallic chassis) is usually applied because of the irregular shape of the coupling strip, at a particular location along the gap, it could be narrowed down to $G_2 = 0.9\,\text{mm}$. In summary, parameters G_1 and G_2 will highly affect the input impedance at the feed point, and thus a typical practice is to introduce an additional capacitance at the feeding point.

3.5 Electromagnetic Exposure and Mitigation

The Radio Frequency (RF) Electromagnetic Field (EMF) exposure for antenna arrays intended for User Equipment (UE) such as smartphone and radio Base Stations (BS) in the 5G mobile communication systems have been a topic of interest, especially for frequency between 10 and 60 GHz [36]. However, in this section, we are more focused on the antenna performance mitigation when the 5G antenna

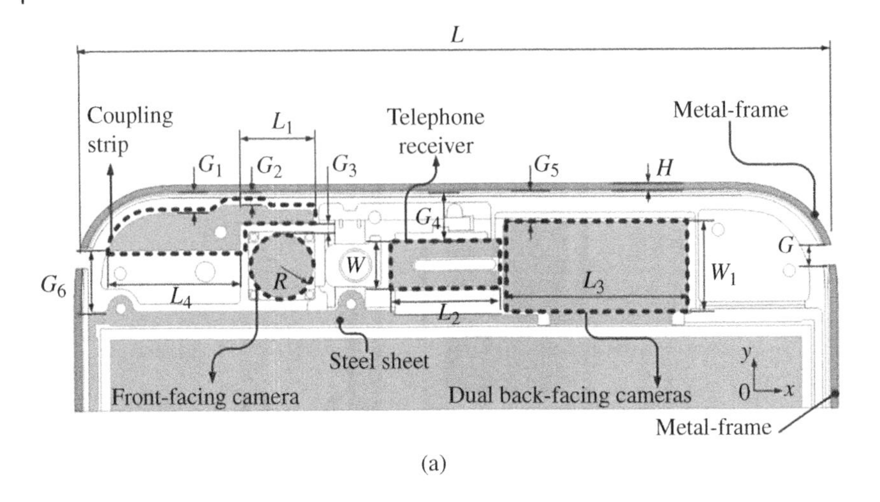

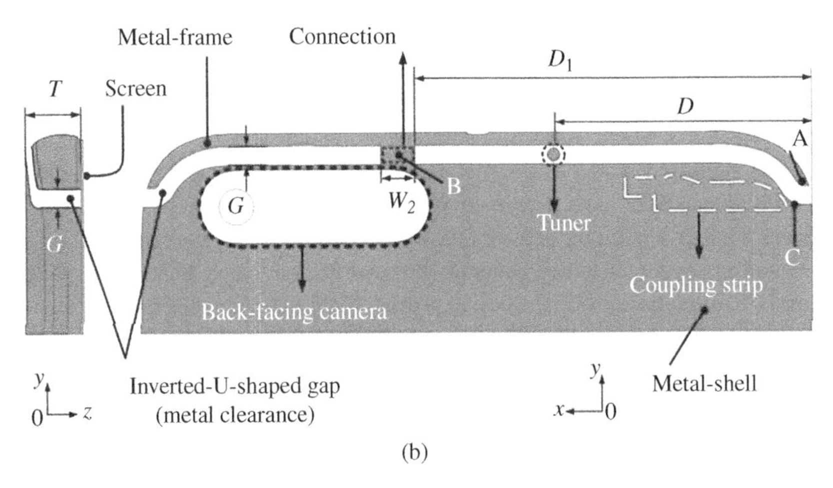

Figure 3.27 Geometry of a practical mobile phone integrated with a 4G LTE antenna with all the necessary components: (a) top section view with metallic chassis and various components, and (b) top and side view of the metallic back panel (with an inverted U-shaped gap). *Source:* From [5]. © 2018 IEEE. Reproduced with permission.

array is placed in the user's hand, rather than studying the EMF exposure, such as electric field strength and Specific Absorption Rate (SAR) distribution in the head by the 5G antenna [37, 38]. The reason for not studying the EMF head model exposed to the 5G antenna array is because there is a misconception that the 5G communication is the same as its 4G counterpart, in which the users will be using the 5G communication system even when it is performing the "Talk mode." Therefore, the authors have to remind the readers that the 5G communication system is mainly used for very large data throughout the transmission of up to gigabytes (>10 Gps) when requested by the UE operating in the 5G "Data mode". If the "Talk mode" is required by the UE at the same time, it is usually the 4G communication system that is stepping in. Therefore, it is unnecessary to study the EMF exposure to the head by the 5G antenna array.

The hand blockage on the EMF radiated by the 5G smartphone is an important issue, as the hand-blocking coverage region may exceed around 50% of the original spherical coverage [39]. Figure 3.28 shows the spherical coverage of a single antenna module and a three-antenna module, and the comparison between them has proved that the three-antenna module has allowed better spherical coverage of 78% by placing the antennas on the right, bottom, and top sections of the smartphone. Even when it is blocked by the user's hand, 60% spherical coverage is still achieved, whereas the single antenna module type has realized only 18% spherical coverage when blocked by the user's hands.

To further comprehend the effects of the user's hand, one can simulate the effects of placing the 5G FR1 antenna array in the user's hand, by referring to Table 3.1 that gives the user's hand permittivity and conductivity of the user's hand at various frequencies (from 3 to 6 GHz). The following Sections 3.5.1–3.5.3 will discuss the effects of a user's hands in single-hand mode (SHM) and dual-hand mode (DHM) when different conventional 5G FR1 antenna array types (monopole, slot, and loop) are applied. From the simulated results, it can be concluded that the user's hand does not have much effects on the isolation between

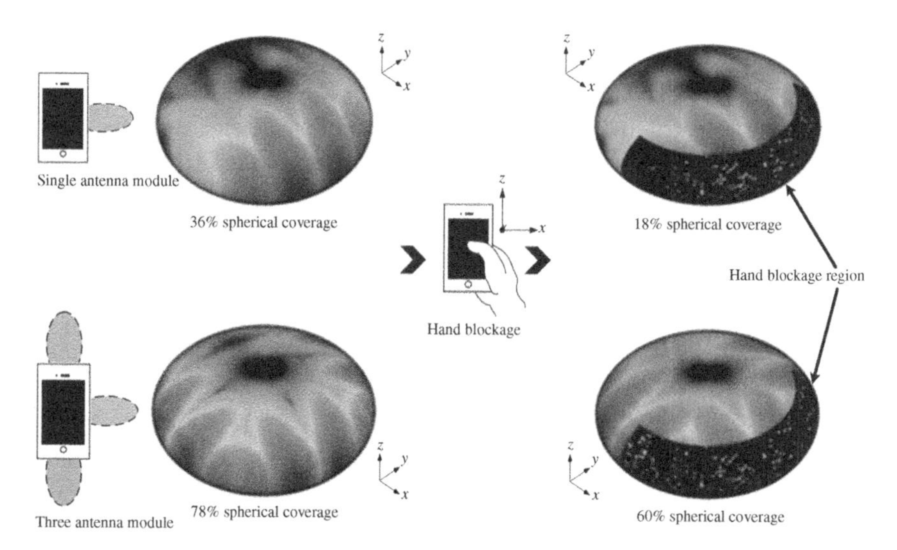

Figure 3.28 Spherical coverage of single antenna module and multiple antenna module. *Source:* From [39]. Qualcomm.

Table 3.1 User's hand parameters in simulation for various frequencies.

Frequency (GHz)	Permittivity	Electrical Conductivity (S/m)
3	24.8	1.61
4	23.5	2.18
5	22.2	2.84
5.2	22.0	2.98
5.4	21.7	3.11
5.6	21.4	3.25
5.8	21.2	3.38
6	20.9	3.52

adjacent elements, due to the fact that the user's hands/fingers are absorbing the EM wave. Thus, less radiation is coupled to the adjacent antenna elements. However, the slot antenna array type has shown much lower total efficiencies (approximately 10%) when the slot element is in close contact with the user's hands (or fingers), and the loop element is slightly above 10%, whereas the monopole element has better total efficiencies of 18%, across the lower C-band (3300–3600 MHz) of the 5G FR1 band.

3.5.1 Monopole Antenna Array User's Hand Effects

An eight-antenna array wideband antenna (formed by eight monopole antenna elements) for 5G FR1 smartphone applications has been reported [40], and its corresponding simulated antenna efficiencies for all antenna elements in the SHM and DHM are shown in Figure 3.29. Here, one can see that in SHM, Ant. 6 and Ant. 7 are in close contact with the user's hand, and thus their simulated antenna efficiencies are approximately 20% (in the 3.5 GHz band). Similar phenomena were also observed for the DHM case, as the two hands are blocking the monopole antenna element at four corners of the smartphone, resulting in mitigation of antenna efficiencies to approximately 17–34% for Ant. 1, Ant. 4, Ant. 5, and Ant. 8.

3.5.2 Slot Antenna Array User's Hand Effects

A more thorough investigation of the eight-antenna array (formed by eight printed double open-end slot antenna elements) in the SHM and DHM scenarios has also been studied [6]. As shown in Figure 3.30, because Ant. 2, Ant. 6, and Ant. 7 are in close contact with the user's fingers in the SHM, the reflection coefficients of these three open slot elements are affected, in which Ant. 6 has demonstrated

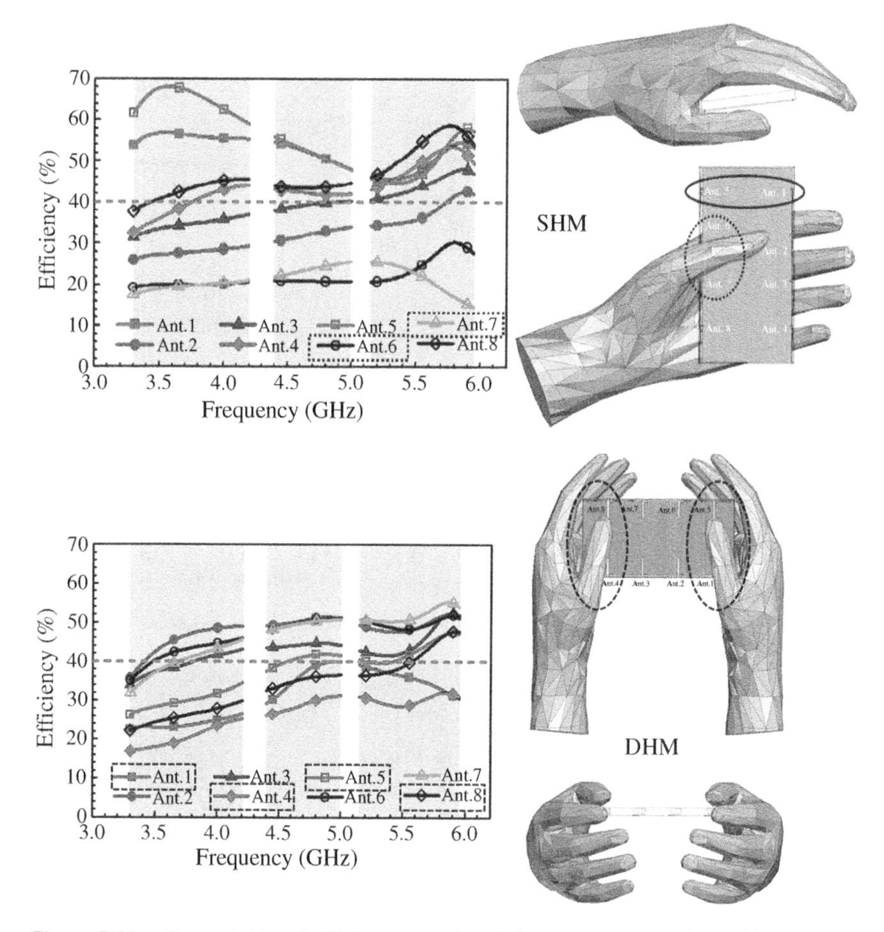

Figure 3.29 Simulated total efficiencies of the eight-antenna array (formed by monopole elements) for 5G FR1 smartphone applications in the SHM and DHM. *Source:* Extended diagrams of [40].

large impedance mismatching. Due to the power absorption by the user's hand tissue, the total efficiencies of Ant. 2, Ant. 6, and Ant. 7 were reduced to <30%, and Ant. 3 has shown the worst total efficiency of approximately 10% because it is in full contact with the user's palm. In comparison, as the other four open-end slot antennas element (Ant. 1, Ant. 4, Ant. 5, and Ant. 8) are located farther away from the user's fingers, minor variations in the resonant modes were observed, and they have shown better total efficiencies of between 35% and 55%.

For the DHM scenario, because the two right-side corner antenna elements (Ant. 1 and Ant. 5) were covered by the user's thumbs, their resonant modes were shifted to the higher frequency spectrum with a slight impedance mismatch.

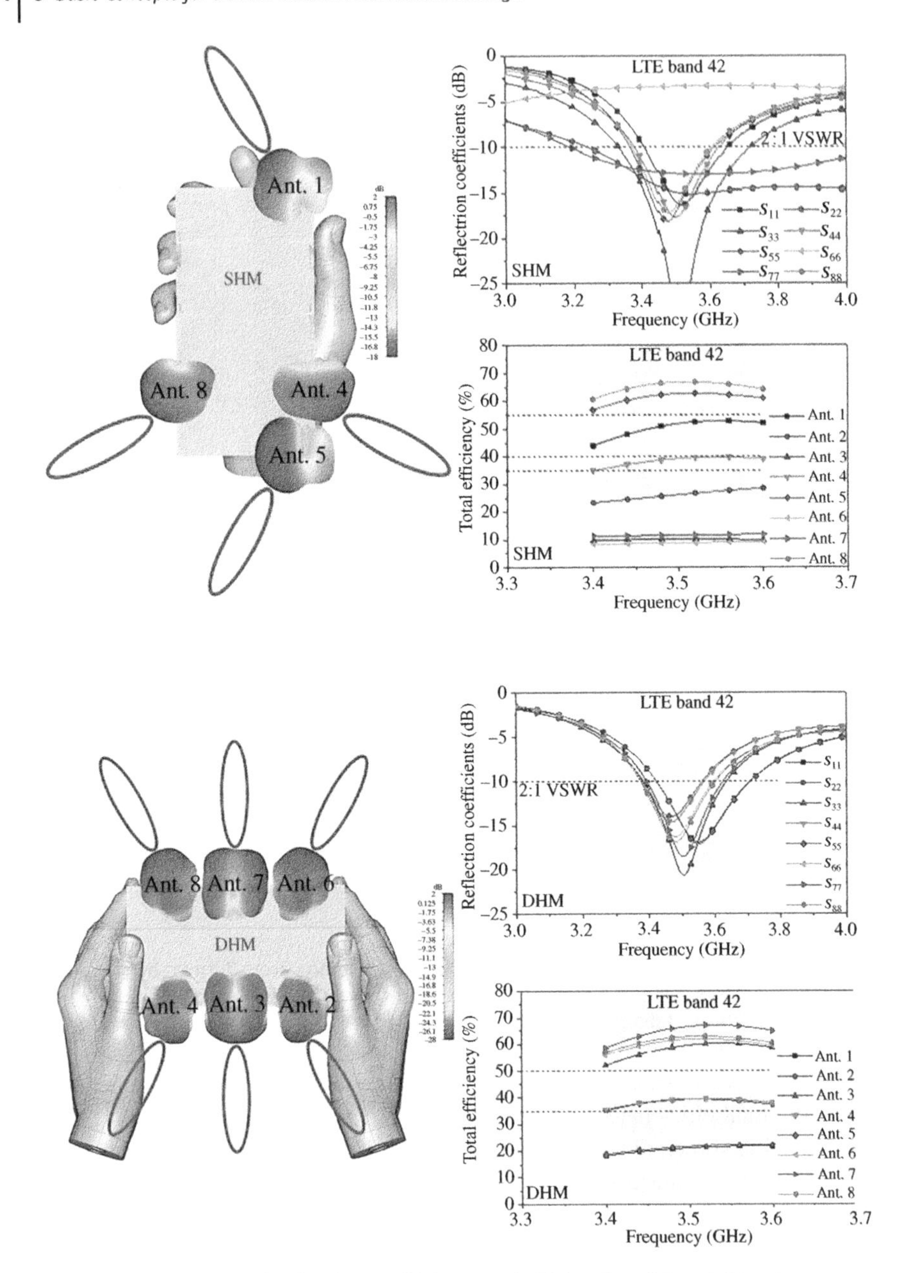

Figure 3.30 Simulated reflection coefficients and efficiencies of the eight-antenna array (formed by open slot elements) for 5G FR1 smartphone applications in the SHM and DHM. *Source:* From [6]. ©2019 IEEE. Reproduced with permission.

Furthermore, the total efficiencies of Ant. 1 and Ant. 5 have been largely mitigated to <25%. As Ant. 2 and Ant. 4 were near the two user's little fingers (left and right), poorer impedance matchings were also observed, but they have exhibited slightly better total efficiencies of >35%. As expected, because Ant. 3, Ant. 6, Ant.7, and Ant. 8 are located much farther away from the user's hands, their corresponding resonant characteristics are nearly the same as before, and they have radiated much acceptable total efficiencies of >50%.

3.5.3 Loop Antenna Array User's Hand Effects

Figure 3.31 shows the simulated reflection coefficients and total efficiencies of a 5G FR1 eight-antenna array (formed by using loop antenna elements) reported in [41]. In the SHM scenario, as the user's fingers are directly in contact with Ant. 1, Ant. 7, and Ant. 8 at the lower section of the smartphone, very poor impedance matching was observed. As for Ant. 2, as it is located behind the finger and was not in direct contact with the finger, slight impedance mismatching was observed. Nevertheless, the isolations between all ports are not affected. Once again, the simulated total efficiencies of Ant. 1, Ant. 2, Ant. 7, and Ant. 8 were lower than 35% because of the EM absorption effects from the user's hands, whereas Ant. 3, Ant. 4, Ant. 5, and Ant. 6 have exhibited total efficiencies of >50%.

For the DHM scenario, Ant. 1 to Ant. 4 are placed close to the user's fingers (not in direct contact), whereas Ant. 5 to Ant. 8 are located farther away from the user's fingers. As Ant. 1, Ant. 2, Ant. 7, and Ant. 8 are not physically touched by the user's fingers, their simulated reflection coefficients (on one hand only) are only slightly affected, and the eight-antenna MIMO array can still cover the 5G FR1 band of 3400–3600 MHz. Similarly, their corresponding isolations are only slightly affected. Due to the absorption effects from the user's fingers, the total efficiencies of the loop elements have different degrees of mitigation, but they are still >30%.

3.6 Conclusion

In this chapter, we covered many different areas in the design of the 5G FR1 antenna elements for smartphone applications. We begin by considering the other passive/active elements (including the WiFi, GPS, Bluetooth, and 4G LTE antennas, etc.) located in a smartphone before the allocation of the 5G FR1 MIMO antenna array, followed by showing the various effects contributed by the smartphone hardware, such as chassis, battery, and LCD screen. The various typical 5G FR1 antenna elements (monopole, slot, loop, PIFA/IFA, etc.) selected for the smartphone applications were presented, and their feeding mechanism to achieve the desired frequency and impedance matching were addressed as well. A very

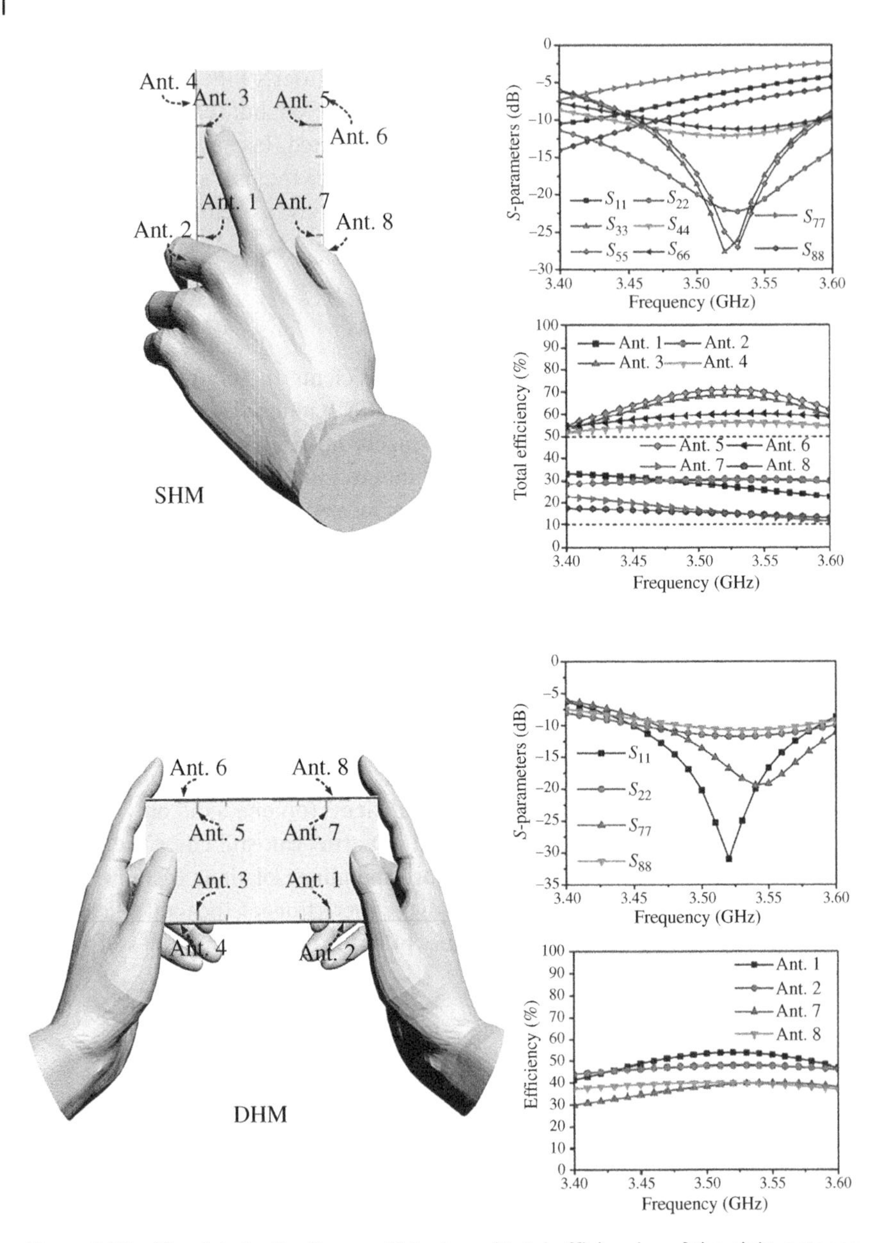

Figure 3.31 Simulated reflection coefficients and total efficiencies of the eight-antenna array (formed by loop elements) for 5G FR1 smartphone applications in the SHM and DHM. *Source:* From [41]. ©2019 IEEE. Reproduced with permission.

detailed description of the effects of the smartphone chassis (metallic or plastic), and how to overcome these effects were shown. Finally, the exposure of the 5G FR1 antenna array to the user's hands (or fingers) was fully investigated, showing the mitigation of total efficiencies of three typical array antenna structures (monopole, slot, and loop).

References

1 (Online) https://www.itu.int/net/pressoffice/press_releases/2015/56.aspx.

2 C. You, D. Jung, M. Song, and K. L. Wong, "Advanced coupled-fed MIMO antennas for next generation 5G smartphones," *In Int. Symp. Antenna Propag.*, Busan, South Korea, 2018.

3 (Online) https://zh.ifixit.com/Device/Samsung_Galaxy_S20_Ultra.

4 W. Hong, "Solving the 5G Mobile antenna puzzle: assessing future directions for the 5G mobile antenna paradigm shift," *IEEE Microwave Mag.*, vol. 18, no. 7, pp. 86–102, Nov.-Dec. 2017.

5 C. Z. Han, G. L. Huang, T. Yuan, W. Hong, and C. Y. D. Sim, "A frequency-reconfigurable tuner-loaded coupled-fed frame-antenna for all-metal-shell handsets," *IEEE Access*, vol. 6, pp. 64041–64049, Nov. 2018.

6 Y. Li, C. Y. D. Sim, Y. Luo, and G. Yang, "High-isolation 3.5 GHz eight-antenna MIMO array using balanced open-slot antenna element for 5G smartphones," *IEEE Trans. Antennas Propag.*, vol. 67, no. 6, pp. 3820–3830, Jun. 2019.

7 Y. Yan, Y. L. Ban, G. Wu, and C. Y. D. Sim, "Dual-loop antenna with band-stop matching circuit for WWAN/LTE full metal-rimmed smartphone application," *IET Microw. Antennas Propag.*, vol. 10, no. 15, pp. 1715–1720, Dec. 2016.

8 J. Y. Lu, H. J. Chang, and K. L. Wong, "10-antenna array in the smartphone for the 3.6-GHz MIMO operation," *in 2015 IEEE International Symposium on Antennas and Propagation & USNC/URSI National Radio Science Meeting*, Vancouver, BC, 2015, pp. 1220–1221.

9 K. L. Wong and J.Y. Lu, "3.6-GHz 10-antenna array for MIMO operation in the smartphone," *Microwave Opt. Technol. Lett.*, vol. 57, no. 7, pp. 1699–1704, Apr. 2015.

10 K. L. Wong, C. Y. Tsai, J. Y. Lu, D. M. Chian, and W. Y. Li, "Compact eight MIMO antennas for 5G smartphones and their MIMO capacity verification," *in 2016 URSI Asia-Pacific Radio Science Conference*, South Korea, pp. 1054–1056, 2016.

11 M. Y. Li, Y. L. Ban, Z. Q. Xu, C. Y. D. Sim, K. Kang, and Z. F. Yu, "Eight-port orthogonally dual-polarized antenna array for 5G smartphone applications," *IEEE Trans. Antennas Propag.*, vol. 64, no. 9, pp. 3820–3830, Sept. 2016.

12 Y. Li, C. Y. D. Sim, and G. Yang, "Multiband 10-antenna array for Sub-6 GHz MIMO applications in 5-G smartphones," *IEEE Access*, vol. 6, pp. 28041–28053, May 2018.

13 Y. Li, C. Y. D. Sim, Y. Luo, and G. Yang, "12-port 5G massive MIMO antenna array in sub-6 GHz mobile handset for LTE bands 42/43/46 applications," *IEEE Access*, vol. 6, pp. 344–354, Oct. 2017.
14 Y. L. Ban, C. Li, C. Y. D. Sim, G. Wang, and K. L. Wong, "4G/5G multiple antennas for future multi-mode smartphone applications," *IEEE Access*, vol. 4, pp. 2981–2988, Jun. 2016.
15 X. Shi, M. Zhang, S. Xu, D. Liu, H. Wen, and J. Wang, "Dual-band 8-element MIMO antenna with short neutral line for 5G mobile handset", *2017 11th European conference on Antennas and Propagation (EUCAP)*, 18 May 2017.
16 J. Guo, L. Cui, C. Li, and B. Sun, "Side-edge frame printed 8 port dual-band antenna array for 5G smartphone applications," *IEEE Trans. Antennas Propag.*, vol. 66, no. 12, pp. 7412–7417, Dec. 2018.
17 M. Wang, B. Xu, Y. Li, Y. Luo, H. Zou, and G. Yang, "Multiband multiple-input multiple-output antenna with high isolation for future 5G smartphone applications," *Int. J. RF Microwave Comput. Aided Eng.*," vol. 29, no. 7, e21758, Jul. 2019.
18 A. Zhao and Z. Ren, "Multiple-input and multiple-output antenna system with self-isolated antenna element for fifth-generation mobile terminals," *Microwave Opt. Technol. Lett.*, vol. 61, no. 1, pp. 20–27, Jan. 2019.
19 A. Zhao and Z. Ren, "Size reduction of self-isolated MIMO antenna system for 5G mobile phone applications," *IEEE Antennas Wireless Propag. Lett.*, vol. 18, no. 1, pp. 152–156, Jan. 2019.
20 Y. Liu, Y. Lu, Y. Zhang, and S. Gong, "MIMO antenna array for 5G smartphone applications," *13th European Conference on Antennas & Propagation 2019 (EuCAP 2019)*, Krakow, Apr. 2019.
21 Z. Ren, A. Zhao, and S. Wu, "MIMO antenna with compact decoupled antenna pairs for 5G mobile terminals," *IEEE Antennas Wireless Propag. Lett.*, vol. 18, no.7, pp. 1367–1371, May 2019.
22 A. Zhao and Z. Ren, "5G MIMO antenna system for mobile terminals," *in 2019 IEEE International Symp. Antennas Propag. USNC-URSI Radio Science Meeting*, Atlanta, Georgia, USA, 2019.
23 R. Li, Z. Mo, H. Sun, X. Sun, and G. Du, "A low-profile and high-isolated MIMO antenna for 5G mobile terminal," *Micromachines*, vol. 11, pp. 1–12, Oct. 2018.
24 H. Piao, Y. Jin and L. Qu, "A compact and straightforward self-decoupled MIMO antenna system for 5G applications," *IEEE Access*, vol. 8, pp. 129236–129245, Jul. 2020.
25 W. Jiang, B. Liu, Y. Cui, and W. Hu, "High-isolation eight-element MIMO array for 5G smartphone applications," *IEEE Access*, vol. 7, pp. 34104–34112, Mar. 2019.
26 X. Zhao, S. P. Yeo, and L. C. Ong, "Decoupling of inverted-F antennas with high-order modes of ground plane for 5G mobile MIMO platform," *IEEE Trans. Antennas Propag.*, vol. 66, no. 9, pp. 4485–4495, Jun. 2018.
27 W. Hu, X. Liu, S. Gao, L. H. Wen, L. Qian, T. Feng, R. Xu, P. Fei, and Y. Liu, "Dual-band ten-element MIMO array based on dual-mode IFAs for 5G terminal applications," *IEEE Access*, vol. 7, pp. 178476–178485, 2017.

28 M. Y. Li, C. Li, Y. L. Ban, and K. Kang, "Multiple antennas for future 4G/5G smartphone applications," *in 2016 IEEE MTT-S International Microwave Workshop Series on Advanced Materials and Processes for RF and THz Applications (IMWS-AMP)*, Chengdu, China, Oct. 2016.

29 K. L. Wong, C. Y. Tsai, and J. Y. Lu, "Two asymmetrically mirrored gap-coupled loop antennas as a compact building block for eight-antenna MIMO array in the future smartphone," *IEEE Trans. Antennas Propag.*, vol. 65, no. 4, pp. 1765–1778, Apr. 2017.

30 K. L. Wong, B. W. Lin, and W. Y. Li, "Dual-band dual inverted-F/loop antennas as a compact decoupled building block for forming eight 3.5/5.8-GHz MIMO antennas in the future smartphone," *Microwave Opt. Technol. Lett.*, vol. 59, no. 11, pp. 2715–2721, 2017.

31 Q. Chen, H. Lin, J. Wang, L. Ge, Y. Li, T. Pei, and C.Y.D. Sim, "Single ring slot based antennas for metal-rimmed 4G/5G smartphones," *IEEE Trans. Antennas Propag.*, vol. 67, no. 3, pp. 1476–1487, Mar. 2019.

32 H. D. Chen, Y. C. Tsai, C. Y. D. Sim and C. Kuo, "Broadband eight-antenna array design for sub-6 GHz 5G NR bands metal-frame smartphone applications," *IEEE Antennas Wireless Propag. Lett.*, vol. 19, no. 7, pp. 1078–1082, Jul. 2020.

33 X. Zhang, Y. Li, W. Wang, and W. Shen, "Ultra-wideband 8-port MIMO antenna array for 5G metal-frame smartphones," *IEEE Access*, vol. 7, pp. 72273–72282, May 2019.

34 Q. Cai, Y. Li, X. Zhang, and W. Shen, "Wideband MIMO antenna array covering 3.3–7.1 GHz for 5G metal-rimmed smartphone applications," *IEEE Access*, vol. 7, pp. 72273–72282, 2019.

35 (Online), Youtube: https://www.youtube.com/watch?v=RD188LlRBGM.

36 B. Thors, D. Colombi, Z. Ying, T. Bolin, and C. Tornevik, "Exposure to RF EMF rrom array antennas in 5G mobile communication equipment," *IEEE Access*, vol. 4, pp. 7469–7478, 2016.

37 K. Y. Yazdandoost and I. Laakso, "EMF exposure analysis for a compact multi-band 5G antenna," *Prog. Electromagn. Res.*, vol. 68, pp. 193–201, 2018.

38 Y. Li and M. Lu, "Study on SAR distribution of electromagnetic exposure of 5G mobile antenna in human brain," *J. Appl. Sci. Eng.*, vol. 23, no. 2, pp. 279–287, 2020.

39 (Online) https://www.qualcomm.com/media/documents/files/5g-nr-mmwave-deployment-strategy-presentation.pdf.

40 C. Y. D. Sim, H. Y. Liu, and C. J. Huang, "Wideband MIMO antenna array design for future mobile devices operating in the 5G NR frequency bands n77/n78/n79 and LTE band 46," *IEEE Antennas Wireless Propag. Lett.*, vol. 19, no. 1, pp. 74–78, Jan. 2020.

41 A. Ren, Y. Liu, H. W. Yu, Y. Jia, C. Y. D. Sim, and Y. Xu, "A high-isolation building block using stable current nulls for 5G smartphone applications," *IEEE Access*, vol. 7, pp. 170419–170429, 2019.

4

Multi-Band 5G FR1 Band Mobile Antenna Design

As mentioned in Chapters 1–3, the 5G New Radio (NR) Band has initially identified their middle band (or C-band) as NR band n77 (3300–4200 MHz), band n78 (3300–3800 MHz), and band n79 (4400–5000 MHz) in the 3GPP (Third-Generation Partnership Project) Release-15 NR specifications [1]. However, the application of the unlicensed spectrum has also received increasing attention. In the new 3GPP Release-16 specifications, it has also been working on a new radio access technology, known as the 5G NR Unlicensed (5G NR-U) band. This 5G NR-U band aims at extending the previous 5G NR band n77/n78/n79 to the unlicensed bands between 5925 and 7125 MHz (called 5G NR-U band n96). The aim is to study the coexistence between Wi-Fi and 5G NR-U systems, and this 5G NR-U band can supposing yield higher data throughput and lower latency than the Wi-Fi (802.11 ac).

In addition to the U-NII (Unlicensed National Information Infrastructure), tech giants such as Qualcomm, Intel, Google, and Apple have also actively promoted this license-free 6-GHz frequency band. Once the specification of 5G NR-U band n96 is officially finalized, the spectrum resources available for mobile communication can be further improved. Notably, with the development of the unlicensed frequency LTE band (Long-Term Evolution in Unlicensed Spectrum, LTE-U), Licensed-Assisted Access (LAA), and LTE-WLAN Aggregation (LWA), combined with the unlicensed frequency band, the application of the Heterogeneous Networks and other technologies to improve data transmission capacity will eventually become a trend. Therefore, this 5G NR-U band has also extended downward to the previous LTE-U band 46 (5150–5925 MHz), and it is now called 5G NR-U band n46 (5150–5925 MHz).

Because of the tremendous changes in the 5G NR spectrum from the initial lower narrow C-band (3300–3600 MHz) to now (from 3300 to 7125 MHz), there is

Microwave and Millimeter-Wave Antenna Design for 5G Smartphone Applications,
First Edition. Wonbin Hong and Chow-Yen-Desmond Sim.

Published 2023 by John Wiley & Sons, Inc.

an urgent need to design 5G FR1 antenna for smartphone applications operating in these 5G NR band n77/n78/n79, 5G NR-U band n46, and 5G NR-U band n96. This chapter will introduce planar antenna design methods that can yield multi-band and wideband 5G operations. Different novel techniques that combine other antenna technologies (called a hybrid antenna), as shown in Figure 3.23, would be introduced. After designing a single antenna element, the array expansion strategies are essential in determining good array performances, and a brief description of choosing different placement methods and techniques to achieve that is discussed. As the 5G FR1 antenna will be eventually collocated with other antenna placements in the smartphone, the coexistence design with 3G/4G antennas, or even 5G mmWave FR2 antennas, is a piece of vital knowledge to the smartphone industry. Lastly, as aforementioned, we will discuss the various 5G FR1 antenna design techniques. Besides achieving the ability to cover the 5G NR band n77/n78/n79, it can also cover up to 5G NR-U bands n46 and n96.

4.1 Planar Antenna Design Topologies

The planar antenna designs for achieving the lower spectrum of the 5G NR band n77 or partial 5G NR band n78 were reviewed in Chapter 3. In this section, we will begin by showing the various planar antenna designs that apply the same antenna structure (such as monopole, loop, or slot) that can achieve dual band, multi-band, or even wideband operations for 5G NR band n77/78 (3300–4200 MHz) and n79 (4400–5000 MHz). For those that have applied hybrid antenna structures (such as the combination of two different antenna structures, for example, monopole and slot) will be discussed in Section 4.2.

4.1.1 Dual Band and Wideband Loop Antenna

Figure 4.1 shows the conceptual diagram of transforming a single band loop antenna that can only cover the lower spectrum of C-band into a dual band type that fits for covering partial 5G NR band n77/n78 and band n79. By observing the coupled-fed loop antenna in [2] that excites only a single band, one can alter the design to a dual band type by loading a similar loop (but with a different loop size) coupled to the other side of the T-shaped coupled-fed structure [3, 4]. As for the coupled-fed and coupled-ground design in [5], besides replicating another loop on its side, the idea of [6] was to replicate another set of the double loop (with different size) at its bottom with the coupled-fed structure and coupled ground structure loaded between the two sets of double loops (upper and lower). Nevertheless, the dual band loop design methods in [3] have excited low-band (3.34–3.84 GHz) and high-band (4.66–5.51 GHz) in the larger loop and smaller loop, while the

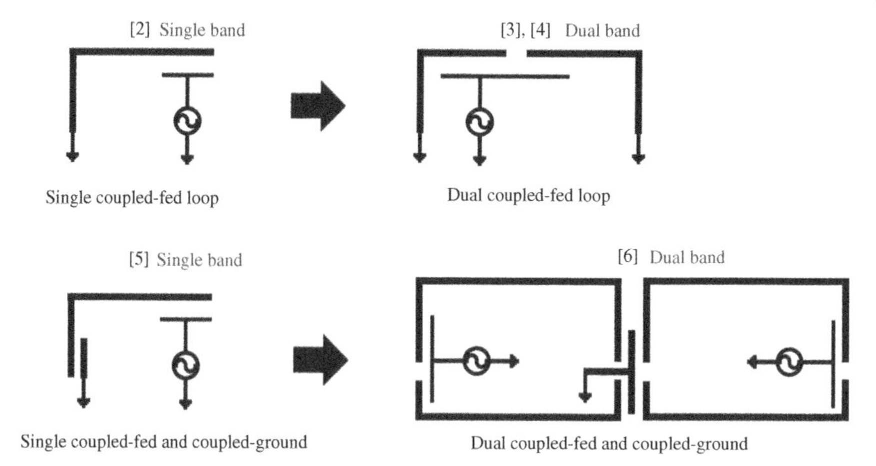

Figure 4.1 Conceptual feeding mechanism diagram for transforming a single band single loop antenna type to a dual band dual loop antenna type. Reference numbers are included for each design.

larger loop and smaller loop in [4], respectively, generate a low-band (2.4–2.5 GHz) and high-band (5.1–5.9 GHz) that can cover the 5G NR band n46. As for the dual band double loop design in [6] that is an expansion from [5], it generated low-band (3.4–3.6 GHz) and high-band (4.8–5 GHz) that cover only the 5G China's operations. Thus, the above loop design structures cannot fulfill the 5G NR band n77/n78/n79 spectrum requirement.

To avoid applying such complicated coupling methods as in [3, 4, 6], with loop expansion that results in increasing the entire loop antenna size, a modified wide-band coupled loop antenna that requires only a single loop structure was introduced to generate wideband operation for 5G NR band n77/n78/n79 [7]. Instead of replicating another similar loop or expanding the loop into multiple sets, it has adopted a unique coupled method that breaks the traditional loop into multiple smaller sections and coupled them (front and back) together by printing these smaller sections on the two sides of the substrate. Figure 4.2 shows the detailed configuration of this modified coupled antenna, along with its eight-antenna array design. Figure 4.3 shows the conceptual structure of the modified coupled loop antenna element with different configurations and their corresponding reflection coefficients. Figure 4.4 further shows the current distributions of the traditional loop element and the modified coupled loop antenna element to provide a better perspective on how to achieve wideband operation. As expected, for a traditional loop, it will yield three separated resonance modes at 3.17 GHz (0.5λ), 4.73 GHz (1.0λ), and 5.99 GHz (1.5λ). Thus, only partial 5G NR band n77/n78/n79 spectrums are satisfied. To further excite a new resonance mode between the

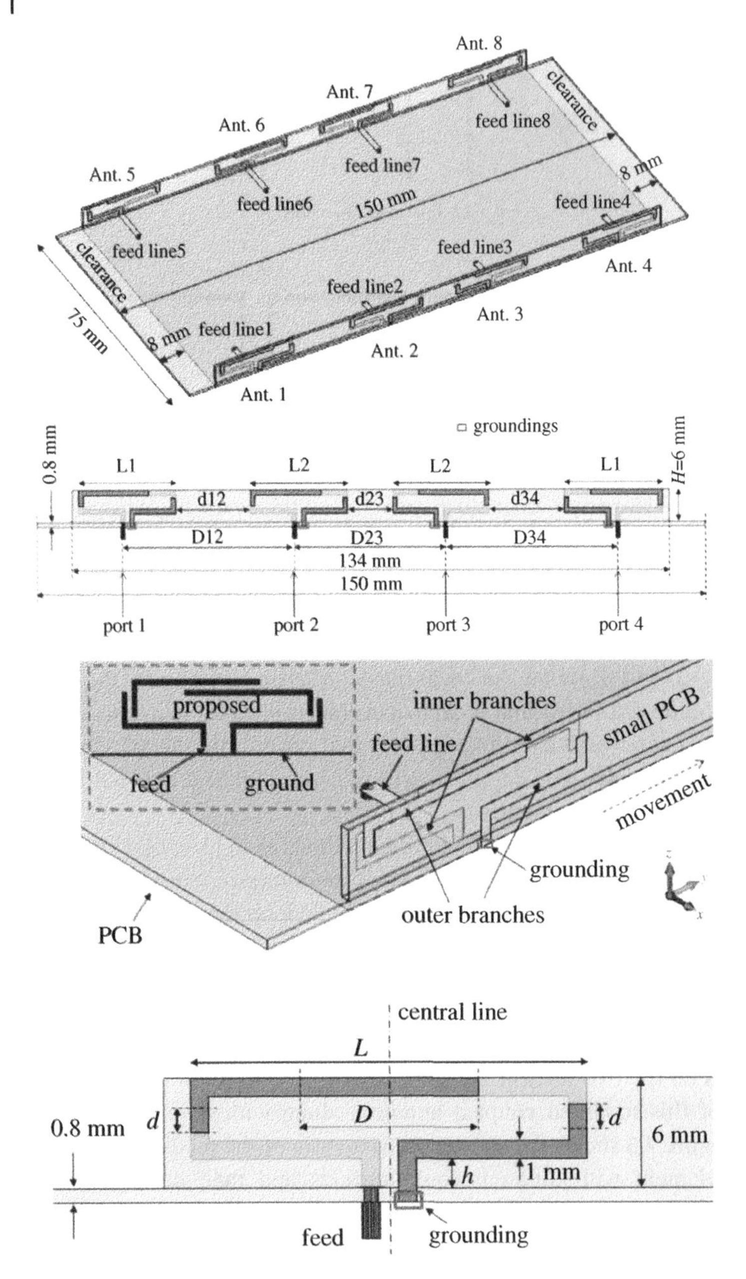

Figure 4.2 Layout of the wideband eight-antenna array, and the configuration of the modified coupled loop antenna element printed on the two sides of the substrate. *Source:* From [7]. ©2019 IEEE. Reproduced with permission.

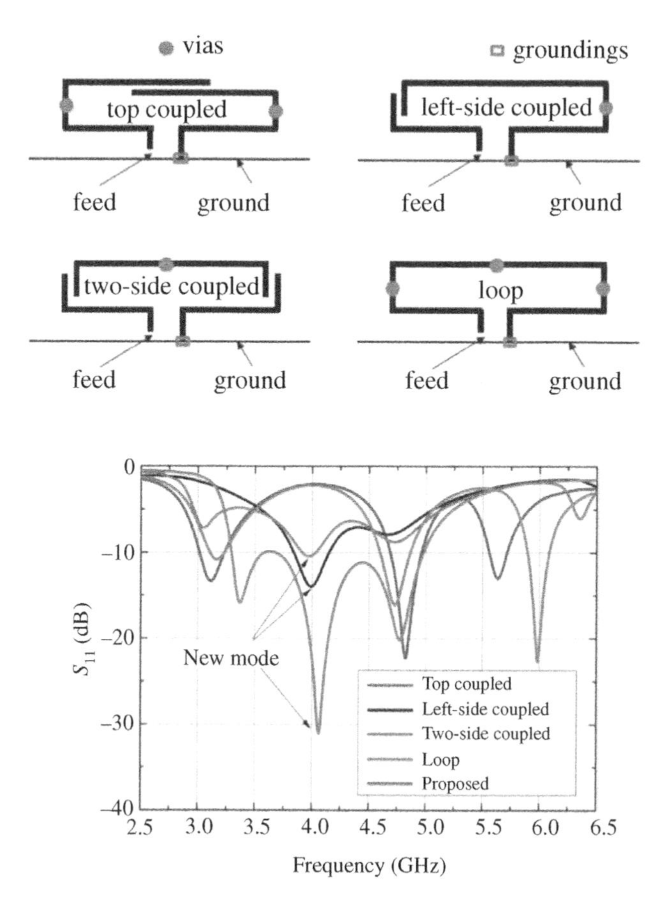

Figure 4.3 Conceptual structure of the loop antenna element with different configurations, and its corresponding simulated reflection. *Source:* From [7]. ©2019 IEEE. Reproduced with permission.

fundamental mode (0.5λ) and second higher-order mode (1.0λ), the technique of coupling multiple smaller sections (front and back) of the loop structure was applied, as shown in Figures 4.2 and 4.3. Due to the coupled effects, a new resonance mode (0.75λ) at 4.06 GHz is excited, not to mention that the resonance modes at 0.5λ and 1.0λ are shifted to 3.37 and 4.78 GHz, respectively. Therefore, by observing Figure 4.4, the combination of these three resonance modes was able to yield desirable 6-dB impedance bandwidth that covers the 5G NR band n77/n78/n79 (3300–5000 MHz). Besides demonstrating good simulation and measured results, the measured isolation of the eight-antenna array type was also better than 14.5 dB, showing that the antenna has self-isolation capability, which will be discussed further in Chapter 5.

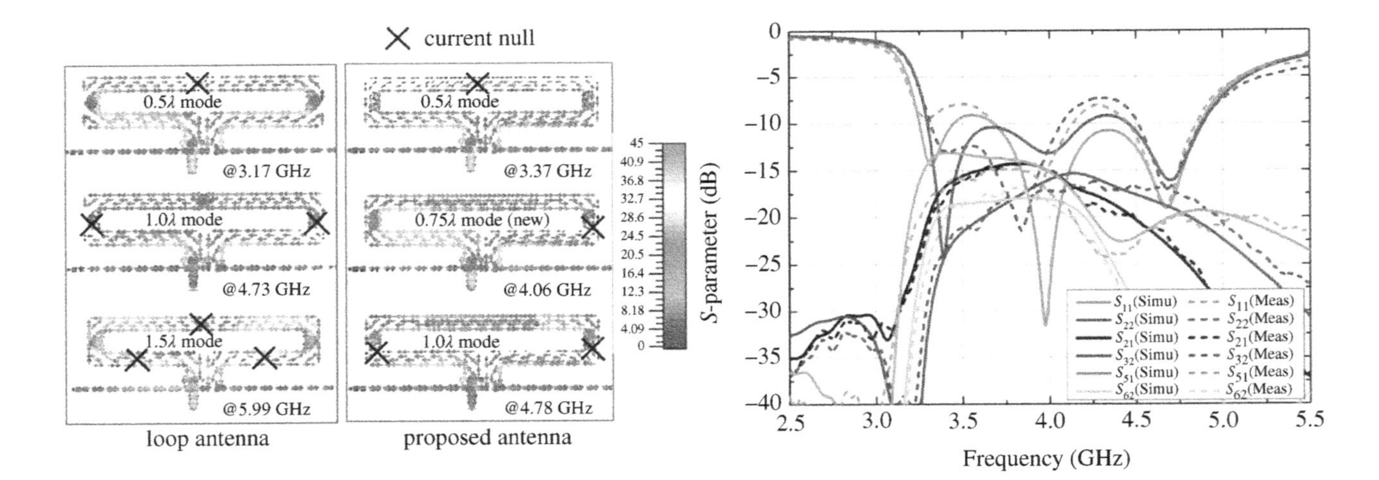

Figure 4.4 The current distributions of the first three modes (traditional loop and coupled loop antenna element), and their corresponding simulated/measured reflection coefficients and isolations. *Source:* From [7]. ©2019 IEEE. Reproduced with permission.

4.1.2 Dual Band and Wideband Monopole Antenna

Figure 4.5 shows the conceptual diagram of transforming a single band monopole antenna [8] that can only cover the lower spectrum of C band into a dual band type that fits for covering partial 5G NR band n77/n78 and band n79 [9–11]. As shown explicitly in [9], the design of such dual band monopole was realized by extending another monopole branch (or strip) so that the longer strip (strip 1) will excite the low-band resonance, and the shorter strip is responsible for the high-band resonance. Notably, the work reported in [12] has further extended more branches (strips), and thus it can cover more operational bands within the 5G FR1 band. However, [9–11] can only excite narrow bandwidths (for low-band and high-band) that can cover partial 5G NR band n78 and n79, respectively. Even though [12] (with multiple branches) was able to cover 3100–3850 MHz for the low-band and 4800–6000 MHz for the high-band, it still cannot satisfy the entire 5G NR band n77/n78 (3300–4200 MHz).

As mentioned earlier, it is rather challenging to apply the monopole antenna structure (even with extended branches as in [9–12]) to yield wideband operation for 5G NR band n77/n78/n79 because the monopole antenna is a quarter-wavelength antenna that usually yields narrow bandwidth. Nevertheless, one dual monopole structure work, in particular, has managed to cover the entire 5G NR band n77/n78 (3300–4200 MHz) and partial (4800–5000 MHz) of band n79 (4400–5000 MHz) for smartphone applications, as reported in [13]. Figure 4.6 shows the geometry and detailed dimensions of the eight-antenna array formed by using a double branch monopole and ground-connected T-Shaped Decoupling Stub (TSDS).

In [13], Ant. 1 and Ant. 2 are considered as a set of dual double branch antenna that has their longer monopole branches facing each other with the TSDS loaded

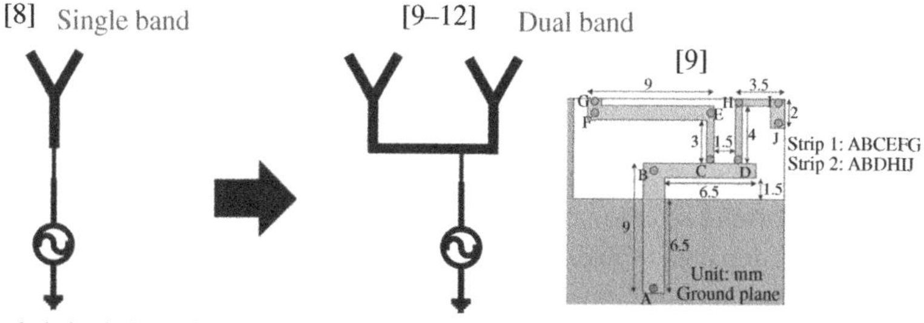

Figure 4.5 The conceptual diagrams of forming a single branch monopole (single band) to a double branch monopole (dual band) with a single feeding structure. An example of a dual band monopole design in [9] was included. Reference numbers are included for each design. *Source:* From [9]. ©2017 IEEE. Reproduced with permission.

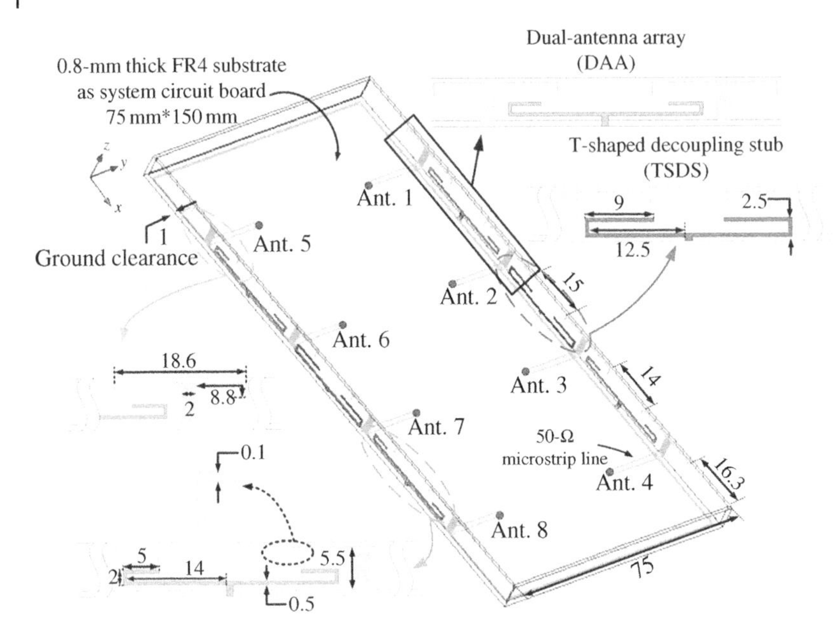

Figure 4.6 The geometry and dimensions of the proposed eight-antenna array with double branch monopole and ground-connected T-shaped decoupling stub (TSDS). *Source:* From [13]. ©2019 IEEE. Reproduced with permission.

between them. Here, branch 1 and branch 2 are extended from the feeding branch with a length of 11.5 and 8.8 mm, respectively, which are approximately quarter-wavelength at 3.5 and 4.9 GHz. Therefore, a dual band operation can be realized by branch 1 and branch 2. However, from Figure 4.7a, the 6-dB impedance bandwidth of branch 1 was (3200–3900 MHz), which is unable to cover the entire 5G NR band n77/n78 in the low-band, and it is only when tuning the width W_0 of branch 2 to a very narrow width of 0.1 mm that the high-band can cover the desired 4800–5000 MHz for China's 5G FR1 band. To further improve the low-band in Figure 4.7a, a tuning technique was applied in [13], in which the ground-connected T-shaped decoupling stub (TSDS), besides acting as a decoupling element, it is also used to slightly coupled to the main radiating branch, allowing it to act as a tuning stub to branch 1 so as to increase the low-band operational bandwidth, as depicted in Figure 4.7b. By comparing Figure 4.7a with Figure 4.7b, one can see that the 6-dB impedance bandwidth of both the S_{11} and S_{22} has been improved to cover the 5G NR band n77/n78. Therefore, it is concluded that if a double branch monopole is to be applied to cover a wider operation for the 5G FR1 band, it is a good practice to insert a tuning stub into the main monopole radiator. The decoupling technique applied in this work will be further explained in Chapter 5.

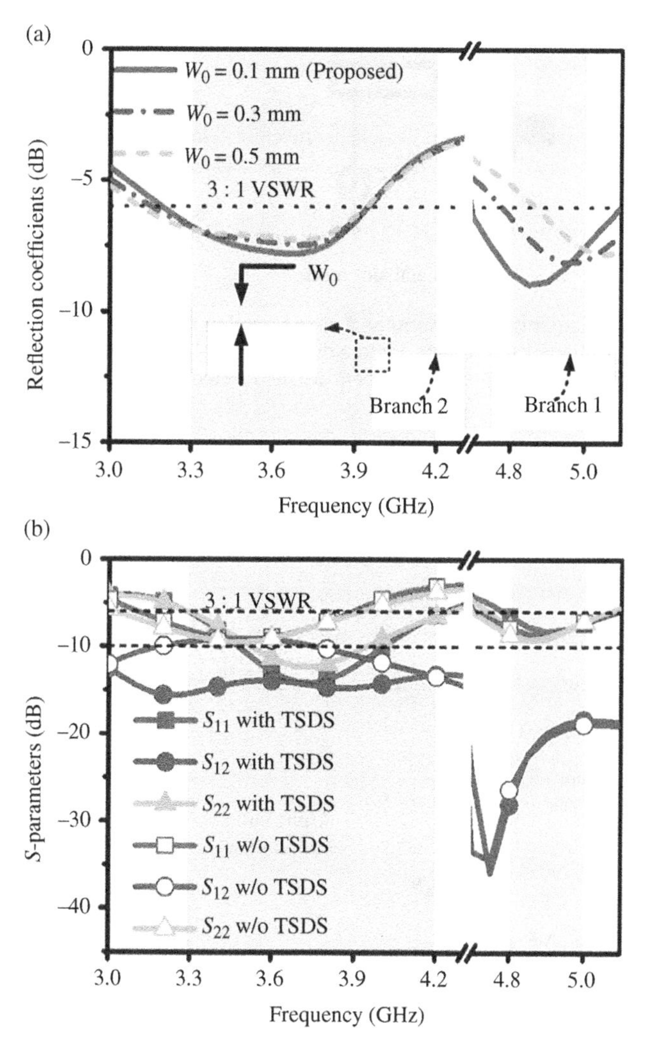

Figure 4.7 (a) The reflection coefficient of Ant. 1 (a double branch monopole) with different width W_0. (b) The reflection coefficients of Ant. 1 and Ant. 2 (with/without TSDS) and isolation between Ant. 1 and Ant. 2 (with/without TSDS). *Source:* From [13]. ©2019 IEEE. Reproduced with permission.

4.1.3 Dual Band and Wideband Slot Antenna

As shown in Chapter 3, it is learnt that an open-end slot antenna with an L-shaped feeding structure can only yield a single slot mode [9], as depicted in Figure 4.8. However, from the work reported in [14], if the open slot position can be meticulously positioned to form an unequal slot length on the two sides of the slot

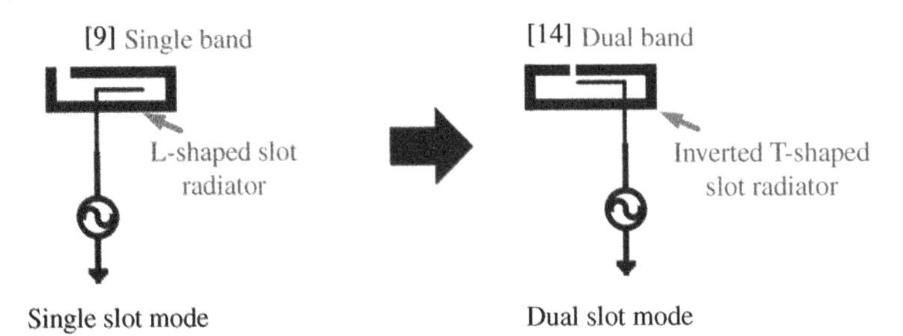

Figure 4.8 The conceptual diagrams of transforming a single band open-end slot coupled-fed antenna (with L-shaped slot radiator) into a dual band dual slot mode (with inverted T-shaped slot radiator) type. Reference numbers are also included for each design.

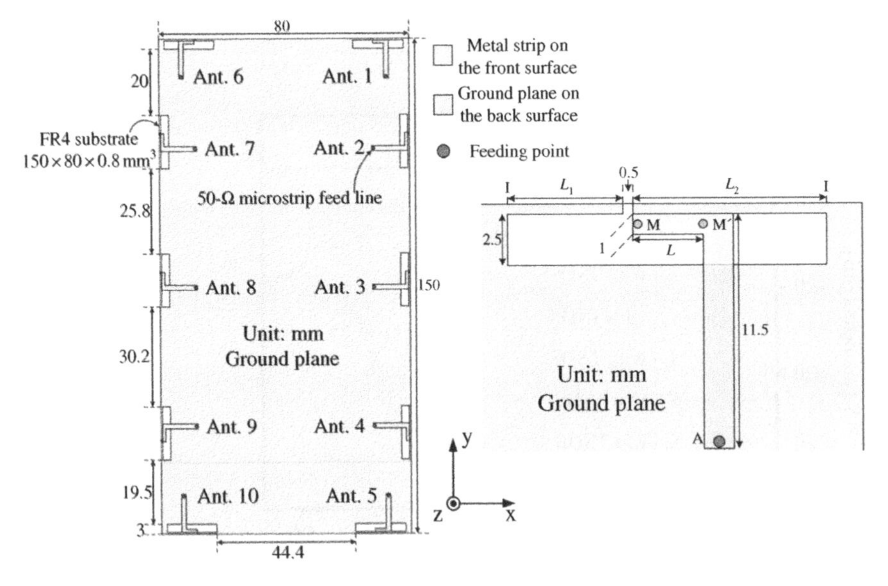

Figure 4.9 The layout of the 10-antenna array element, and the detailed configuration of the dual band inverted T-shaped slot radiator. *Source:* From [14]. ©2018 IEEE. Reproduced with permission.

(transformation from an L-shaped slot to an inverted T-shaped slot), and allowing the open-end position of the L-shaped feeding strip to be located very close to the narrow open slot, a dual band operation (or dual slot mode) can be excited as well.

Figure 4.9 shows the detailed configuration and dimensions of the dual band inverted T-shaped slot radiator, as well as its 10-antenna array layout on the smartphone system ground. As explained in Figure 4.8, the initial L-shaped slot was modified in such a way that it can be denoted as an inverted T-shaped slot with a very short vertical slot section. As for the L-shaped feeding structure, as indicated earlier, it must place meticulously near the narrow open slot so that dual band operation can be realized.

To better comprehend the operating principles of this inverted T-shaped slot radiator, its corresponding surface electric field (*E*-field) distributions at 3500 and 5500 MHz were simulated and plotted in Figure 4.10. As the opening section (narrow open slot) of an open slot antenna usually exhibits maximal impedance, whereas the two closed sections (located on both ends of the slot) have a minimal impedance, confined by this boundary condition, the *E*-field maximum was exhibited at the opening section, and the two closed sections have shown *E*-field minimum. Thus, it signifies that this inverted T-shaped open slot radiator can excite two different quarter-wavelength resonance modes via the L-shaped feeding structure. By further observing the inverted T-shaped slot radiator at 3500 MHz, a strong quarter-wavelength E-field path was distributed along the longer open slot (located on the right section), whereas the shorter open slot (located on the left section) has demonstrated strong quarter-wavelength *E*-field distribution at 5500 MHz.

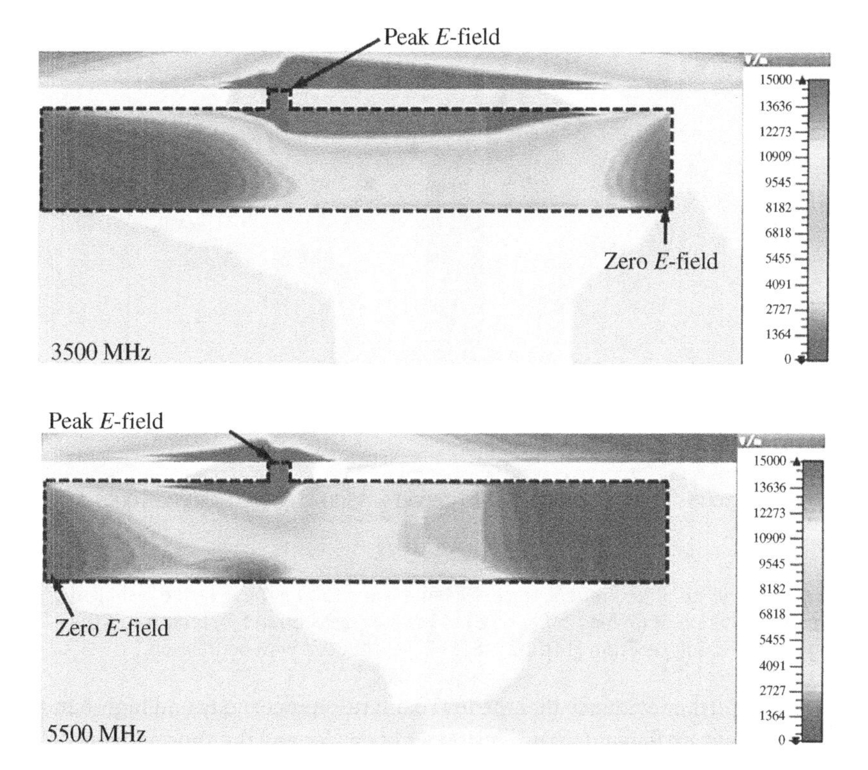

Figure 4.10 The surface electric field (*E*-field) distribution of the inverted T-shaped radiator at 3500 and 5500 MHz. *Source:* From [14]. ©2018 IEEE. Reproduced with permission.

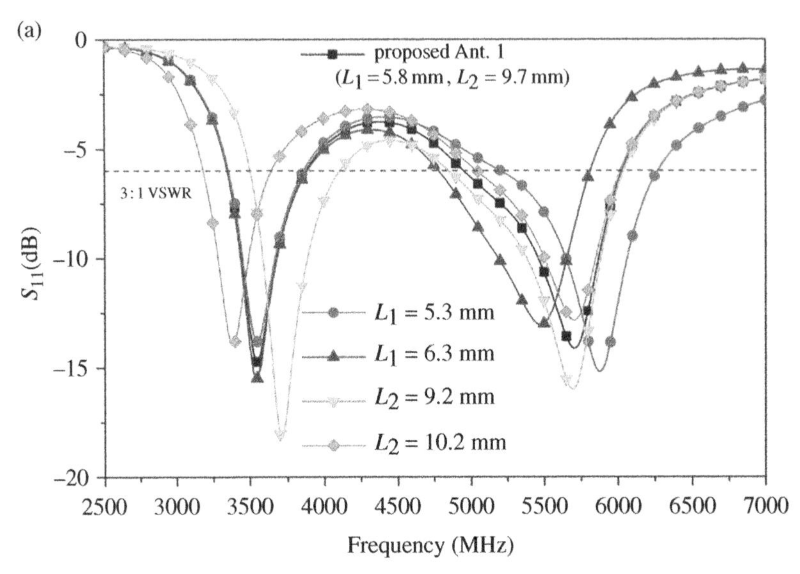

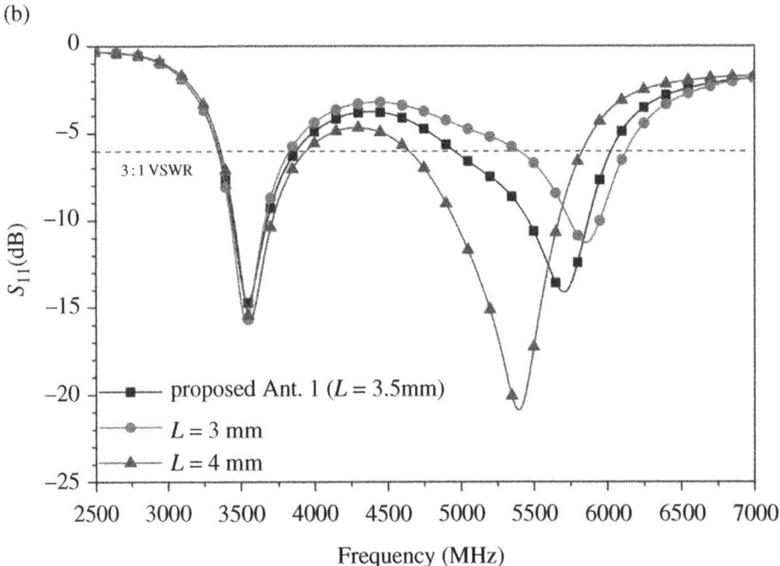

Figure 4.11 The simulated reflection coefficients when tuning the inverted T-shaped radiator as a function of (a) length L_1 and L_2 of the slot radiator, and (b) length L of the feeding structure. *Source:* From [14]. ©2018 IEEE. Reproduced with permission.

To provide further evidence that the low-band resonance mode and high-band resonance mode are dependent on the longer open slot and the shorter open slot sections of the inverted T-shaped slot radiator, respectively, Figure 4.11a shows their corresponding simulated reflection coefficients when tuning the lengths

(L_1 and L_2) of the slot radiator. Here, decreasing L_1 and L_2 will allow the low-band and high-band resonances to shift to the higher frequency spectrum, and vice versa. As shown in Figure 4.11b, it is realized that tuning the feeding strip length (L) will mainly affect the shifting and matching of the high-band resonance. Therefore, it is concluded that tuning these three vital parameters (L_1, L_2, and L) can realize desirable dual band impedance bandwidth and matching. Nevertheless, the above slot antenna design can only excite dual band operation with measured stacked 6-dB impedance bandwidths of 11.54% (3396–3812 MHz) and 31.08% (4810–6580 MHz), respectively, and it still cannot fulfill the desired wideband operation for 5G NR band n77/n78/n79 (3300–5000 MHz).

To fulfill the 5G NR band n77/n78/n79 spectrum, as well as providing good decoupling (isolation) between adjacent slot antenna element of better than 10.8 dB, [15] has adopted a very unique integrated slot antenna pairs technique that combines the ground/bottom slot mode (excited by loading a horizontal slot on the system ground) and the chassis/top slot mode (excited by cutting an open slot to the metallic chassis and loading a connecting line over it, which forms a T-shaped slot) so that the two different slot modes (top and bottom slot modes) can be integrated and forms a much wider operational bandwidth.

Figure 4.12 shows the design evolution of the integrated slot antenna pair that stemmed from the design of two closely spaced IFA (with a gap of 1 mm). As shown in Figure 4.12a, broad 6-dB impedance bandwidth of 14.8% (3750–4350 MHz) was realized after loading a decoupling element (lump inductance L), as depicted in Figure 4.12d. To accommodate the metallic chassis of the smartphone and improve the operational bandwidth, as shown in Figure 4.12b, the two-ports IFA was modified and co-designed with the metallic chassis of the smartphone, and one can see in Figure 4.12e that the 6-dB impedance bandwidth was enhanced tremendously (from 3300 MHz to over 5000 MHz) because the IFA mode was replaced by the slot mode generated via the open slot (loaded at the chassis) and the horizontal slot (loaded at the system ground), not to mention that the good impedance matching across the higher frequency spectrum was realized by loading a lump capacitance C into the feeding point of each port. However, by observing further into Figure 4.12e, the isolation (decoupling) level of such design by loading a decoupling element (lump inductance) into the open section of the slot radiator was undesirable because it can only yield 10-dB isolation levels across the 3500–4000 MHz band. Therefore, as shown in Figure 4.12c,f, the lumped inductance was substituted by a metal-connecting line loaded across the open section of the slot radiator, which results in constructing an integrated slot antenna pair that has abundant 6-dB impedance bandwidth of >41.0% covering the 5G NR band n77/n78/n79 (3300–5000 MHz).

To avoid further confusion, the detailed theories regarding the excitation of the common mode (CM) and differential mode (DM) of [15] are not discussed in here. By simply observing Figure 4.13a,b, which show the CM E-field distributions at the lower (3.4 GHz) and higher (4.9 GHz) frequency, respectively, in the low-band, the

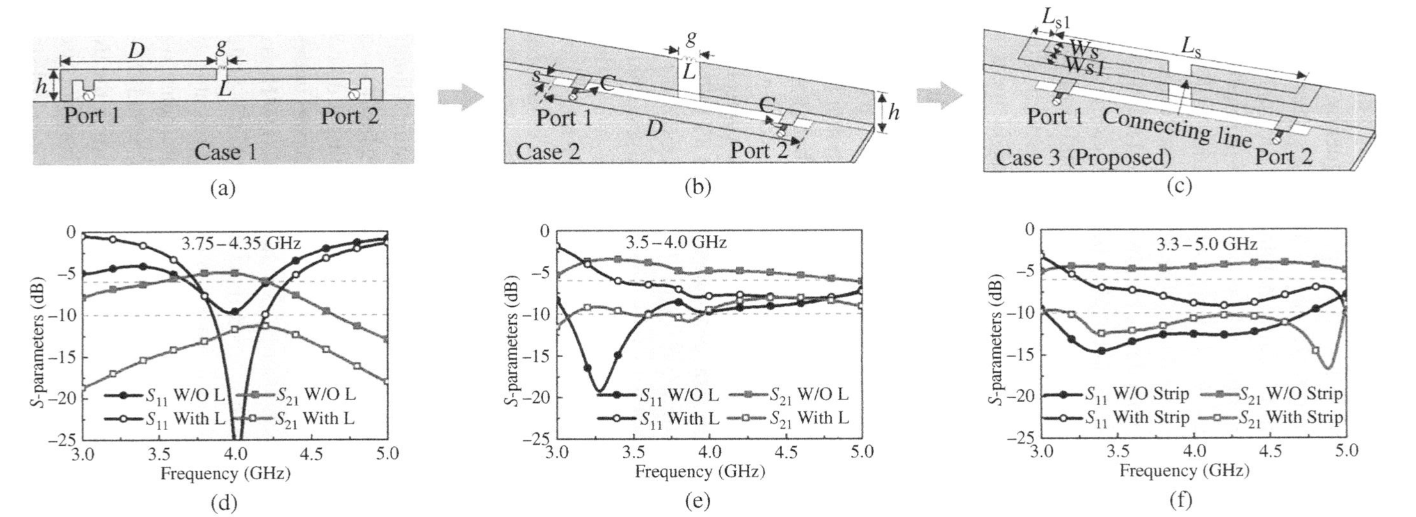

Figure 4.12 The design evolution of the integrated slot antenna pair. (a) Case 1: Two IFA designs decoupled by an inserted inductor (L = 21 nH). (b) Case 2: Two open slot antennas decoupled by an inserted inductor (L = 16 nH). (c) Case 3: Two open slot antennas decoupled by a connecting line. (d)–(f) Simulated S-parameters of Cases 1–3. *Source:* From [15]. ©2021 IEEE. Reproduced with permission.

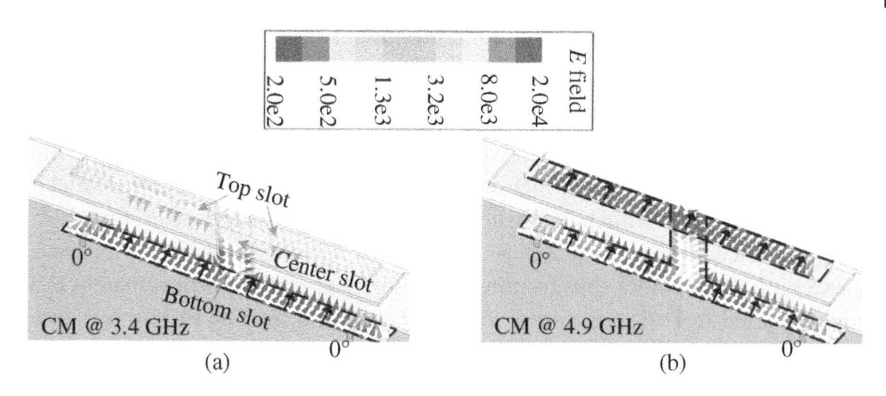

Figure 4.13 The common mode (CM) *E*-field distributions. (a) at 3.4 GHz. (b) at 4.9 GHz. *Source:* From [15]. ©2021 IEEE. Reproduced with permission.

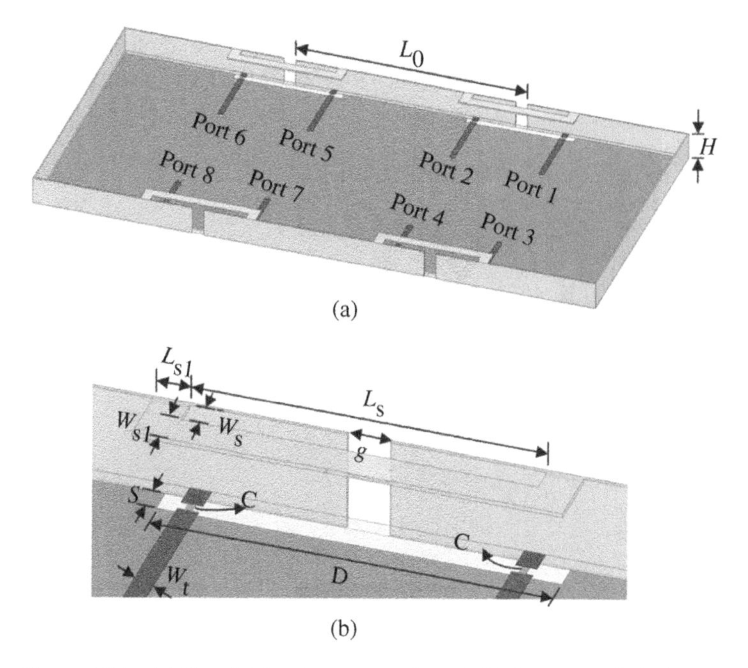

Figure 4.14 (a) The eight-antenna array layout. (b) Geometry of the integrated slot antenna pair loaded with two lump capacitances for impedance matching. *Source:* From [15]. ©2021 IEEE. Reproduced with permission.

bottom slot is exciting a "balanced slot mode," while in the high-band, the top slot is excited because of the strong electrical coupling between the bottom and top slots. The final layout of the eight-antenna array formed by four sets of integrated slot antenna pair is shown in Figure 4.14a, and the final geometry of the integrated slot antenna pair loaded with two lump capacitances is shown in Figure 4.14b.

4.1.4 Dual Band PIFA and Dual Band Monopole

In Chapter 3, we learned that a conventional planar inverted-F antenna (PIFA) could only yield a single quarter-wavelength mode [16] because it has a single branch linked to the feeding point while one of its open ends is connected to the shorting point, as depicted in Figure 4.15. However, from work reported in [17, 18], if an additional branch of dissimilar length is also extended from the PIFA, the two unequal-length branches will indeed yield two different quarter-wavelength PIFA modes thus forming a dual-band operation. Unfortunately, the work reported in [17] only covers <3 GHz bands, and the one in [18] has only covered the (3400–3600 MHz) and (5725–5785 MHz) bands. Even though the idea of applying multiple branches of PIFA can be applied to the 5G smartphone, the PIFA's height (profile) must be carefully evaluated.

The monopole antenna usually excites a quarter-wavelength mode at the fundamental mode or half-wavelength mode at 2nd higher-order mode, as reported in [19]. To yield dual band operations for 5G FR1 smartphone applications at (3400–3600, 4550–4750 MHz) and (3300–3600, 4800–5000 MHz), the technique of

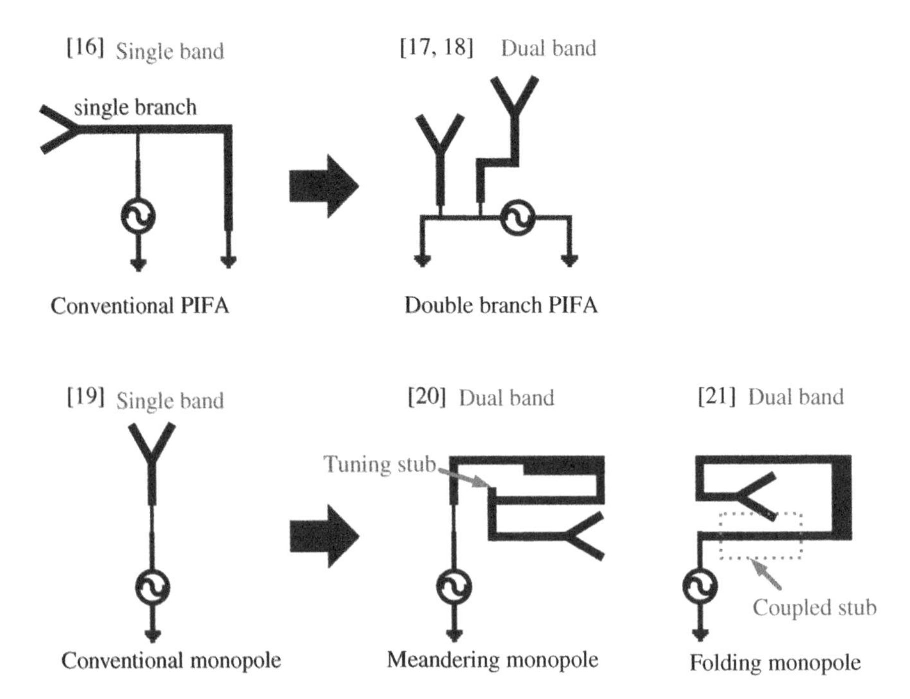

Figure 4.15 The conceptual diagrams of transforming a single band conventional PIFA (with a single branch) into a dual band PIFA (with double branch) type. Reference numbers are included for each design.

applying the meandering/folding monopole structure was reported in [20, 21], respectively, as shown in Figure 4.15. Notably, during the process of bending and folding the conventional monopole structure, certain sections of the monopole structure may require different width lengths, and a tuning stub [20] or coupled stub [21] may be applied for achieving better impedance matching.

4.2 Hybrid Antenna Design Topologies

In the previous section, the antenna types applied in every MIMO array are of the same kind; thus, each antenna element has exhibited very similar radiation characteristics. However, in recent years, the hybrid antenna designs for 5G FR1 smartphone applications have been a topic of interest because different antenna type (such as loop, slot, monopole, and PIFA) has its own unique radiation characteristic performance. If they are separately applied or integrated as part of the eight-antenna MIMO array, the radiation characteristics, isolation (decoupling) between adjacent antenna elements, and even MIMO performances such as the Envelope Correlation Coefficient (ECC) may be distinctly improved. In this section, the "separated," "integrated," and "tightly arranged" hybrid antennas for 5G FR1 smartphone applications are mainly categorized as follows:

1) Separated slot and monopole
2) Integrated and tightly arranged slot and monopole
3) Tightly arranged loop and monopole
4) Tightly arranged loop and slot
5) Integrated loop and IFA

In the following sub-sections, because the description of radiation patterns is very tedious, only Section 4.2.1 will discuss the detailed radiation characteristics of each reported work. As for the rest of the sub-sections, we will mainly focus on the design of achieving the hybrid antenna type.

4.2.1 Separated Slot and Monopole

The technique of separately loading the slot radiator and monopole radiator into the same system ground of the 5G smartphone is discussed in this sub-section. From the open literature, one of the earliest works that apply both the slot radiator and monopole radiator was reported in [22], as shown in Chapter 3, Figure 3.15. As shown in Figure 4.16, because of its unique distributions along the two longer side edges of the system ground, in which all four coupled strip radiators are collocated in the middle section, whereas the four open slot radiators are placed at four corners of the system ground, good radiation diversity was achieved.

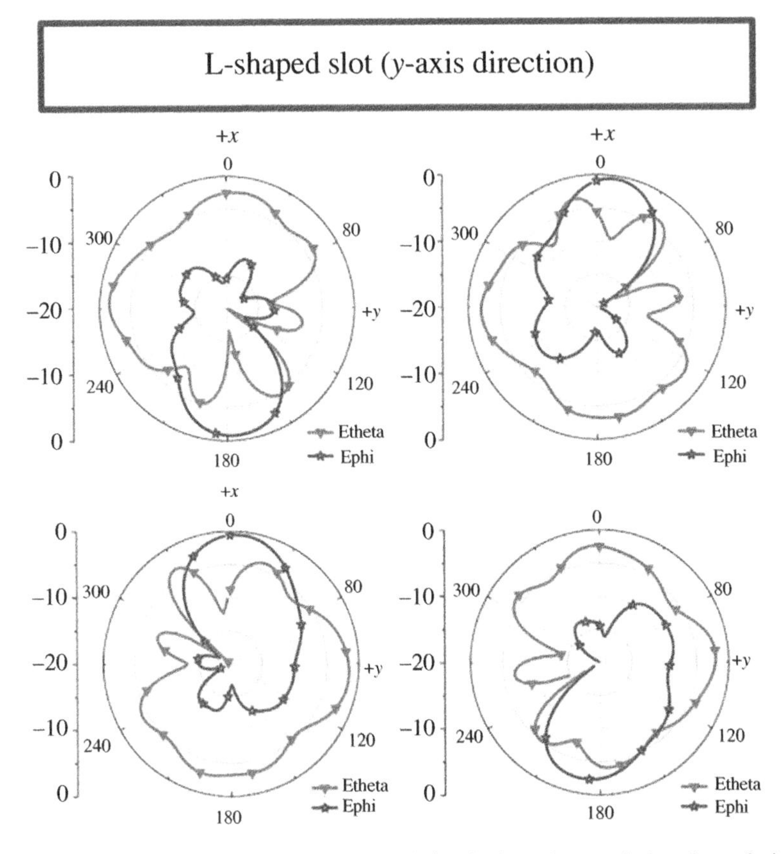

Figure 4.16 The layout of the separated slot (L-shaped open slot) and coupled strip radiators (C-shaped monopole) on the 5G smartphone ground system, and the conceptual Maximum Radiation Orientation (MRO) of all radiators with radiation plots. *Source:* From [22]. ©2016 IEEE. Reproduced with permission.

As each pair of coupled strip radiators are collocated on the same edge (Ants. 5 and Ant. 8, Ant. 6, and Ant. 7), and each antenna pair has their respective open ends pointing at each other, as shown in Figure 4.16, the main radiation beams of the four coupled strip radiators are pointing outward, which signifies that the MROs of Ant. 6 and Ant. 7 are pointing toward the −ve x-axis direction, and the MROs of Ant. 5 and Ant. 8 are pointing toward the +ve x-axis direction. As the four L-shaped open slot radiators (Ant. 1 to Ant. 4) have their respective open slot section placed near each corner of the system ground, they have also exhibited similar phenomena of main radiation beams pointing outward, but the MROs of Ant. 1 and Ant. 2 are in the −ve y-axis direction, and the MRO of Ant. 3 and Ant. 4 are in the +ve y-axis direction. Because all the antenna elements (radiators) have

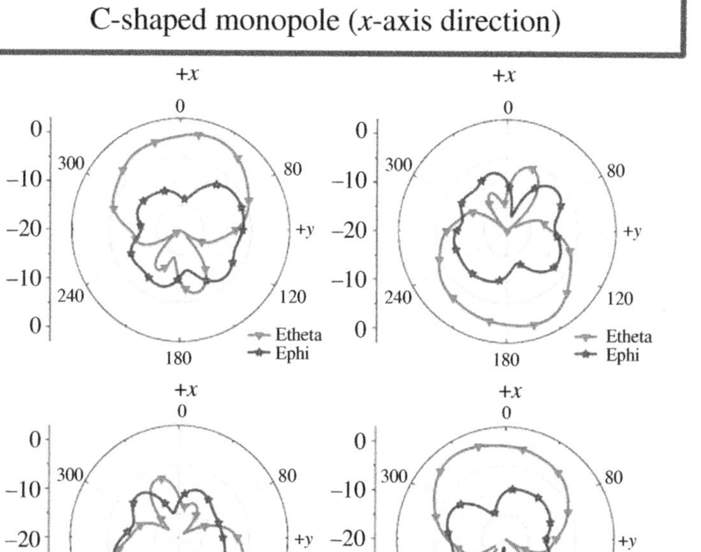

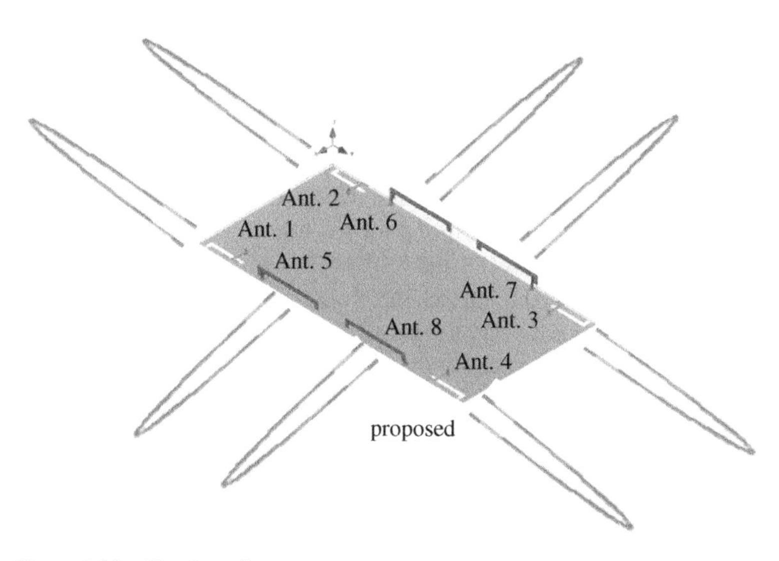

Figure 4.16 (Continued)

their respective MROs pointing outward (not toward each other or not pointing toward the system ground) with good radiation diversity, therefore, desirable Mean Effective Gain (MEG) with differences of MEGs <1 dB and ECC < 0.15 across the bands of interest were calculated, which signifies that the MIMO performances of the antenna array are stable.

As shown in Figure 4.17, [9] has reported hybrid 12-antenna elements that can form an 8-antenna MIMO array (Ants. 1, 2, 3, 5, 7, 8, 10, 12) for LTE bands 42/43 and a 6-antenna MIMO array (Ants. 1, 2, 4, 6, 9, 11) for LTE band 46. Here, Ant. 1 and Ant. 2 are dual branch monopole radiators (called DM) with dual band operation in the 3.5 and 5.5 GHz, Ants. 3, 5, 7, 8, 10, 12 are open slot radiators with a longer slot length (called LS) achieving 3.6 GHz slot mode, and Ants. 4, 6, 9, 11 are other open slot radiators with a shorter slot length (called SS) achieving 5.0 GHz slot mode. The positioning of these monopole radiators (DM) and slot radiators (SS and LS) are very vital in this work, as their exciting main lobes are dependent on it, as shown in Figures 4.18 and 4.19.

Figure 4.18 shows the measured theta-polarized and phi-polarized gain patterns plotted in the *xy* plane of the MIMO array at 3600 MHz (for the eight-antenna MIMO array). Only the results of Ants. 1 to 7 are given for brevity. Here, the two DMs (Ant. 1 and Ant. 2) have shown strong radiation in the phi = 45° and phi = 315° directions, thus achieving good radiation diversity. In addition to that, the main lobes of LS (Ant. 3, Ant. 5, and Ant. 7) have also shown strong radiation in the +ve *x*-axis direction, which can be observed at near phi = 315°, phi = 0°, and phi = 45°, respectively.

Figure 4.19 shows the measured theta-polarized and phi-polarized gain patterns plotted in the *xy* plane of the MIMO array at 5500 MHz (for the six-antenna MIMO array). In this figure, the two DMs (Ant. 1 and Ant. 2) have shown maximum radiation oriented in −ve *y*-direction and +ve *y*-direction, and the two SS (Ant. 4 and Ant. 6) are radiating nearly toward phi = 45° and phi = 315° directions, respectively.

Summarizing the radiation patterns excited by the hybrid antennas (slot and monopole radiators) applied in [9], the two dual branch monopole radiators (DM) and the open slot antennas (LS and SS) have shown distinctive and complementary patterns with MRO scattered, which is a good feature for obtaining good spatial/radiation diversity performances. Furthermore, both DM and SS have exhibited relatively weaker radiation at near −ve *x*-axis direction, which can be justified by the high isolation levels (>12 dB) between the 2 DM and 10 open slot radiators that are collocated along the left side edge and right side edge of the system ground. Further work in [19] has also reported the slot and monopole radiators for a 10-antenna MIMO array, as the slot antenna and monopole structures are very similar to those reported in Chapter 3. The results are not shown here.

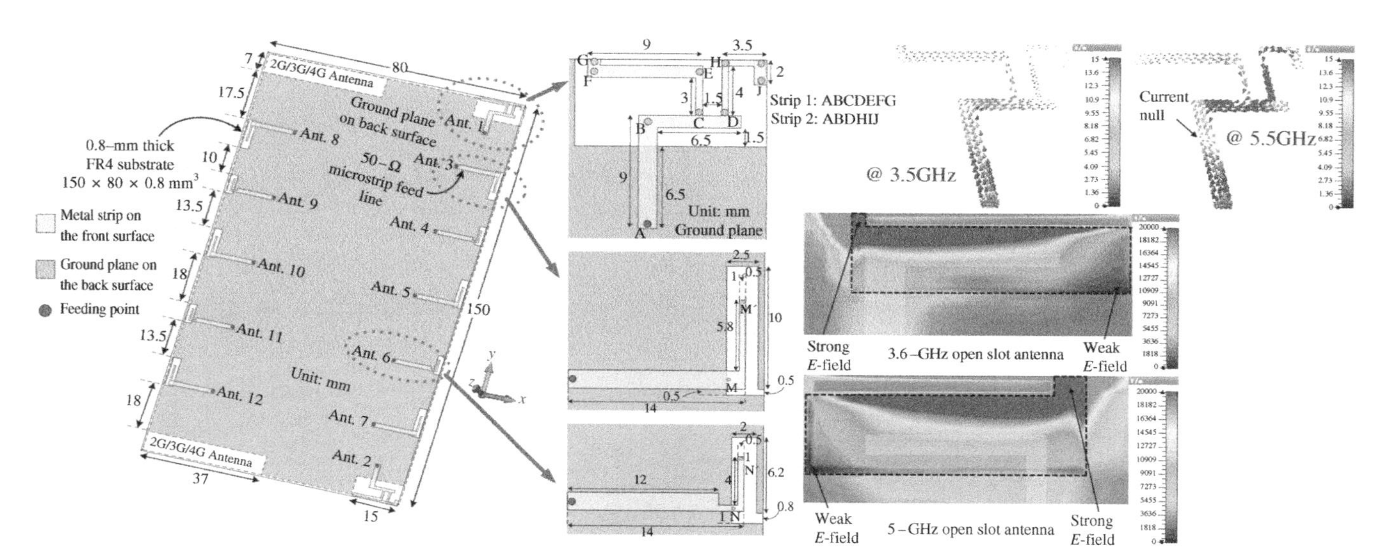

Figure 4.17 The layout of the separated slot (L-shaped open slot) and dual branch monopole radiators on the 5G smartphone ground system with their respective current and *E*-field distributions. *Source:* From [9]. ©2017 IEEE. Reproduced with permission.

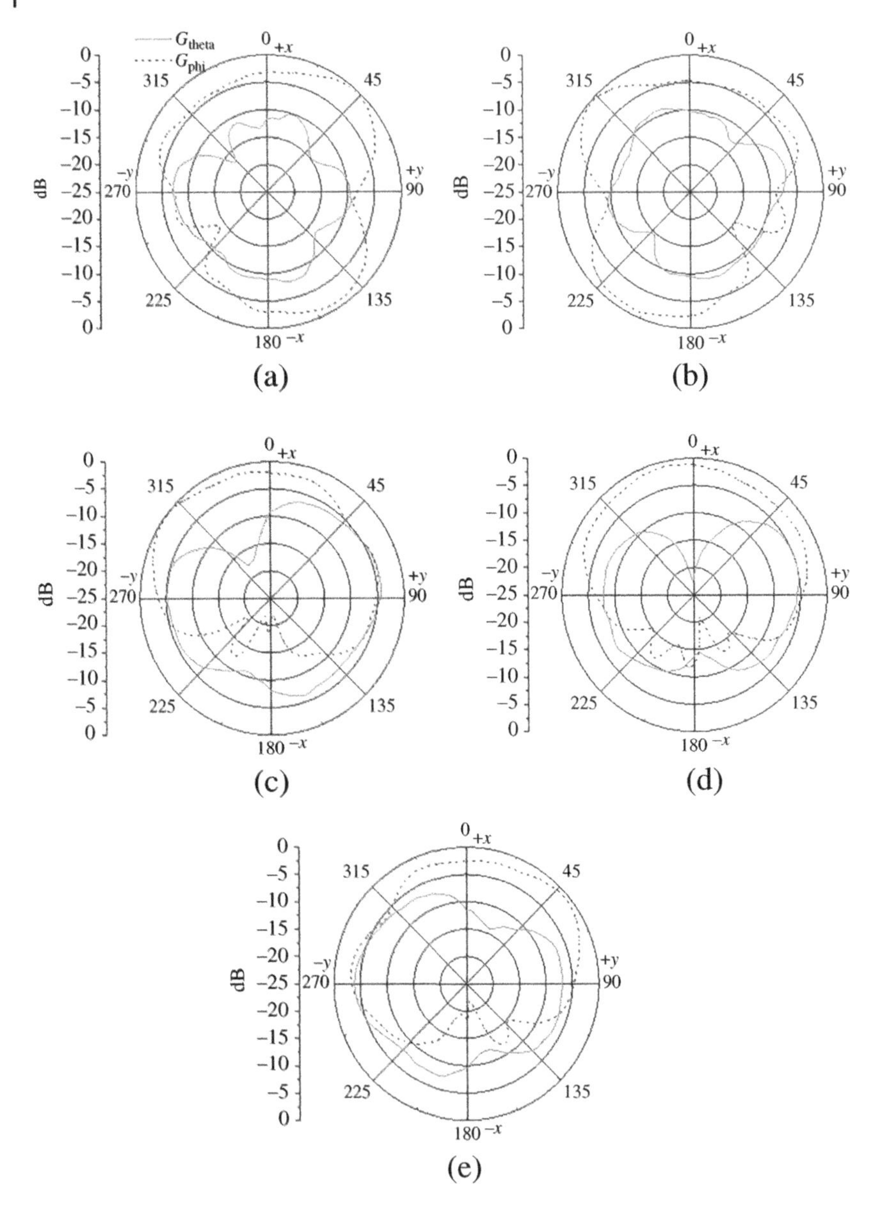

Figure 4.18 The measured radiation patterns of the hybrid antenna in [9] at 3600 MHz. (a) DM Ant. 1, (b) DM Ant. 2, (c) LS Ant. 3, (d) LS Ant. 5, and (e) LS Ant. 7. *Source:* From [9]. ©2017 IEEE. Reproduced with permission.

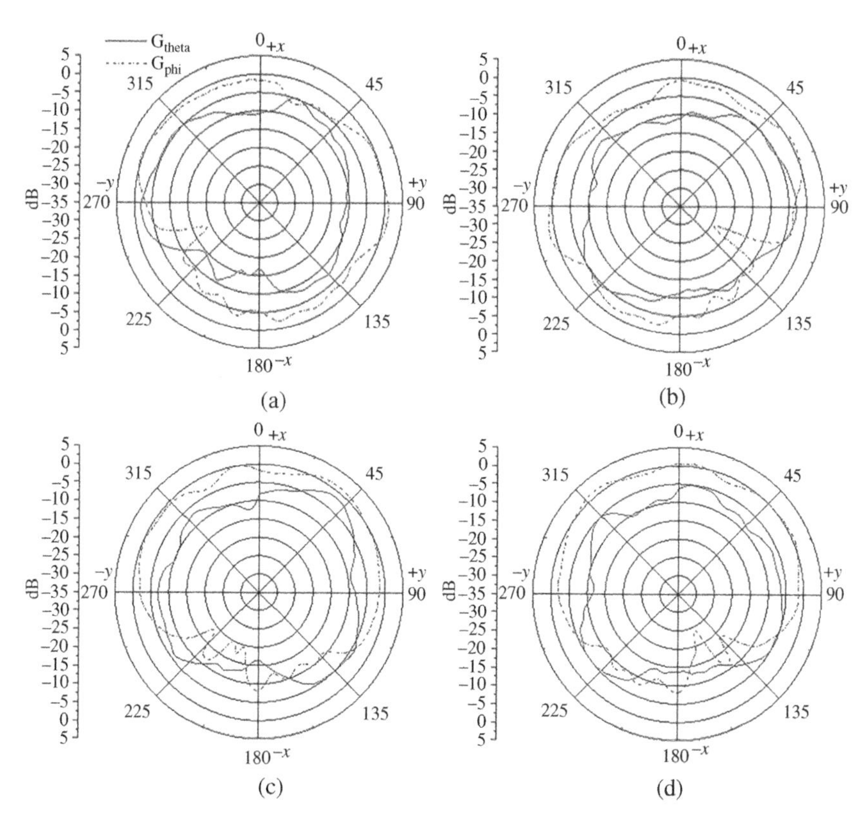

Figure 4.19 The measured radiation patterns of the hybrid antenna in [9] at 5500 MHz. (a) DM Ant.1, (b) DM Ant. 2, (c) SS Ant. 4, and (d) SS Ant. 6. *Source:* From [9]. ©2017 IEEE. Reproduced with permission.

4.2.2 Integrated and Tightly Arranged Slot and Monopole

Figure 4.20 shows an example of an integrated slot and monopole radiator that forms a dual band eight-antenna array for 5G smartphone applications. Each hybrid antenna is composed of an L-shaped open slot radiator and a U-shaped monopole radiator, in which the slot radiator is realized by etching an L-shaped slot onto the system ground and the standing monopole radiator is printed on the smartphone chassis. Thus, low-band (3500 MHz) and high-band (5350 MHz) operations can be excited separately, and their respective resonance mode can be tuned separately without affecting the other one.

By further observing Figure 4.20, the feeding port was fed at the right bottom corner of the monopole radiator, which allows the excitation of the low-band resonance (at 3500 MHz) by the slot radiator (with quarter-wavelength slot mode),

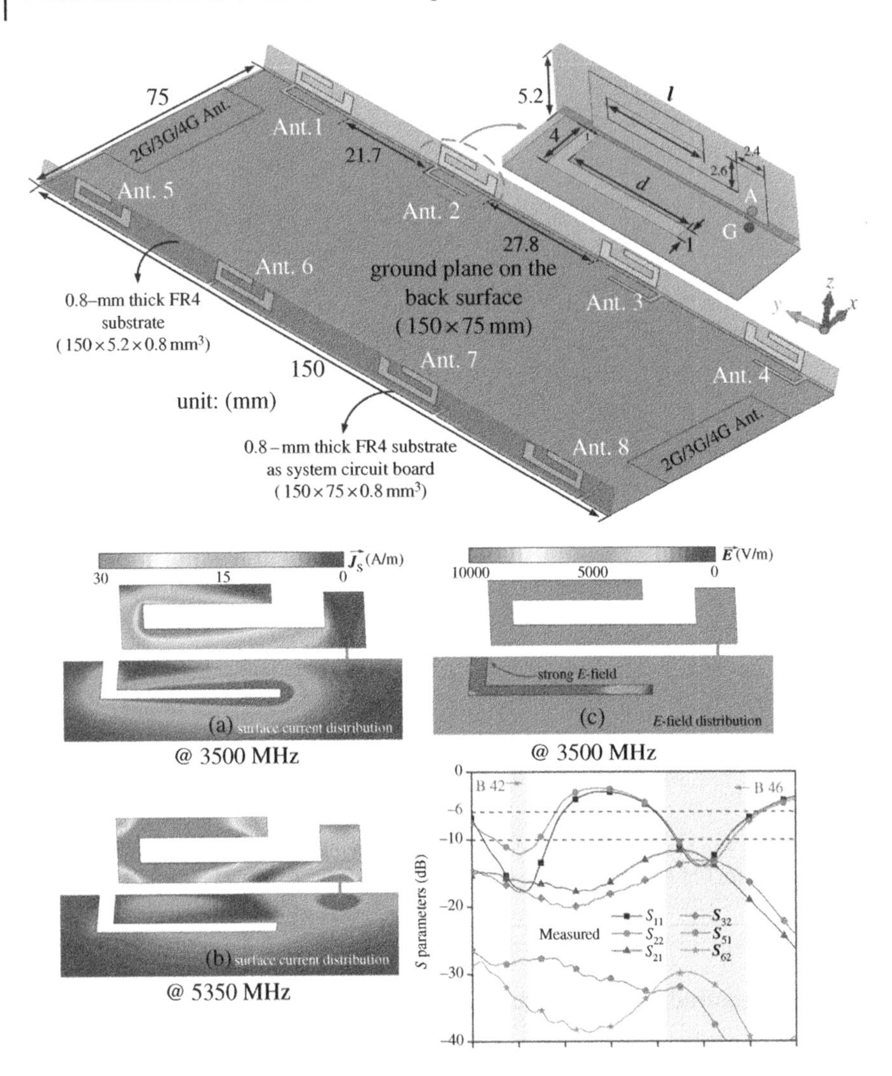

Figure 4.20 The eight-antenna MIMO array layout and configuration of the integrated slot and monopole radiator. The current distributions are at 3500 and 5350 MHz, and electric field distribution plot is at 3500 MHz. The measured *S*-parameters are included. *Source:* From [23]. ©2018 Wiley. Reprinted with permission.

and the high-band resonance (at 5350 MHz) excited by the monopole radiator (with half-wavelength higher-order mode). From the measured *S*-parameters, such integrated slot and monopole antenna have stacked 6-dB impedance bandwidths that can cover the 5G NR band n78 (3300–3800 MHz) and 5G NR band n46 (or LTE band 46, 5150–5925 MHz).

As mentioned earlier in Section 4.21, during the process of designing the eight-antenna MIMO array, the placement of each hybrid antenna must be meticulously collocated along the two longer side edges of the system ground so that good pattern diversity can be realized, leading to improved MIMO performances. By simply observing the measured radiation patterns for Ant. 1 to Ant. 4 (collocated on one of the longer side edges) at the two resonances, as shown in Figure 4.21, good MROs are plotted in the *x*-axis direction, which signifies that the main beam radiation is pointing outward from the system ground (or smartphone chassis), and thus good pattern diversity was realized.

As far as the authors are concerned, it is rather difficult to realize a very wideband operation for an integrated-type antenna, especially when only one

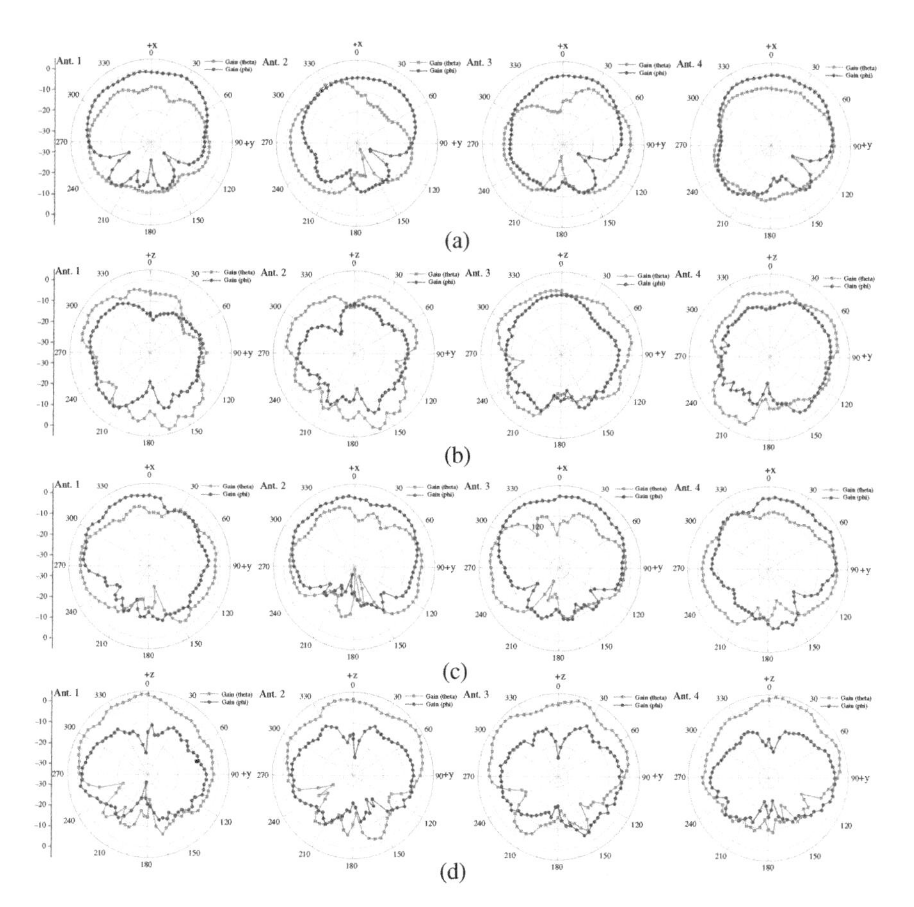

Figure 4.21 Measured radiation patterns of the integrated slot and monopole eight-antenna MIMO array. (a) 3.5 GHz, *xy*-plane. (b) 3.5 GHz, *yz*-plane. (c) 5.5 GHz, *xy*-plane. (d) 5.5 GHz, *yz*-plane. *Source:* From [23]. ©2018 Wiley. Reprinted with permission.

feeding port is applied to excite the two radiators (slot and monopole) simultaneously. Therefore, a new arrangement technique was recently promoted, and it is called the "tightly arranged" technique, in which the two radiators are closely arranged in a certain position, and they can electrically complement each other, thus achieving wide operational bandwidths as well as improved isolation and radiation characteristics. By applying this tightly arranged technique to a hybrid antenna type, two feeding ports are now required (instead of one). Thus, the hybrid antenna is also analogous to an antenna pair.

As shown in Figure 4.22, a wideband tightly arranged hybrid slot and monopole radiator that forms an antenna pair for metallic chassis 5G FR1 smartphone applications were reported in [24]. To achieve wideband operation, the monopole radiator was a single port feeding type (port 1), whereas the slot radiator (analogous to an inverted π-shaped slot) was fed by a differential port-feeding type (port 2+ and port 2−) via a balun chip. Therefore, it can excite a monopole mode at 3500 MHz and a slot mode at 4500 MHz. By combining these two modes, the measured stacked 6-dB impedance bandwidth of each antenna pair can yield wide bandwidth of >41% that can cover the 5G NR band n77/n78/n79 (3300–5000 MHz), and the simulated result has also indicated very high isolation of up to 21 dB between the two different radiators (of the antenna pair).

Figure 4.23 shows the design evolution procedure of the work reported in [24], which stems from the tightly arranged coupled-fed loop radiator and monopole radiator design reported in [25]. The designed structure for each case (Cases 0–4) was explained very clearly with the simulated S-parameter results. To further comprehend the excitation of these two resonance modes at 3500 and 4500 MHz, the simulated current distributions of the tightly arranged hybrid antenna pair excited by the two ports (port 1 and port 2) at 3500 and 4500 MHz, and their respective 3-D radiation patterns are plotted in Figure 4.24. In this figure, at 3500 MHz, the hybrid antenna pair can realize an x-polarized monopole mode and y-polarized dipole mode when excited by port 1 and port 2, respectively. At 4500 MHz, the slot mode excites an x-polarized radiation field, and the open slot mode excites a y-polarized radiation field. Therefore, at low-band resonance, the monopole mode and dipole mode polarization are orthogonal with each other, and at high-band resonance, the slot mode and open slot mode are also orthogonal with each other; thus, both low and high-bands can yield very high isolation.

4.2.3 Tightly Arranged Loop and Monopole

As aforementioned, the integrated hybrid antenna type usually has one feeding port that fed into the main radiator, followed by coupling the energy to the other radiator that is integrated into it. As for the tightly arranged hybrid antenna type, it usually will have two separate feeding ports for each radiator, but the two

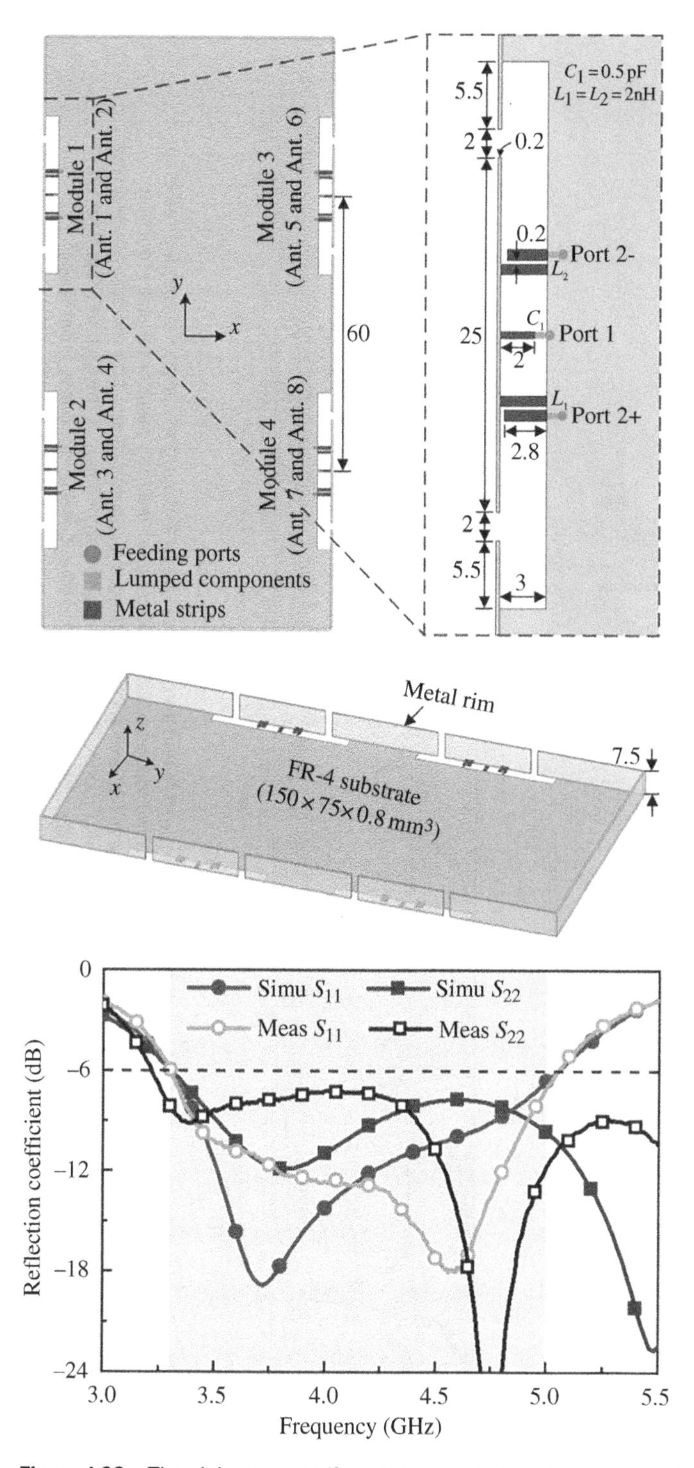

Figure 4.22 The eight-antenna (four-antenna pairs) MIMO array layout and configuration of the tightly arranged slot and monopole radiators. The measured reflection coefficients are also included. *Source:* From [24]. ©2020 IEEE. Reproduced with permission.

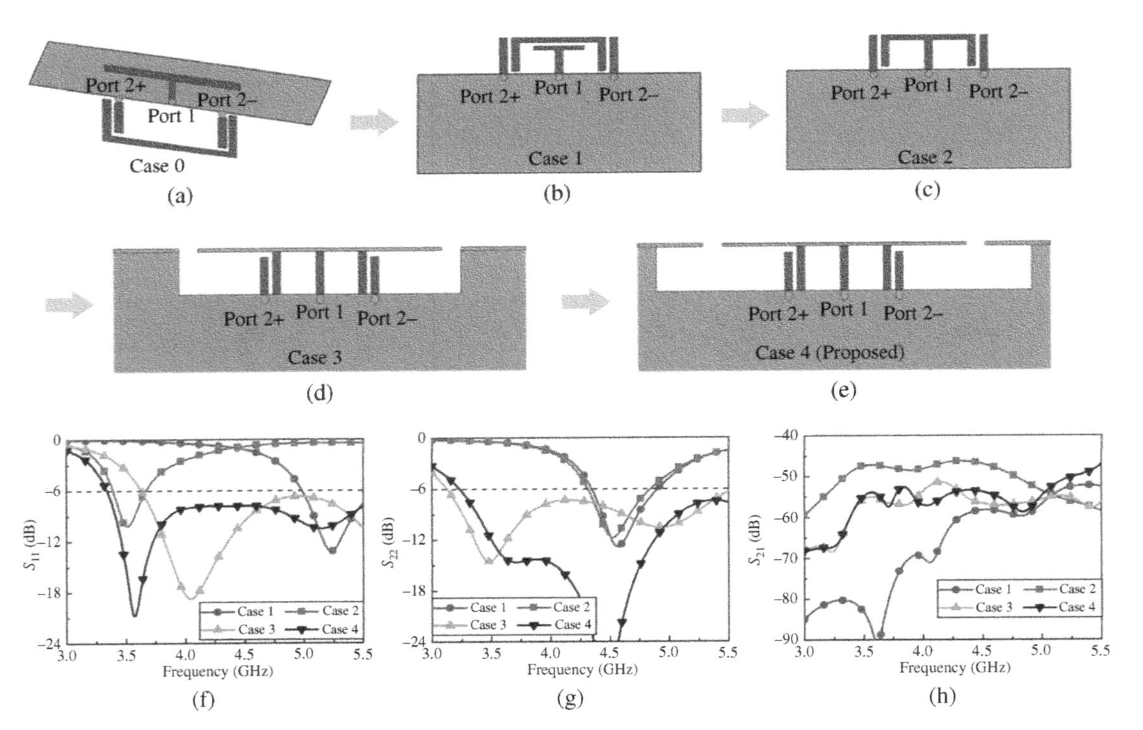

Figure 4.23 The design evolution procedure of the tightly arranged hybrid antenna pair. (a) Case 0: A tightly arranged coupled-fed loop radiator and monopole radiator in [25]. (b) Case 1: Arranging the monopole radiator and coupled-fed loop radiator for size reduction. (c) Case 2: Merging the two different radiators as a shared radiator. (d) Case 3: Constructing the shared radiator into a metal-rimmed type and the horizontal strip was replaced by the metal rim. (e) Case 4 (Proposed): Lengthening the etched slot in the system ground to allow new orthogonal modes. (f)–(h) Comparison of S_{11}, S_{22}, and S_{21} of the antenna pairs proposed in Cases 1–4. *Source:* From [24]. ©2020 IEEE. Reproduced with permission.

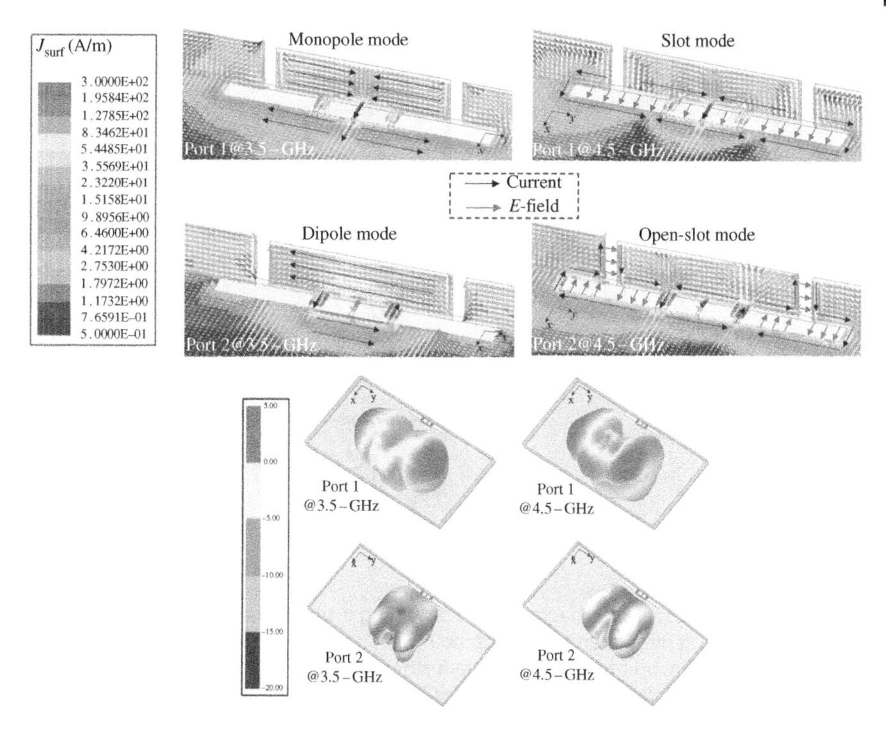

Figure 4.24 The simulated current distributions of the tightly arranged hybrid antenna pair excited by the two ports (port 1 and port 2) at 3.5 and 4.5 GHz. The simulated 3-D radiation patterns of the tightly arranged hybrid antenna pair are included. *Source:* From [24]. ©2020 IEEE. Reproduced with permission.

different radiator types are closely arranged with each other. Figure 4.25 shows the conceptual diagram of transforming the work in [26] from a single port coupled-fed loop radiator into a single port dual radiator type [27, 28]. The main design concept is to transform the previous non-radiating coupled-fed structure of [26] into a monopole radiator, and the energy from this monopole radiator was then coupled to the loop radiator, which, in turn, formed an integrated hybrid antenna pair. Notably, both [27, 28] have exhibited desirable isolation of larger than 17.9 and 14.4 dB, respectively, and their stacked 6-dB impedance bandwidths can well cover the 5G NR band n77/n78 (3300–4200 MHz). As for the work in [29], due to the tightly arranged positions of the two radiators, complemented radiation patterns are realized, and thus desirable isolation of >25 dB can be achieved between the two radiators. However, besides requiring an additional matching circuit for improving the impedance matching of the radiators, [29] has only achieved narrow 6-dB impedance bandwidth of 4800–5000 MHz (partial of 5G NR band n79).

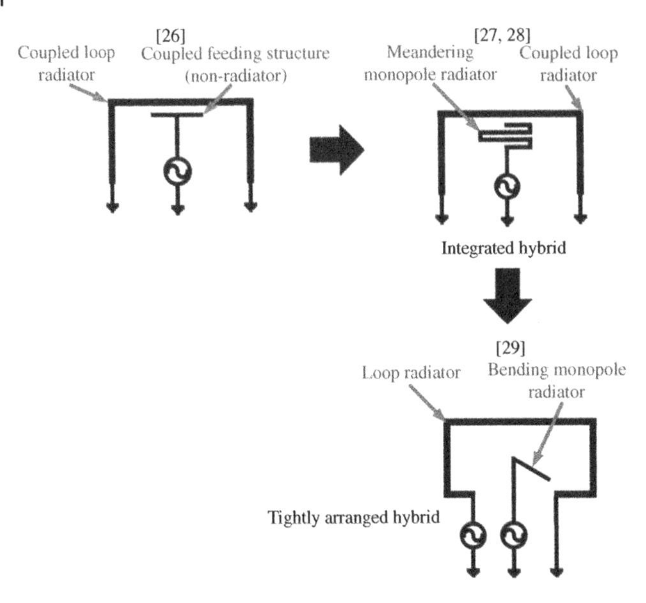

Figure 4.25 The conceptual diagram of transforming a single port coupled-fed loop radiator [26] into an integrated hybrid antenna (with loop and monopole radiators) [27, 28]. This integrated hybrid technique can be further developed into a tightly arranged hybrid antenna with dual antenna pair excited by two different ports [29]. Reference numbers are also included for each design.

Even though [29] is a simple, tightly arranged loop and monopole antenna design, the main problem is the protruded monopole may hinder its ability to fit into an actual 5G smartphone. Thus, another tightly arranged antenna with loop and monopole radiators [25] is introduced in this sub-section. As it is a printed type without any required matching circuit, it can be more easily fitted into a 5G smartphone.

Figure 4.26 shows the configuration of the four-antenna array, as each two-antenna pair (formed by the tightly arranged loop and monopole radiators) is col-located on the two longer side edges of the system ground [25]. By further observing the conceptual diagram, the feeding mechanism of the loop radiator (or Ant. 1) is a differential port-feeding type (similar to [24]), as Port 1+ and Port 1− are connected to a coupled strip used to feed the loop radiator. As for the monopole radiator, it is a simple microstrip line fed bent T-shaped monopole radiator (or Ant. 2). To realize the above feeding mechanisms for the two radiators, the system ground (printed on a substrate) was located between the two radiators. Therefore, due to the tightly arranged positions of the loop and T-shaped monopole radiators in [25], it has also shown complementary radiation patterns (orthogonal mode) and high isolation between the two different radiator types.

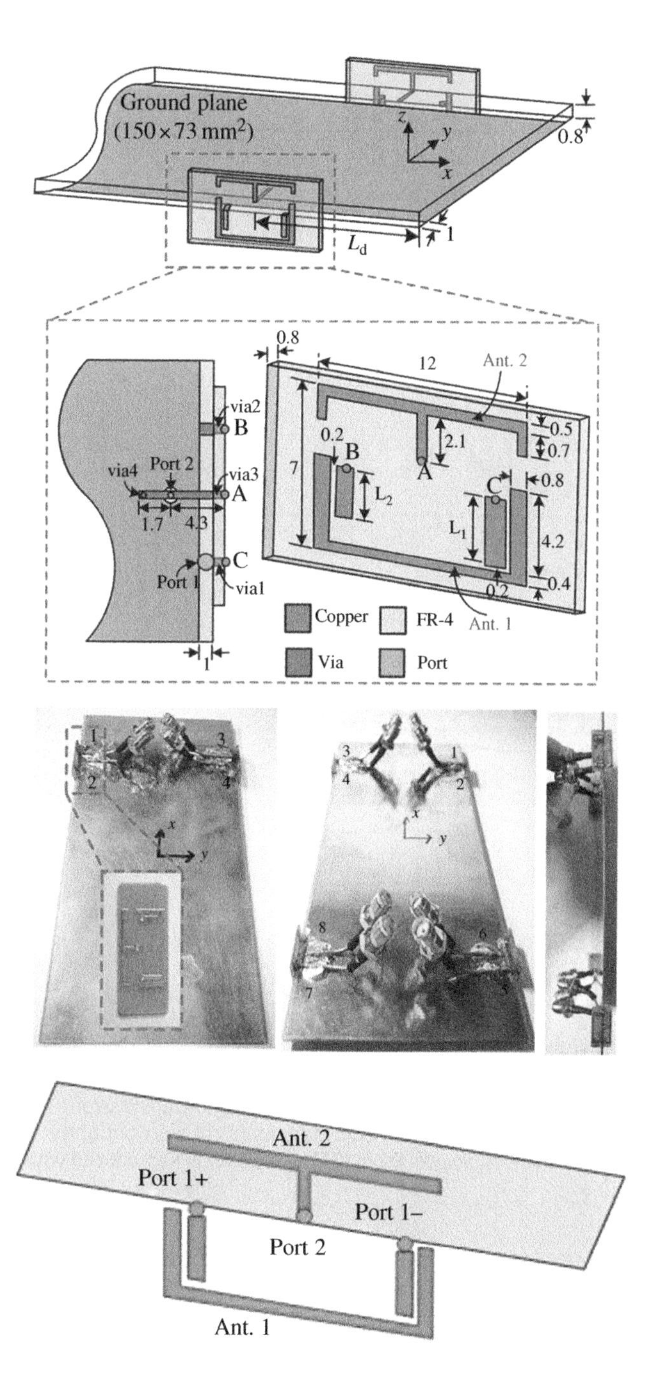

Figure 4.26 The four-antenna (two-antenna pairs) MIMO array layout and configuration of the tightly arranged loop and monopole radiators. The actual fabricated four-antenna and eight-antenna prototypes, and conceptual diagram of the tightly arranged antenna pair are included. *Source:* From [25]. ©2018 IEEE. Reproduced with permission.

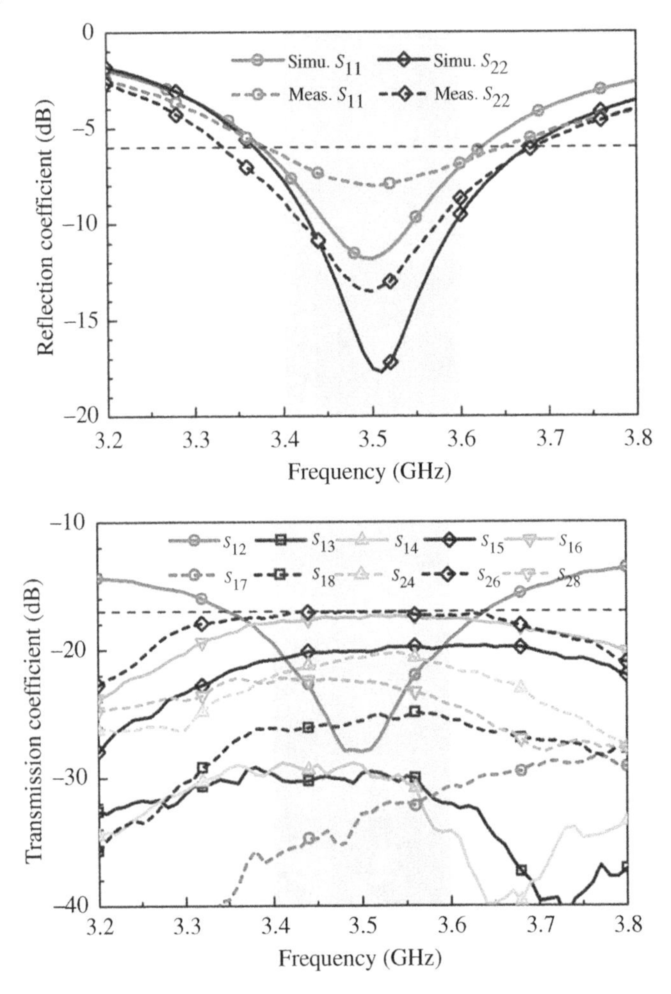

Figure 4.27 The measured and simulated reflection coefficients and transmission coefficients of the eight-antenna MIMO array formed by four identical pairs of tightly arranged loop and monopole radiators. *Source:* From [25]. ©2018 IEEE. Reproduced with permission.

Figure 4.27 shows the measured and simulated reflection coefficients and transmission coefficients of the eight-antenna MIMO array formed by four identical pairs of tightly arranged antenna pairs, and their respective layout positions on the system ground can be seen in Figure 4.26. Similar to [29], narrow bandwidths were realized by the tightly arranged loop and monopole radiators, and its measured stacked 6-dB impedance bandwidth (3400–3600 MHz) was only able to cover partial 5G NR band n77/n78. As shown in Figure 4.28, due to its complementary

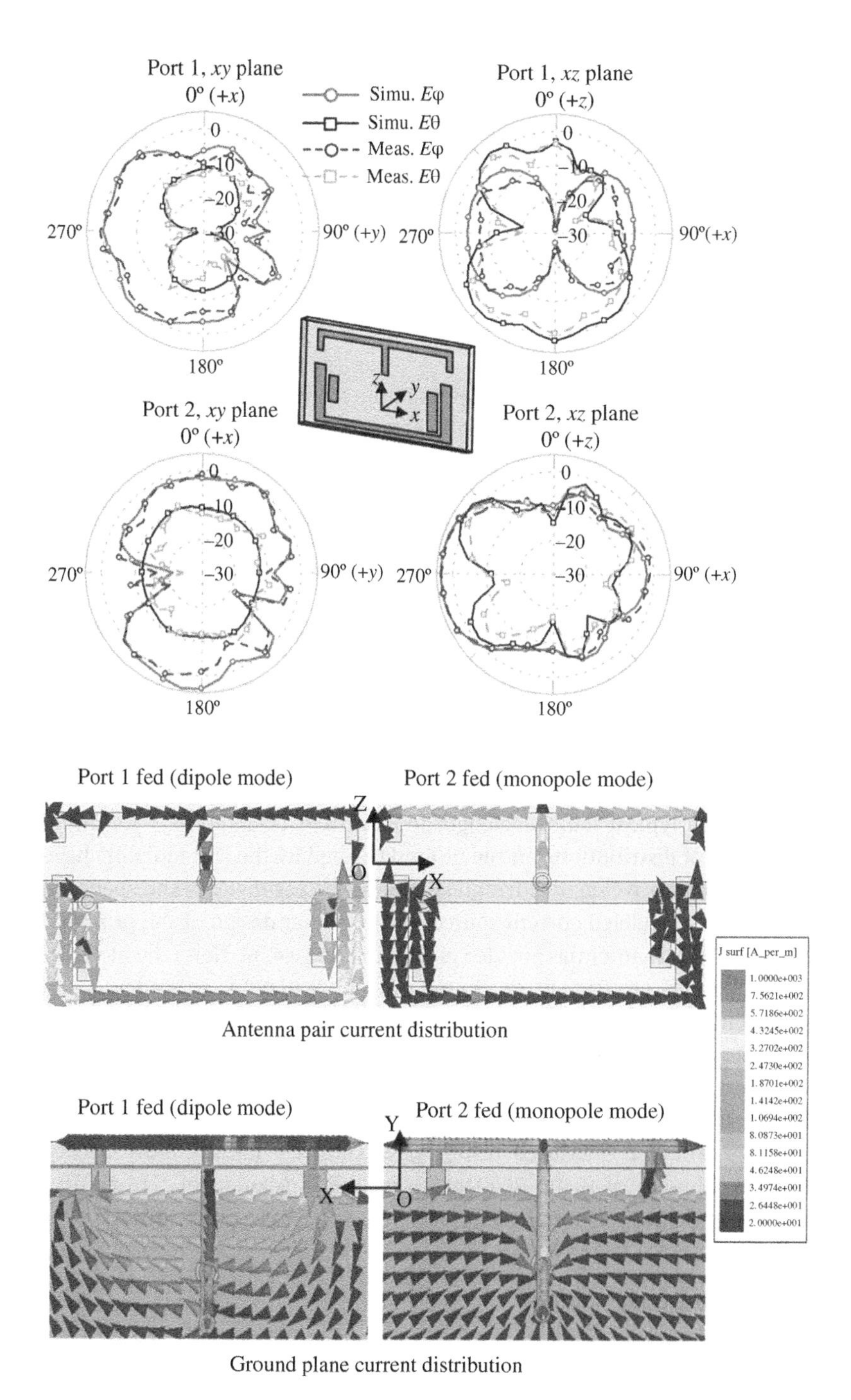

Figure 4.28 The measured and simulated radiation patterns of the tightly arranged antenna pairs at 3.5 GHz. Vector current distributions of the orthogonal-mode antenna pairs and ground plane at 3.5 GHz. *Source:* From [25]. ©2018 IEEE. Reproduced with permission.

radiation characteristics (orthogonal mode at different principal planes) by feeding the two radiators separately at 3.5 GHz, it has also achieved a good isolation level of better than 17 dB across the bands of interest, as shown in Figure 4.27. Nevertheless, to improve the operational bandwidth of this work to 5G NR band n77/n78/n79 (3300–5000 MHz), please see Figure 4.23 [24].

4.2.4 Tightly Arranged Loop and Slot

To realize a tightly arranged loop and slot radiators for 5G FR1 smartphone applications, Figure 4.29 shows a simple tightly arranged method, in which the half-wavelength loop radiator was placed to stand at the longer side edge of the system ground, and an open-end T-shaped slot radiator was loaded into the ground, right underneath and between the feeding port (Port-1) and shorted ends of the loop. On the other hand, the feeding port (Port-2) of the open-end T-shaped slot was located at one end of the T-slot junction, and a capacitor (Cr) was also loaded into the open section of the T-shaped slot for easy tuning of the desired resonance frequency. By arranging the loop radiator and slot radiator in such a way, a self-decoupled mode can be excited, as the loop and slot radiation characteristics will generate complementary radiation patterns (orthogonal mode) and high isolation between the two different radiator types.

Figure 4.30 shows the excitation of the half-wavelength loop mode excited by Port-1, which is a typical half-wavelength loop current orientation. By exciting Port-2, the current distributions on the ground (loaded by the slot radiator) have demonstrated a half-wavelength current path along the perimeter of the open-end T-shaped slot. The modeled current sources of the planar design of the proposed self-decoupled MIMO antennas are also plotted in Figure 4.30. Here, the modeled array of two parallel electric-current sources in the xy-plane is equivalent to an electric-current element (J_1) directed to the x-axis direction. Accordingly, the z-directed magnetic-current element M_2 and the x-directed electric-current element (J_1) have also exhibited orthogonal characteristics. Figure 4.31 shows the fabricated eight-antenna MIMO array formed by the four tightly arranged hybrid antenna pairs, and the measured results have shown a narrow 6-dB stacked impedance bandwidth of (3400–3600 MHz) and high isolation of >16 dB. If a higher or lower spectrum is required, tuning capacitor Cr is a good option.

4.2.5 Integrated Loop and IFA

The advantages of implementing an integrated loop and IFA hybrid antenna type as an antenna element for the MIMO antenna array of the 5G smartphone are compact size, simple structure, and ease in achieving dual band and wideband operations. By observing the conceptual diagrams shown in Figure 4.32, one can

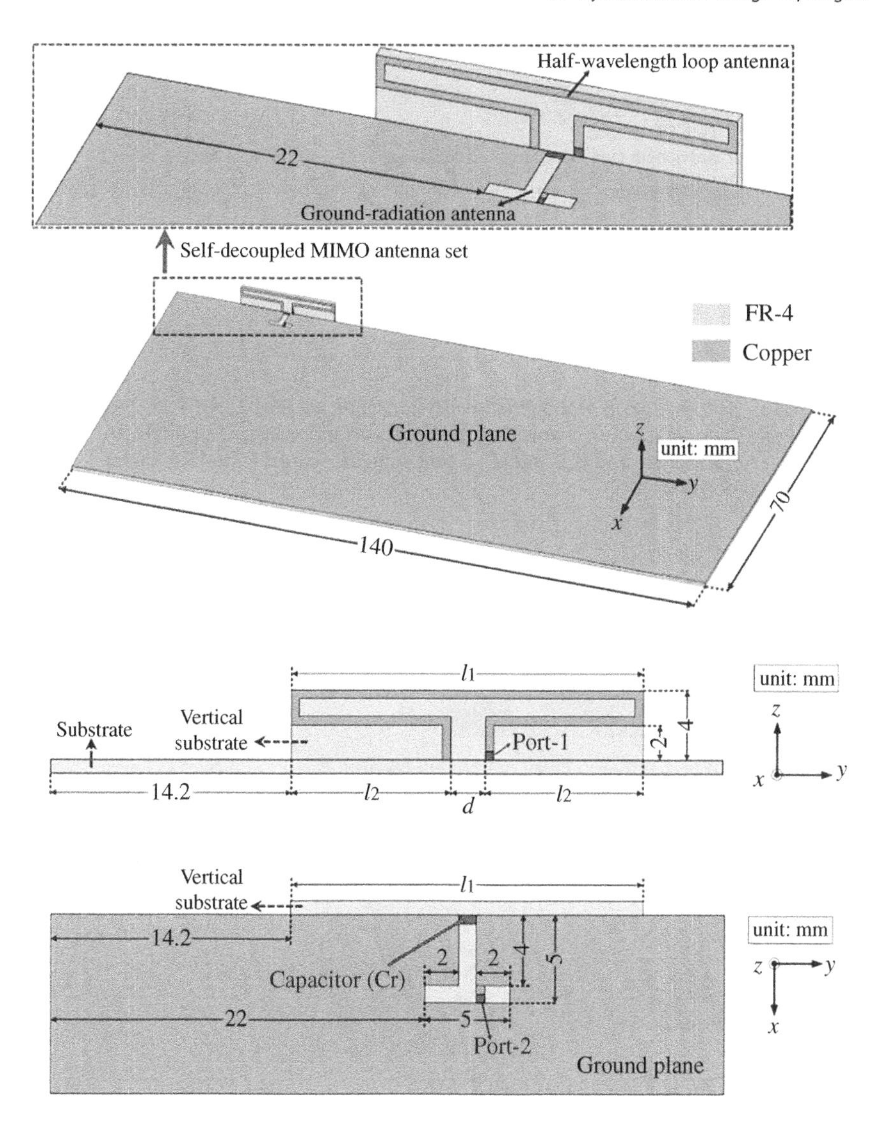

Figure 4.29 The configuration of the tightly arranged loop and slot radiators. *Source:* From [30]. Under a Creative Commons License.

see that the entire integrated loop and IFA hybrid antenna design stemmed from a conventional single coupled-fed loop antenna that has been discussed in Chapter 3, Figure 3.23. Various design concepts that extend this single loop antenna structure into a hybrid antenna type have been reported in [31–33], and they are also included in this figure. By observing [31, 32], the top horizontal

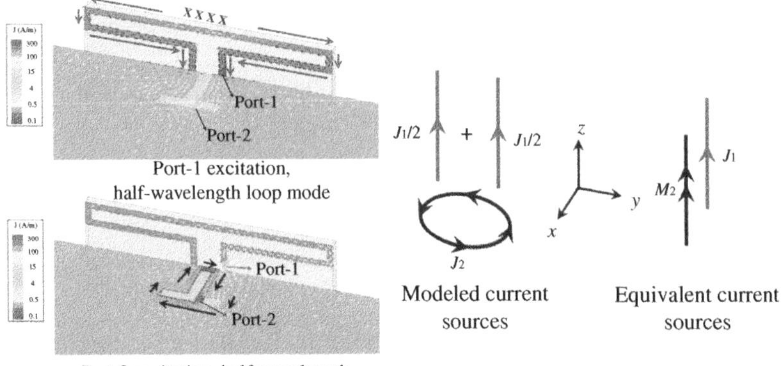

Figure 4.30 The excitation of loop mode and slot mode via Port-1 and Port-2 excitation, respectively. The operation mechanism of the tightly arranged antenna pairs, with modeled current sources and equivalent current sources. *Source:* From [30]. Under a Creative Commons License.

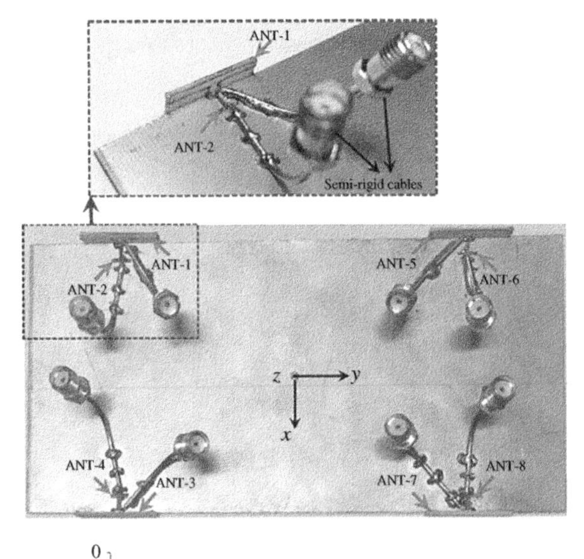

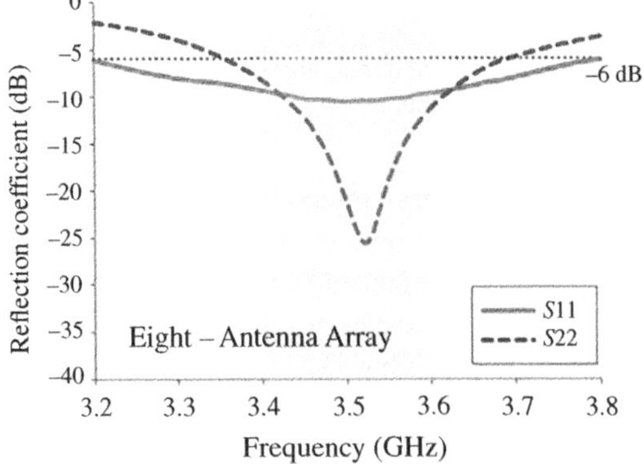

Figure 4.31 The fabricated eight-antenna MIMO array formed by four tightly arranged antenna pairs. The measured reflection coefficients and transmission coefficients of the eight-antenna MIMO array. The effects of tuning capacitor (Cr) loaded at the open section of the T-shaped slot radiator. *Source:* From [30]. Under a Creative Commons License.

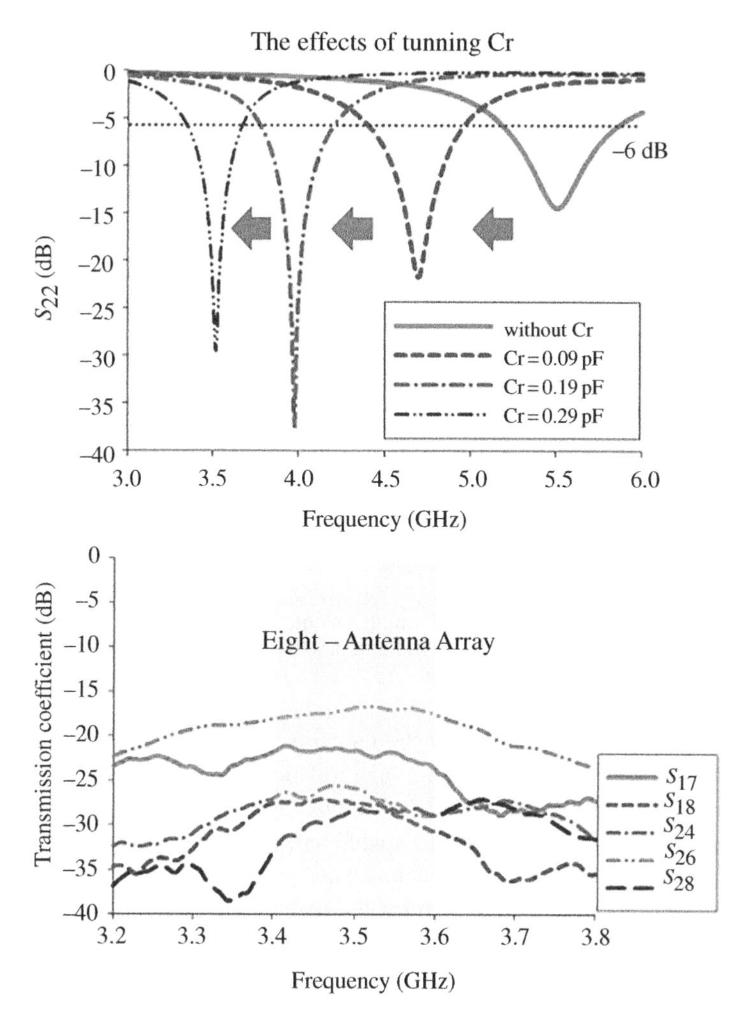

Figure 4.31 (Continued)

section of the initial loop structure was extended to the left and advanced to the right with a bend, respectively, so that an additional quarter-wavelength IFA mode can be generated besides the original quarter-wavelength loop mode. Here, [31] was later constructed as a building block type, and it has successfully shown a compact size decoupled 3.5/5.8 GHz (3400–3600/5725–5875 MHz) dual band integrated loop/IFA hybrid antenna, and similar results were observed in [32]. As for [33], due to the meandering feeding structure, the two separately excited modes (loop mode and IFA mode) can be merged, and thus a wideband integrated loop/IFA hybrid antenna for 5G NR band n77/n78 was realized.

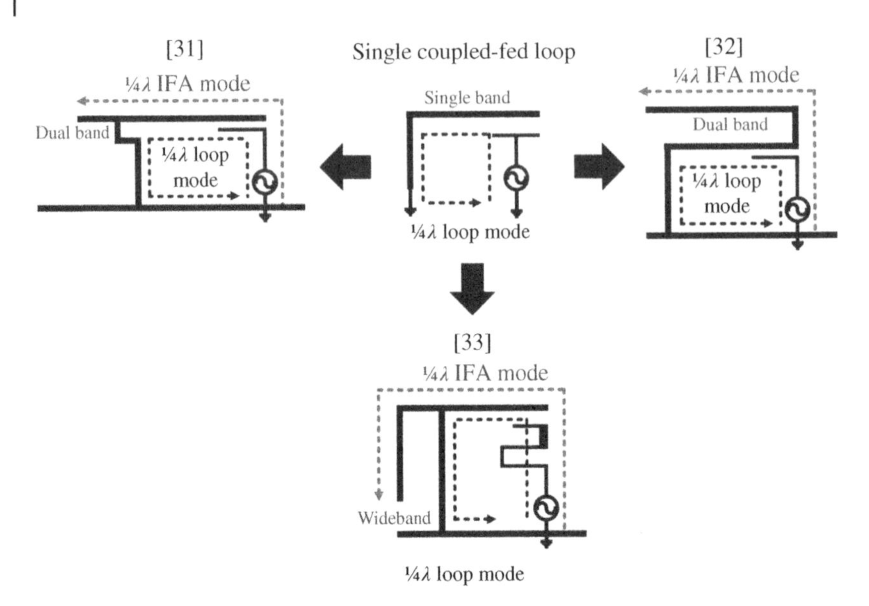

Figure 4.32 The conceptual diagram of transforming a single coupled-fed loop radiator into various dual band/wideband integrated hybrid antenna (with loop and IFA radiators). Reference numbers are included for each design.

In summary, the use of tightly arranged hybrid antenna types can easily achieve good pattern diversity (orthogonal mode) and high isolation (decoupling) between the two different antenna radiators because they are fed by different ports and the nature of the radiation characteristics of the two different radiators have also aided the cause. However, for integrated hybrid antenna type, even though it requires the use of only one feeding port to simultaneously excite two different radiators, it is usually good for realizing dual band operations, and it is rather difficult to apply this integrated hybrid antenna type for achieving wide bandwidth 5G NR band n77/n78/n79 (3300–5000 MHz) operations. Nevertheless, one can also see the work reported in [34] that has applied the integrated loop, slot, and monopole hybrid antenna type, and it can yield triple band operations covering the 5G NR band n78 (3300–3800 MHz), China 5G-Band (4800–5000 MHz), and 5G NR-U n46 (5150–5925 MHz).

4.3 Co-existence with 3G/4G and Millimeter-Wave 5G Antenna Techniques

Since the appearance of various 5G FR1 MIMO antenna array designs operating in the lower section of the C-band (3300–3600 MHz) in 2015, there have been many discussions regarding the integration of 3G/4G antenna with the 5G FR1 antenna array (sub-6 GHz array). However, due to the recent advancement in 5G mmWave antenna technology (especially the Antenna-in-Package, AiP) that is now applied to the 5G smartphone (see Chapter 3, Figure 3.3), the coexistence of 5G FR1 antenna array, 5G FR2 antenna array (mmWave antenna array), and the 3G/4G antennas are increasingly becoming a vital topic for the 5G antenna industry. Previously, in Chapter 3, we have discussed the various antenna placements in a 5G smartphone; however, the exact antenna designs for the coexistence of these 3G/4G antennas, 5G FR1 antennas, and 5G FR2 antennas (including AiP) have not been fully studied. Therefore, this section is mainly categorized as follows:

1) Integrated 4G and 5G sub-6 GHz antennas
2) Integrated 4G and 5G mmWave antennas
3) Miniaturized 5G module of mmWave antennas-in-package integrating non-mmWave antennas (AiPiA)

followed by a short summarized description of future-integrated 5G module and their potential.

4.3.1 Integrated 4G and 5G Sub-6 GHz Antennas

Some of the earliest works that have combined the 4G LTE dual MIMO antenna array with the 5G FR1 eight-antenna MIMO array have been reported in [8, 35], in which the 4G LTE dual antenna array is located at one end of the 5G smartphone system ground section, whereas the 5G FR1 antenna arrays are collocated along the two side edges of the system ground. Figure 4.33 shows the configuration of the integrated 4G LTE and 5G FR1 antenna reported in [8], and its detailed 5G FR1 antenna array design has been shown in Chapter 3, Figure 3.16. The 4G antenna array is an extended version of the one reported in [36] that can cover WWAN/LTE operation bands, and its corresponding S-parameter results are also shown in Figure 4.33. In this design, the feeding strip and shorting strip of the 4G antenna array can excite a fundamental resonant mode at 950 and 850 MHz, respectively, forming a combined dual resonance for GSM 850/900 operations. On the other hand, the shorting strip can generate a higher-order resonance at 1800 MHz, and when combined with the 2600 MHz resonance (introduced by the protruded ground), they can well cover the GSM 1800/1900 and UMTS operations. For brevity, the 5G FR1 antenna array results are not discussed here.

Figure 4.34 shows the configuration of the integrated 4G LTE and 5G FR1 antenna reported in [35], and its 5G FR1 antenna element is a simple PIFA design. Similar to [8], the 4G LTE dual MIMO antenna array of [35] was capable of covering GSM 850/900/1800/1900, UMTS 2100, and LTE 2300/2500 operating bands, and the 5G FR1 eight-antenna MIMO array has realized narrow C-band of (3400–3600 MHz).

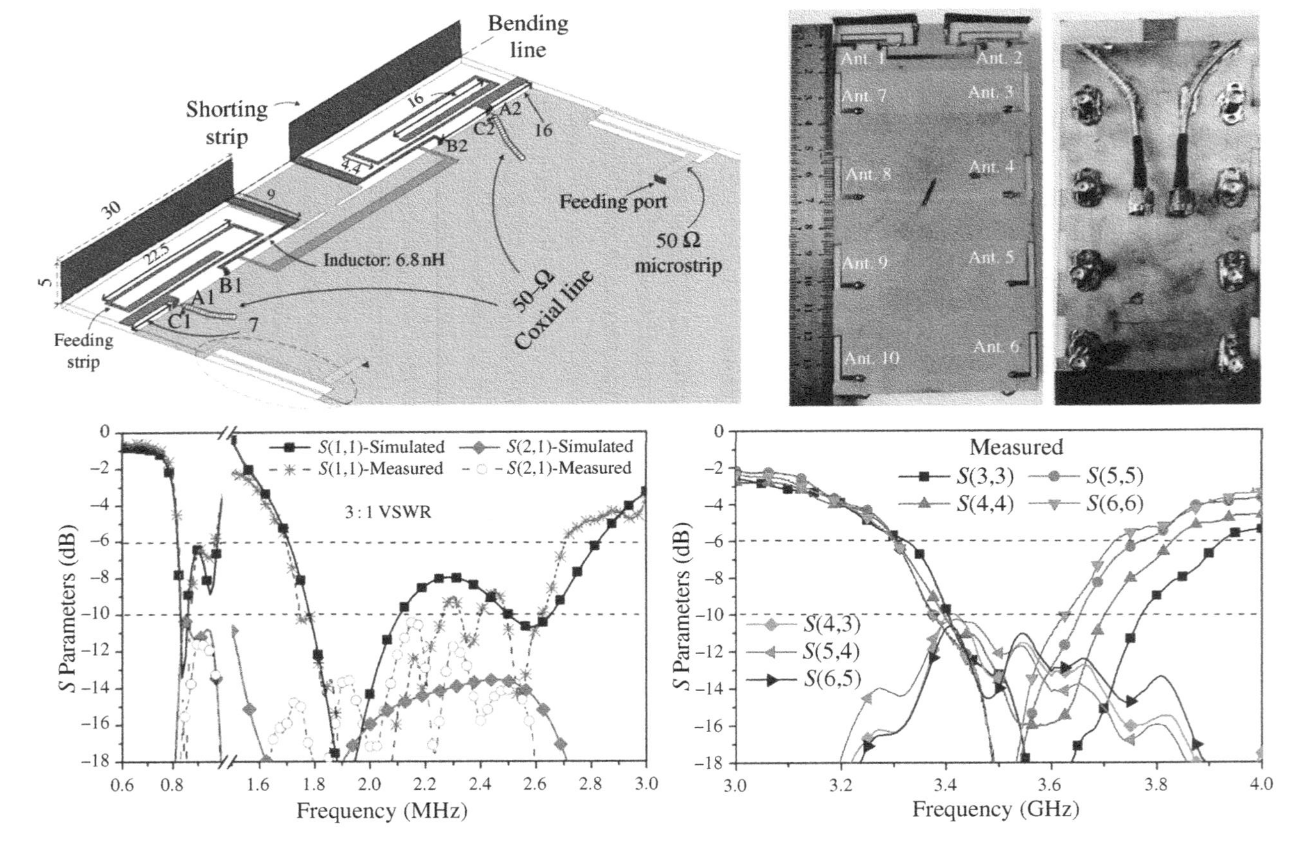

Figure 4.33 The 4G LTE dual MIMO antenna array configuration of [8], and the picture of the integrated 4G LTE MIMO array and 5G FR1 eight-antenna MIMO array. The *S*-parameters of the 4G LTE MIMO array and the 5G FR1 MIMO array are included. *Source:* From [8]. ©2016 IEEE. Reproduced with permission.

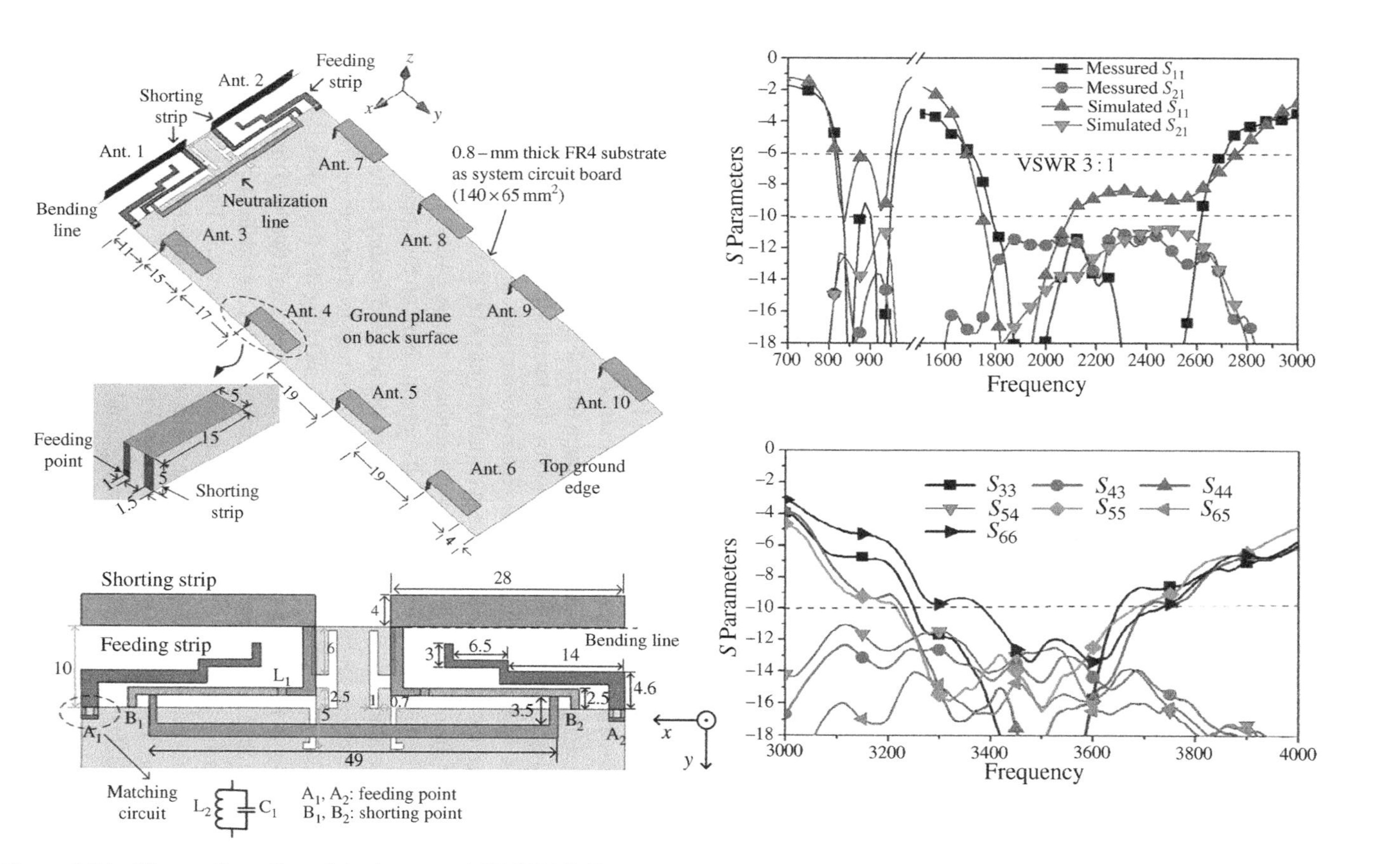

Figure 4.34 The configuration of the integrated 4G LTE MIMO array and 5G FR1 eight-antenna MIMO array of [35]. The S-parameters of the 4G LTE MIMO array and 5G FR1 MIMO array are included. *Source:* From [35]. ©2016 IEEE. Reproduced with permission.

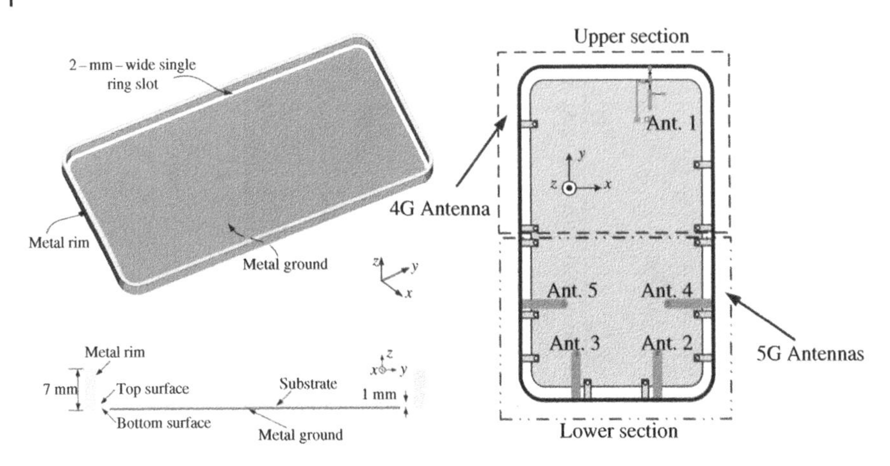

Figure 4.35 The configuration of integrated 4G/5G antennas built on a full metallic chassis 5G smartphone. The upper slot section is to implement the 4G antenna, and the lower slot section is for realizing the 5G four-antenna slot array. *Source:* From [37]. ©2019 IEEE. Reproduced with permission.

As the 4G/5G integrated antennas aforementioned in [8, 35] cannot be applied to 5G smartphones with metallic chassis (or metal-rimmed), a single ring slot-based 4G/5G-integrated antennas for metal-rimmed smartphones were introduced in [37]. Notably, the metallic chassis applied in [37] is a full metallic-rimmed type, which signifies that the metallic chassis has maintained its integrity without any slit nor slot loaded (or cut) into the chassis, as depicted in Figure 4.35. In this figure, the 2 mm-wide single ring slot that was formed by the metallic chassis and the metallic ground can be further modified into two main slot sections, in which the upper slot section is used for the excitation of the 4G slot antenna (Ant. 1), while the lower slot section is for the excitation of the 5G four-antenna slot array (Ant. 2 to Ant. 5).

Figure 4.36a shows the geometry of the 4G antenna built on the upper slot section of the full metallic chassis. Figure 4.36b shows the adopted feeding network in the fabricated prototype, and Figure 4.36c shows the alternate feeding network in a practical design. As depicted in Figure 4.36d, by implementing the three slots (Slot 1, Slot 2, and Slot 3) in the upper slot section, three slot modes and two loop modes can be excited by Slot 1, while two half-wavelength slot modes are excited by Slot 2 and Slot 3. By further observing its measured reflection coefficients in Figure 4.37, for different values of varactor diode when switched to different L values (46 and 5.6 nH), the low-band (820–960 MHz), middle band (1710–2170 MHz), and high-band (2300–2690 MHz) for WWAN/LTE operating bands can be satisfied. As for its four-antenna MIMO slot array design for 5G FR1 band (3400–3600 MHz), they are built on the lower slot section of the full metallic chassis, as depicted in Figure 4.38. The isolation technique applied in this case will be discussed in Chapter 5.

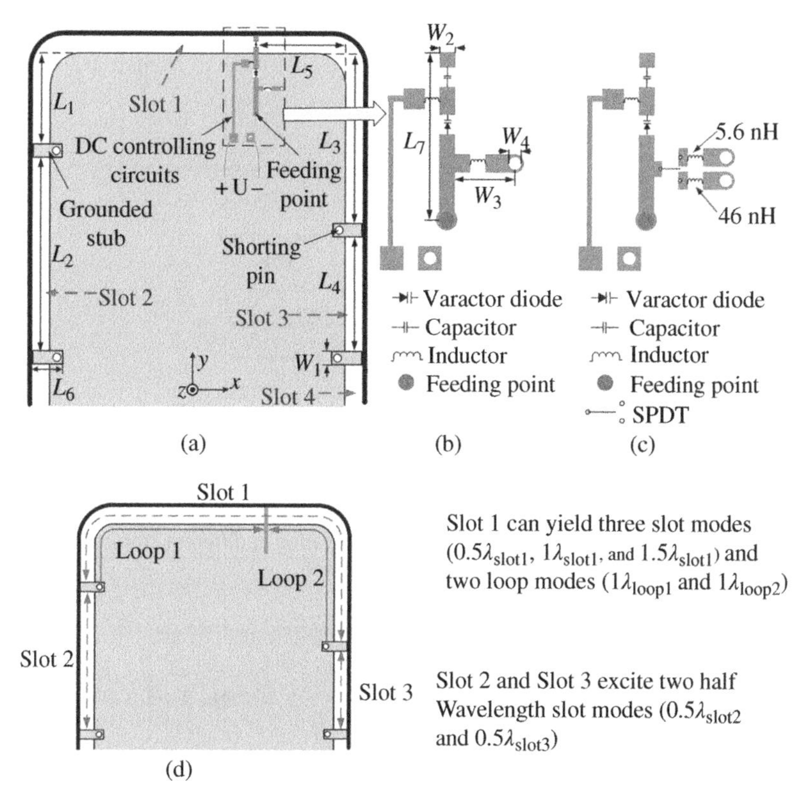

Figure 4.36 (a) The full geometry of the 4G antenna built on the upper slot section of the full metallic chassis. (b) Adopted feeding network in the fabricated prototype. (c) Alternative feeding network in a practical design. (d) Schematic picture of the 4G antenna principle, with one slot (Slot 1), two loops (Loop 1 and Loop 2), and two parasitic slots (Slot 2 and Slot 3). *Source:* From [37]. ©2019 IEEE. Reproduced with permission.

4.3.2 Integrated 4G and 5G mmWave Antennas

To successfully realize the collocation of a 5G mmWave antenna array with a lower frequency spectrum 4G antenna in a 5G smartphone, it is vital to make sure that the radiation pattern and impedance matching of the 4G antenna does not interfere with the 5G mmWave antenna array, and vice versa. As the position of the 4G antenna pairs (main and diversity) are usually constructed at the lower and bottom sections of the smartphone (see Chapter 3, Figure 3.1), therefore, the selection of a desirable position (or multiple positions) for the 5G mmWave antenna array must be meticulously studied. Recently, the mmWave antenna array designs with end-fire radiation have been widely studied by many researchers. Besides the ability to yield higher directivity, mmWave antenna arrays with

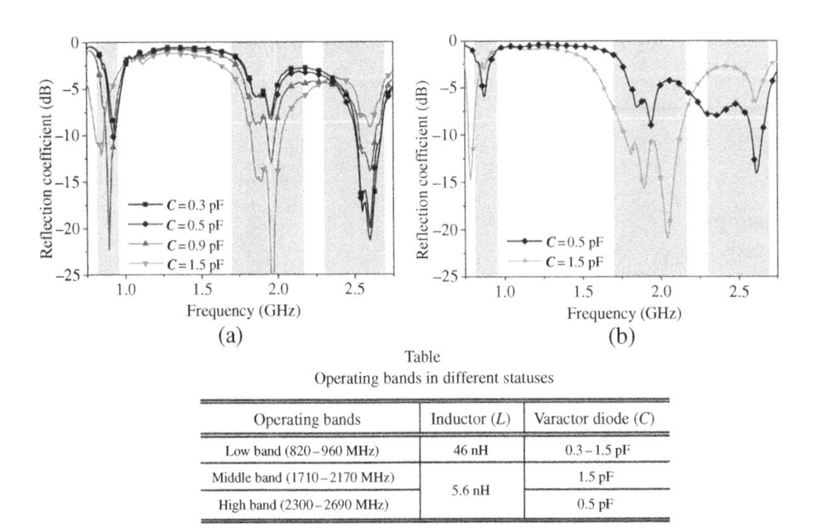

Table
Operating bands in different statuses

Operating bands	Inductor (L)	Varactor diode (C)
Low band (820–960 MHz)	46 nH	0.3–1.5 pF
Middle band (1710–2170 MHz)	5.6 nH	1.5 pF
High band (2300–2690 MHz)		0.5 pF

Figure 4.37 The measured reflection coefficients of the 4G antenna for different values of varactor diode when switched to different L values. (a) $L = 46$ nH. (b) $L = 5.6$ nH. The table for summarizing the 4G operating bands when operating at different statuses. *Source:* From [37]. ©2019 IEEE. Reproduced with permission.

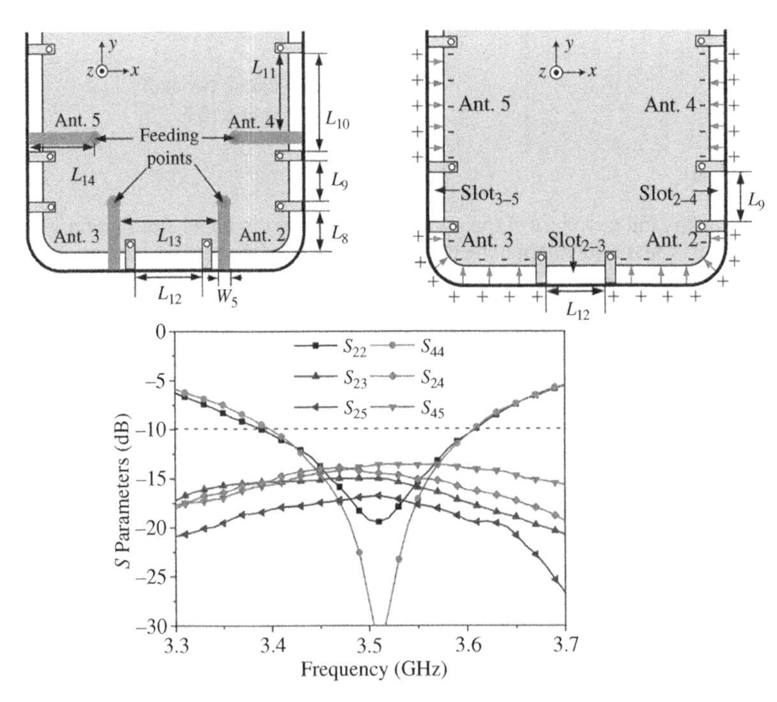

Figure 4.38 The detailed geometry of the 5G MIMO antenna array built on the lower slot section of the full metallic chassis with a simplified conceptual slot radiation diagram in each slot antenna (Ant. 2 to Ant. 5). The measured S-parameters are also included. *Source:* From [37]. ©2019 IEEE. Reproduced with permission.

end-fire radiations are also more immune to user-shadowing (in data mode) [38], as compared with those that exhibit broadside radiation patterns.

At the moment, there are very few works that involve the integration of 4G antennas and 5G mmWave antennas [39–41], and amid these works, [39] has the drawbacks of broadside radiation patterns, fixed radiation pattern beams instead of beam steering, and complex multiple layer fabrications. Furthermore, the two 4G antennas of [39] were collocated and fed at the two upper corners of the smartphone, which is uncommonly applied in actual 5G smartphone antenna design. As explained in [40], because of the limited, finite spacing (small clearance) available at the top or bottom section of a 5G smartphone for allocating the low- and high-frequency antennas together, during the process of collocating the low-frequency 4G planar antenna and 5G mmWave planar antenna array, the following two allocation methods can be considered:

a) The 5G mm-wave planar antenna array is placed in front of the low-frequency 4G planar antenna.
b) The 5G mmWave planar antenna array is placed behind the low-frequency 4G planar antenna.

However, if case (b) is selected, the 4G planar antenna will block the radiation of the 5G mmWave planar antenna array. Thus, [40] has proposed an anti-reflective layer technique so that the low-frequency 4G planar antenna can be transparent (by applying grating strips between the low- and high-frequency antennas) from the 5G mmWave antenna array radiation, allowing the mmWave spectrum radiation to propagate in the end-fire direction through the low-frequency 4G planar antenna with minimum interference. The configuration of this collocated 5G mmWave planar antenna array (folded dipole structure) with dual band 4G PIFA design is depicted in Figure 4.39a,b shows the proposed grating strips that act as the director. The simulated reflection coefficients and gains of the collocated design with and without integrating the grating lobes in Figure 4.39c have shown that the gain has significantly reduced after the grating strips are removed from the design. As shown in Figure 4.39d, the measured 10-dB impedance bandwidth of the 5G mmWave end-fire antenna array was approximately 22–31 GHz (satisfied 5G NR band 257/n258/n261) with isolation better than 13 dB. On the other hand, the 4G PIFA has exhibited 6-dB impedance bandwidths of 740–960 MHz and 1700–2200 MHz, which can cover the 4G LTE bands.

Even though [40] can well cover the 5G mmWave and 4G LTE spectrum, the entire antenna structures are planar type, and they are not applicable to 5G smartphones with metallic chassis. Therefore, [41] has implemented an integrated 4G LTE antenna with a 5G mmWave end-fire array into the metallic chassis of the smartphone. Similar to [40], the aim of this co-design is that the radiation pattern and impedance matching of the two antenna types would not interact with each

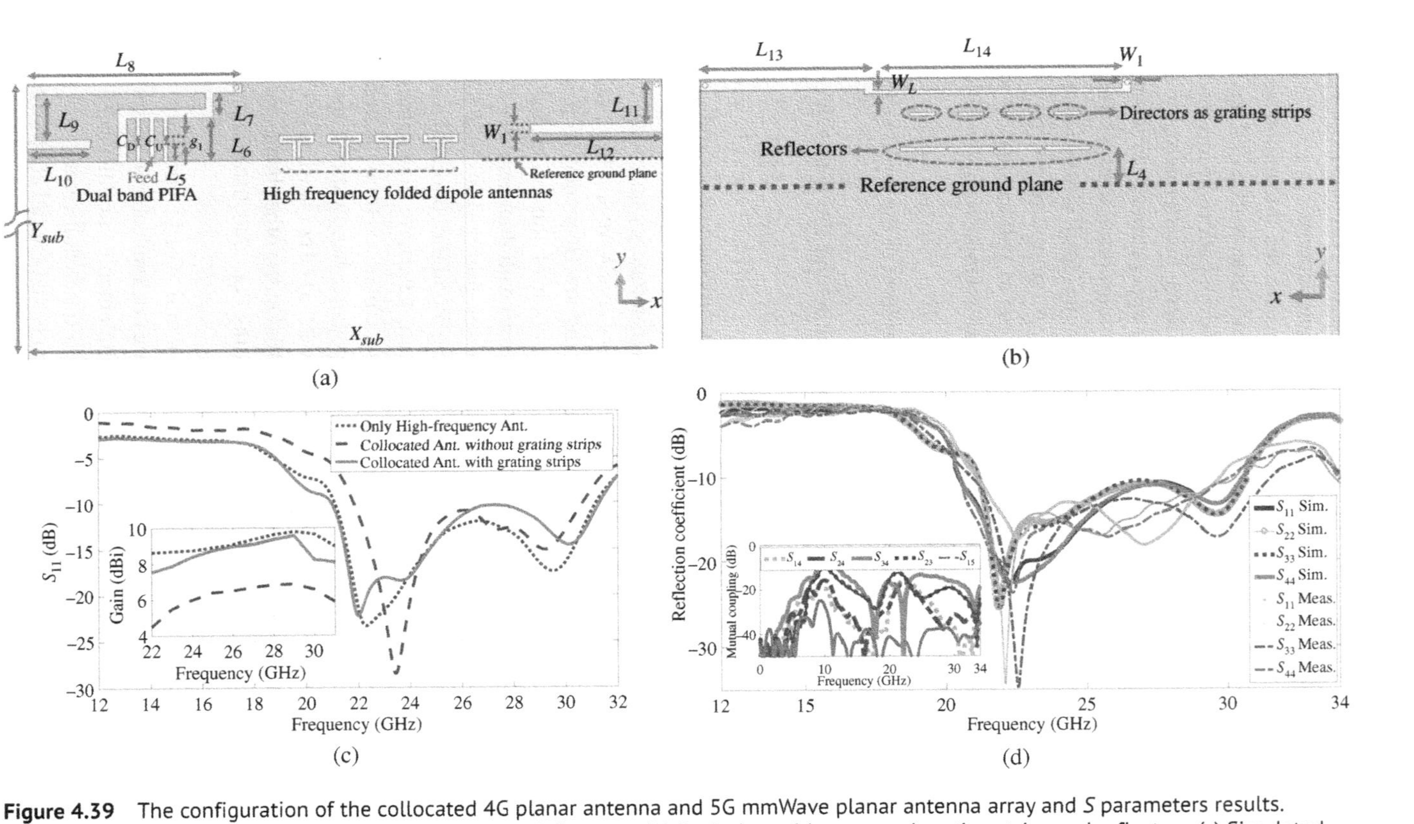

Figure 4.39 The configuration of the collocated 4G planar antenna and 5G mmWave planar antenna array and *S* parameters results. (a) Front view with planar 4G PIFA and 5G mmWave end-fire array. (b) Back view with proposed grating strips and reflectors. (c) Simulated reflection coefficients and total realized gains for three cases. (d) Simulated and measured S-parameters at mmWave spectrum. *Source:* From [40]. ©2019 IEEE. Reproduced with permission.

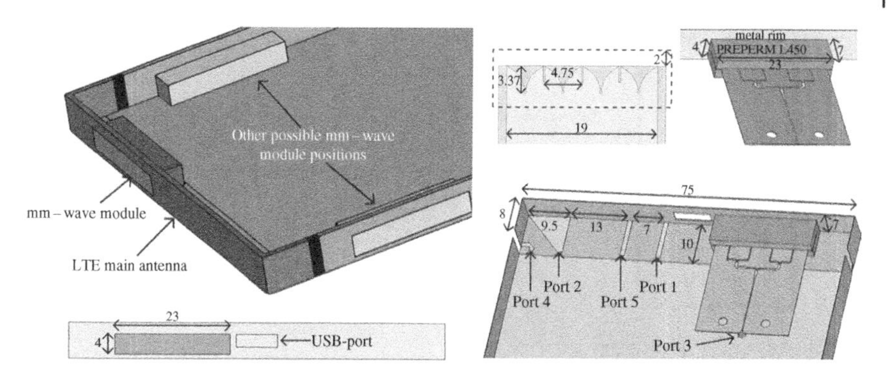

Figure 4.40 The configuration of the collocated 4G LTE antenna and 5G mmWave end-fire antenna array. The design of the 5G mmWave end-fire array (Vivaldi antenna array) was fitted into the plastic-filled window and the 4G LTE antenna constructed via the metallic chassis is included. *Source:* From [41]. Under a Creative Commons License.

other and could, therefore, be designed separately. Figure 4.40 shows the configuration of the 4G LTE main antenna and 5G mmWave end-fire array collocated on the top section of the metallic chassis. Here, the 5G mmWave end-fire array (Vivaldi array) was implemented on a separate PCB (a 0.101-mm thick Rogers RO4350B substrate), and it was enclosed by an RF-optimized injection moldable plastic (PREPERM L450, $\varepsilon_r = 4.5$, $\tan\delta = 0.0005$). To successfully allow the 5G mmWave end-fire array to radiate, a $23 \times 4\,\text{mm}^2$ window was cut into the top surface of the metallic chassis (near the top right corner of the chassis), and a moldable plastic enclosed with the 5G mmWave end-fire array was fitted into this window. Notably, [38] has also indicated that the corner positions of a smartphone chassis can yield the best performance for the 5G mm-wave array system in terms of spatial coverage when user effects are considered.

Figure 4.41 shows the fabricated prototype of the collocated 4G LTE antenna and 5G mmWave end-fire antenna array (Vivaldi array) by using the metallic chassis. To achieve beam steering between 0°, 50°, and 100° from the Vivaldi array, the phases of each feeding port of the four-antenna array Vivaldi are also indicated. From the matching level (reflection coefficients) results in Figure 4.41, good 10-dB impedance matching from 25 to 30 GHz were measured, and its corresponding realized gain patterns have also exhibited a maximum gain of around 5 dBi with desirable beam-steering directions at almost ±40°. On the other hand, the 4G LTE antenna can well cover the LTE low-band (700–960 MHz) and high-band (1710–2690 MHz) operations.

In [42, 43], a smartphone with integrated LTE 4G, 5G sub-6 GHz, and 5G mmWave technologies was briefly described and shown in Figure 4.42. Here, the 4G LTE antenna pairs are constructed by a tunable loop and coupled-fed IFA that can cover

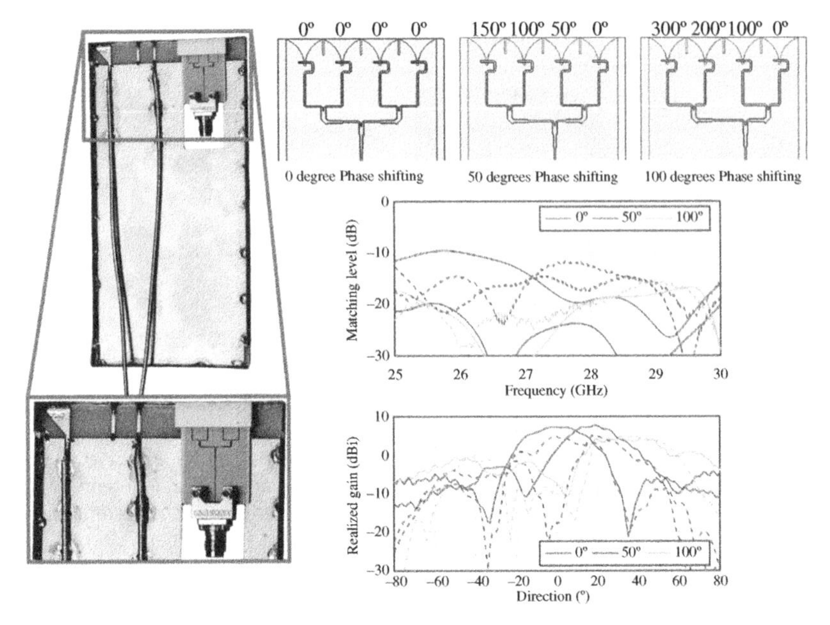

Figure 4.41 The fabricated prototype of the collocated 4G LTE antenna and 5G mmWave end-fire antenna array (Vivaldi antenna array) using the metallic chassis. The reflection coefficient of the Vivaldi antenna array at different phase shifting angles (0, 50, and 100°), and the azimuth plane realized gain patterns of the Vivaldi antenna array at 28 GHz are included. Solid lines: simulations. Dashed lines: measurements. *Source:* From [41]. Under a Creative Commons License.

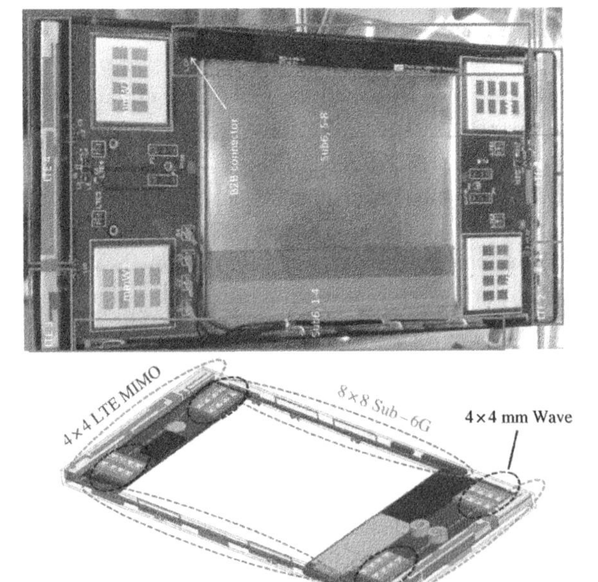

Figure 4.42 The configuration of the collocated 4G LTE antenna pairs, 5G sub-6 GHz 8×8 antenna array, and 4 sets of 5G mmWave 2×4 patch antenna in a smartphone. The two LTE antenna pairs that are located at the top and bottom sections of the smartphone are formed by two different antenna types, tunable loop antenna and coupled-fed IFA. *Source:* Prof. Guangli Yang, Figures obtained with permission.

the LTE low-band (698–960 MHz) and high-band (1710–2690 MHz) operations, with isolation of better than 12 dB. The intended 5G sub-6 GHz eight-antenna array are collocated along the two longer side edges, and they can cover 3400–3800 MHz, while the four sets of 5G mmWave 2×4 patch antenna array that can cover the 5G NR band n258 (24.25–27.5 GHz) are placed next to the 4G LTE antenna pairs.

4.3.3 Miniaturized 5G Module of mmWave Antennas-in-Package Integrating Non-mmWave Antennas (AiPiA)

It was only recently that the idea of integrating the 5G mmWave AiP module with non-mmWave antennas, now known as AiPiA, was proposed by H.C. Huang [44]. The concept of AiPiA is very simple, which is to realize a very miniaturized 5G module that can excite both the 5G mmWave FR2 band and 5G sub-6 GHz FR1 band. Two examples of how to integrate the 5G AiP module with a non-mmWave antenna are shown in Figure 4.43. The left diagram (also appeared in [45]) is a

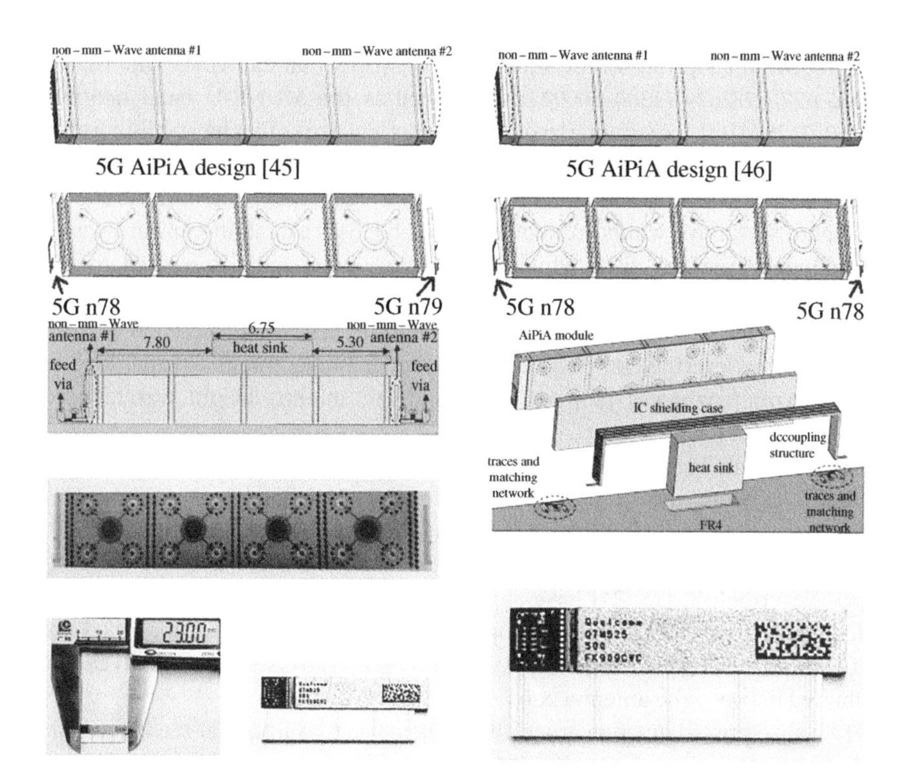

Figure 4.43 5G AiPiA design for mmWave and non-mmWave MIMO applications. *Source:* From [44]. Under a Creative Common License.

5G AiPiA design that has two different 5G sub-6 GHz IFAs built on the two sides edges of the AiP. Their operating frequencies are 5G NR n78 and 5G NR n79. As for the right diagram (also appeared in [46]), the two 5G sub-6 GHz IFAs built on the two sides edges of the AiP are operating at the same 5G n78 bands. Both have shared the same dimensions of ($21.27 \times 4.18 \times 1.36\,mm^3$). To further reduce the dimensions of [45, 46], this work has reported an advanced method to physically remove the two sub-6 GHz IFAs on both side edges and replace them with virtual equivalent ones multitasked by the mm-Wave AiP modules. This design was called the "mmWave AiP as non-mmWave antennas" (AiPaA), and they have also appeared in [47, 48]. The design philosophy of the AiPaA is to reuse the mmWave AiP module to directly function as non-mmWave antennas. As this idea is more on 5G mmWave AiP solutions, we recommend reading [44–48].

4.4 Wideband Antenna Design Topologies Beyond Band n77/n78/n79

Before introducing the recent antenna array works that can cover the 5G NR band n77/n78/n79 (3300–5000 MHz), as well as the 5G NR-U band n46/n96 (5150–7125 MHz), we summarized the previous few works that have fully covered either the 5G NR n77/n78 (3300–4200 MHz) and 5G NR n77/78/n79 bands across the 6-dB impedance bandwidth, respectively, in Figures 4.44 and 4.45. As shown in Figure 4.44, three works are shown, in which [13] is a dual-antenna array with a monopole-type structure, while [27, 28] are of the same integrated hybrid antenna type with a monopole radiator and loop radiator. As [27, 28] have self-decoupled characteristics, their isolation levels are better than 17.9 and 14.4 dB, respectively. Nevertheless, [13] has a very small single antenna planar size of $18.6 \times 5.6\,mm^2$, whereas [28] was bent, and the antenna height was reduced to 4 mm.

As shown in Figure 4.45, [7] has implemented the coupled-loop structurer design to generate another new resonance mode (0.75λ), and by merging this new resonance mode with the two conventional loop resonance modes at 0.5λ and 1.0λ, a wideband operation that can cover the 5G NR band n77/n78/n79 was realized. Nevertheless, [15, 24] have applied the integrated slot antenna pair technique to achieve wide bandwidth operation, and because [24] has applied the differential feeding mechanism to Ant. 2, very good port isolation >21 dB can be achieved between the antenna pairs.

To realize printed antenna arrays that are suitable to collocate at the longer side edges of the 5G smartphone chassis (or frame) with a very wide operational band that can cover the 5G NR n77/n78/n79, as well as 5G NR-U band n46 (5150–5925 MHz), as shown in Figure 4.46, [49] has implemented a compact size ($16 \times 5\,mm^2$)

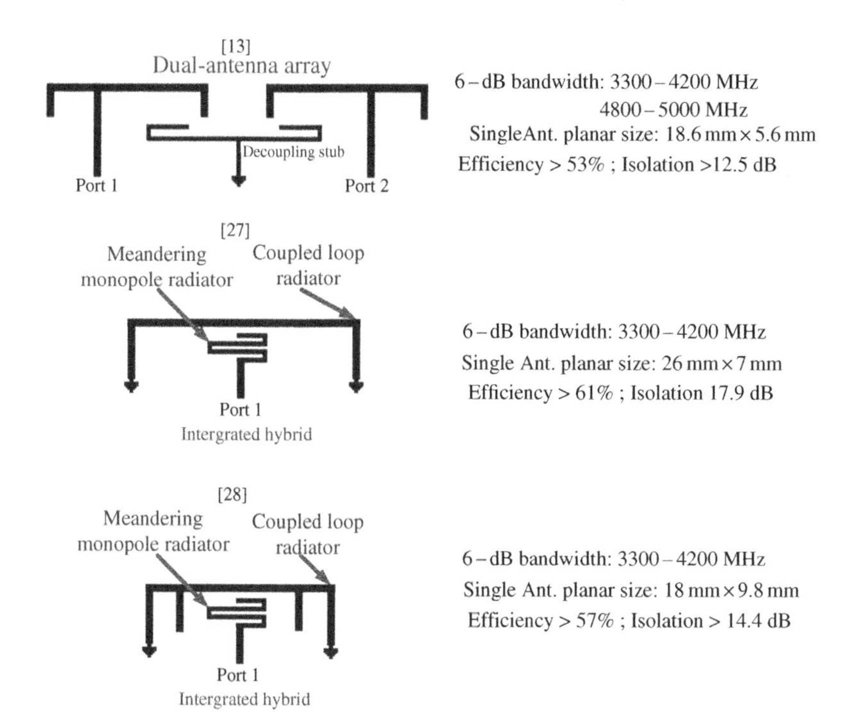

Figure 4.44 Conceptual diagrams and performances of antenna designs that are operating at 5G NR band n77/n78. Reference numbers are included for each design.

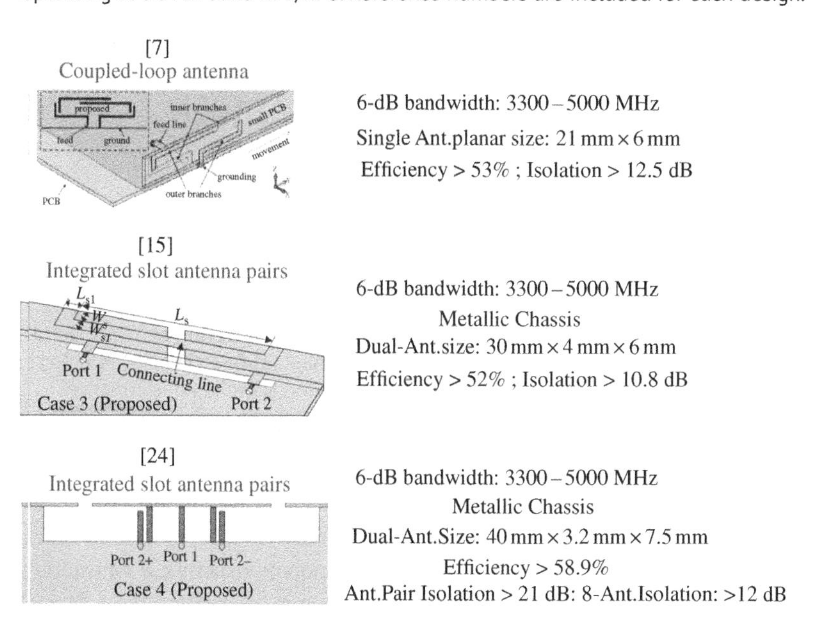

Figure 4.45 Antenna designs that are operating at 5G NR band n77/n78/n79 and their corresponding performances. Reference numbers are also included for each design. *Source:* From [7]. ©2019 IEEE. Reproduced with permission; From [15]. ©2021 IEEE. Reproduced with permission; From [24]. ©2020 IEEE. Reproduced with permission.

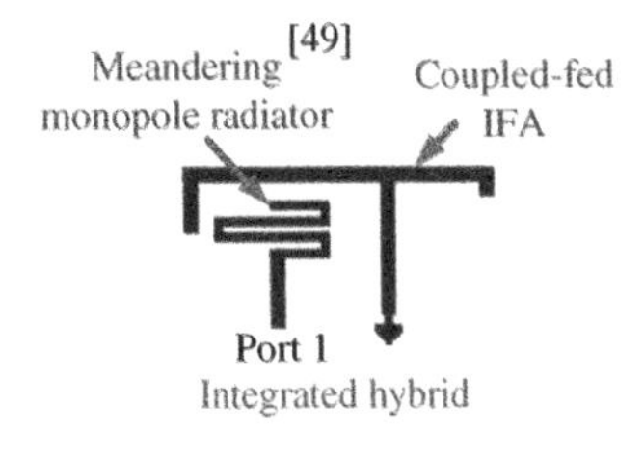

6 – dB bandwidth: 3300 – 6000 MHz
Single Ant. planar size: 16 mm × 5 mm
Efficiency > 56% ; Isolation > 10 dB

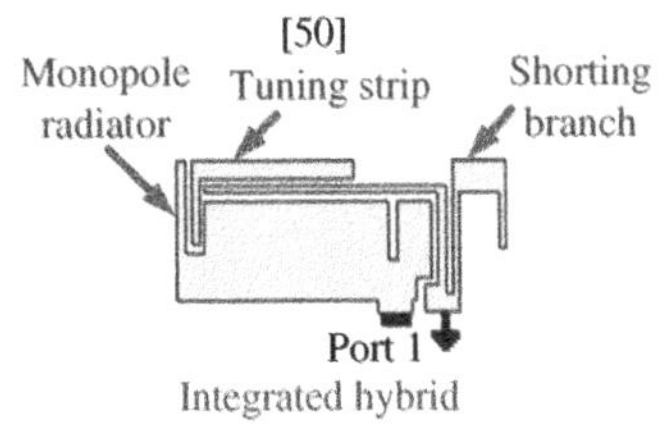

6 – dB bandwidth: 3100 – 6000 MHz
10 – dB bandwidth: 3250 – 5930 MHz
Single Ant. planar size: 13.9 mm × 7 mm
Efficiency > 41% ; Isolation > 10 dB

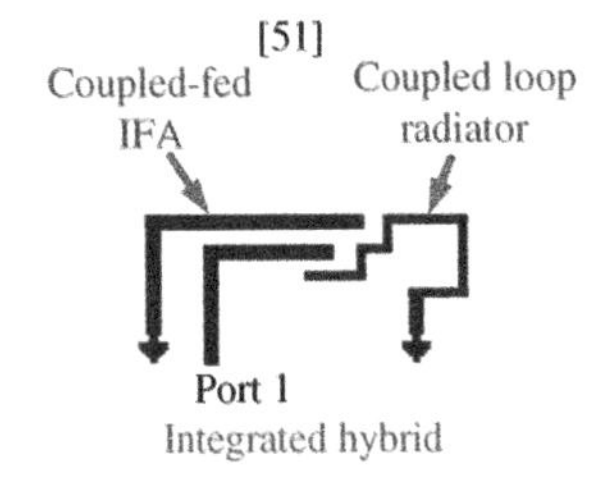

6 – dB bandwidth: 3300 – 6400 MHz
Single Ant. planar size: 20 mm × 6.8mm
Efficiency > 52% ; Isolation > 12 dB

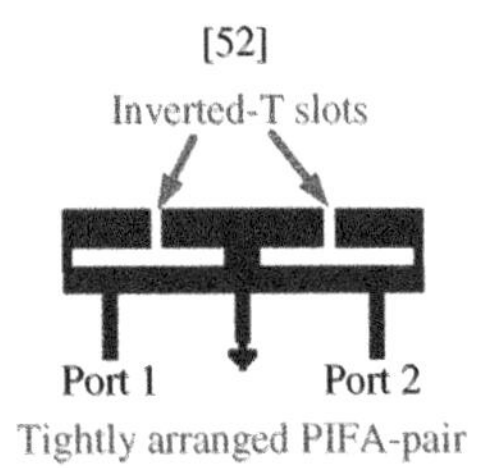

6 – dB bandwidth: 3300 – 7500 MHz
Single Ant. size: 18.6 mm × 7 mm
Dual Ant. pair size: 30 mm × 7 mm
Efficiency > 40% ; Isolation > 10 dB

Figure 4.46 Conceptual diagrams of antenna designs that are operating at 5G NR band n77/n78/n79 (3300–4200 MHz), also including 5G NR-U n46 (5150–5925 MHz) [49–51], as well as 5G NR-U n96 (5925–7125 MHz) [52], and their corresponding performances. Reference numbers are also included for each design. *Source:* From [50]. ©2020 IEEE. Reproduced with permission.

integrated hybrid design, in which the integrated meandering monopole radiator and the coupled-fed IFA can excite three resonance modes covering 6-dB impedance bandwidth of 3300–6000 MHz. In [50], the driven monopole structure is for exciting the 5.2 and 6.05 GHz modes, while the shorting branch is for exciting the 3.54 and 5.8 GHz modes. By merging these resonances, a very good 10-dB impedance bandwidth of 3250–5930 MHz was achieved, and it can cover the 5G NR n77/n78/n79 and

5G NR-U band n46 (3300–5925 MHz). To further achieve good impedance matching, a tuning strip and a narrow slit were loaded into the driven monopole. In [51], the hybrid antenna appeared to have integrated a coupled-fed IFA with a coupled-loop radiator, in which the former has generated a 0.25λ mode at 3529 MHz, and the latter has excited three resonance modes at 4945, 5536, and 6165 MHz. By merging these resonances, a very wide 6-dB impedance bandwidth of 3300–6400 MHz was achieved. But the size of this hybrid antenna was rather large at $20\times6.8\,mm^2$. To further cover the 5G NR-U band n96 (5925–7125 MHz), a tightly arranged PIFA pair was realized by loading two inverted T-shaped slots into the radiator [52].

Besides the above planar antenna array designs collocated along the longer side edges of the smartphone frame (non-metallic type) that can cover the 5G NR band n77/n78/n79 and 5G NR-U band n46/n96, a few antenna array designs [53–56] for 5G smartphone with metallic chassis have also been investigated in recent years. Figure 4.47 shows the work reported in [53], in which the open slot radiator is composed of an asymmetric U-shaped slot (15 mm × 3 mm) and a vertical narrow slot (6 mm × 2 mm) separately loaded into the system ground edge and the metallic chassis, respectively. To excite the antenna and yield good impedance matching, a 50 Ω microstrip feeding line is applied with a tuning stub. Notably, [53] has reported that the two major resonances at approximately 3.3 and 5.3 GHz are mainly excited by the U-shaped slot (path ABDEF) and feeding line (path ABC), respectively, but it is rather peculiar as the feeding line (Path AB) should not be a radiator. Nevertheless, from the measured S-parameters, the stacked 6-dB impedance bandwidth was approximately 3100–6500 MHz, and its corresponding isolation and total efficiency were better than 11 dB and 40%, respectively.

Figure 4.48 shows the work reported in [54], in which the L-shaped open slot radiator is composed of a horizontal narrow slot (9 mm × 2 mm) and a vertical narrow slot (9 mm × 2 mm) separately loaded into the system ground edge and the metallic chassis, respectively. To feed this slot radiator as well as achieve good impedance matching, two tuning stubs are embedded into the 50 Ω microstrip feeding line. As compared with [53], the highlight of this work is that the slot radiator can excite two different slot modes (3.5 and 5.5 GHz) along the same slot path. Furthermore, its measured stacked 10-dB impedance bandwidth was 3270–5920 MHz, and if stacked 6-dB impedance bandwidth is considered, it would be able to cover the entire 5G NR band n77/n78/n79 and 5G NR-U band n46. Its corresponding measured isolation and antenna efficiency were better than 12 dB and 50%, respectively. If an open slot is further loaded into metallic chassis between two adjacent open slot elements, the isolation can be improved to 14.3 dB. Notably, this decoupling technique has also been applied by [55].

Compared with [53, 54], the work reported in [55] has applied a much different slot radiator structure into the metallic chassis, as it is a combination of two open slot radiators, namely inverted T-shaped open slot radiator and inverted C-shaped

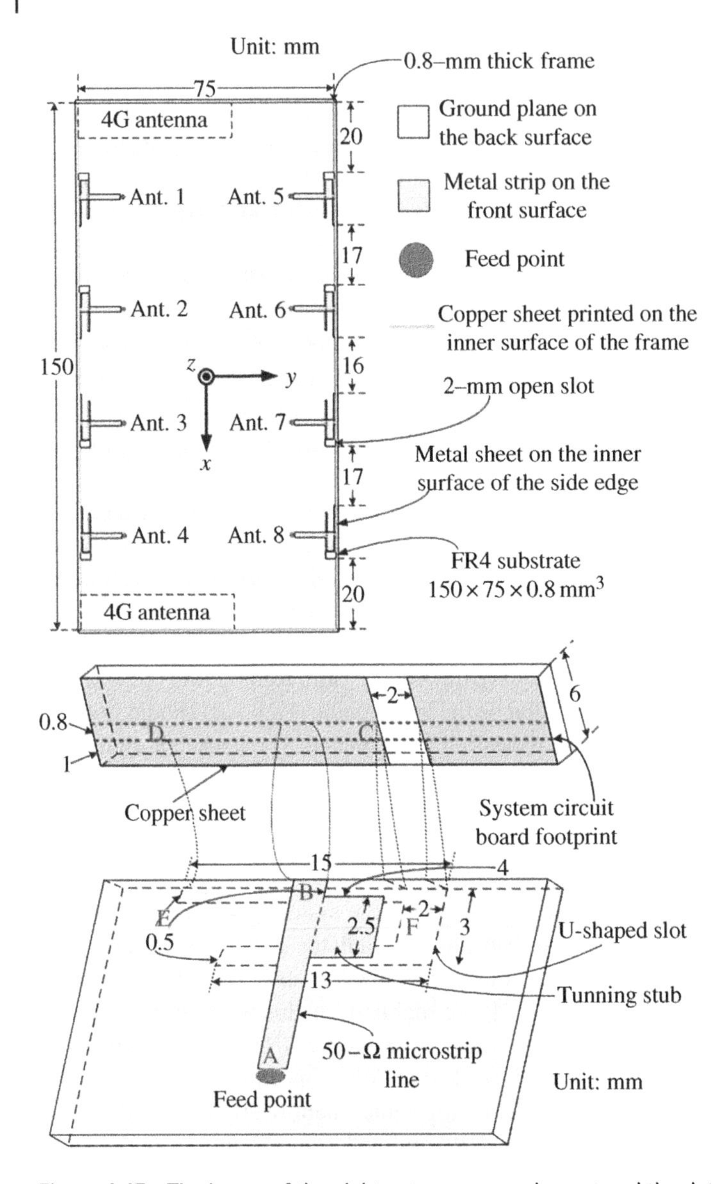

Figure 4.47 The layout of the eight-antenna array element and the detailed configuration of the open slot radiator in [53]. Its corresponding measured S-parameters and simulated current distributions are also included. *Source:* From [53]. ©2019 IEEE. Reproduced with permission.

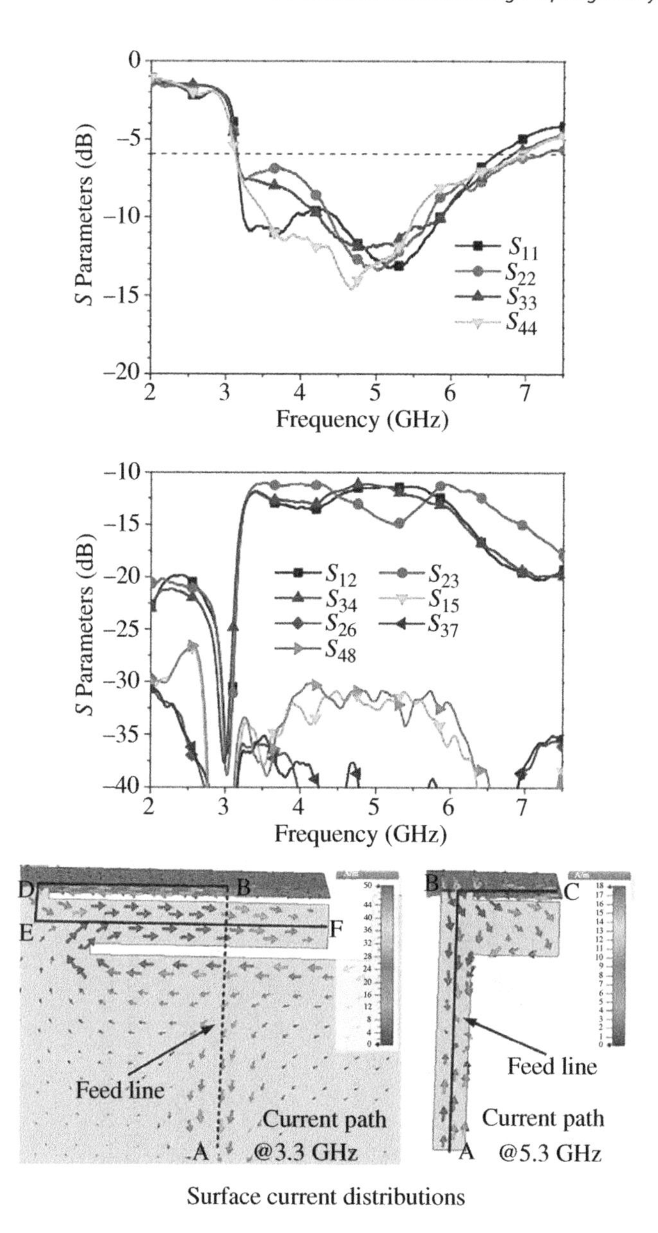

Figure 4.47 (Continued)

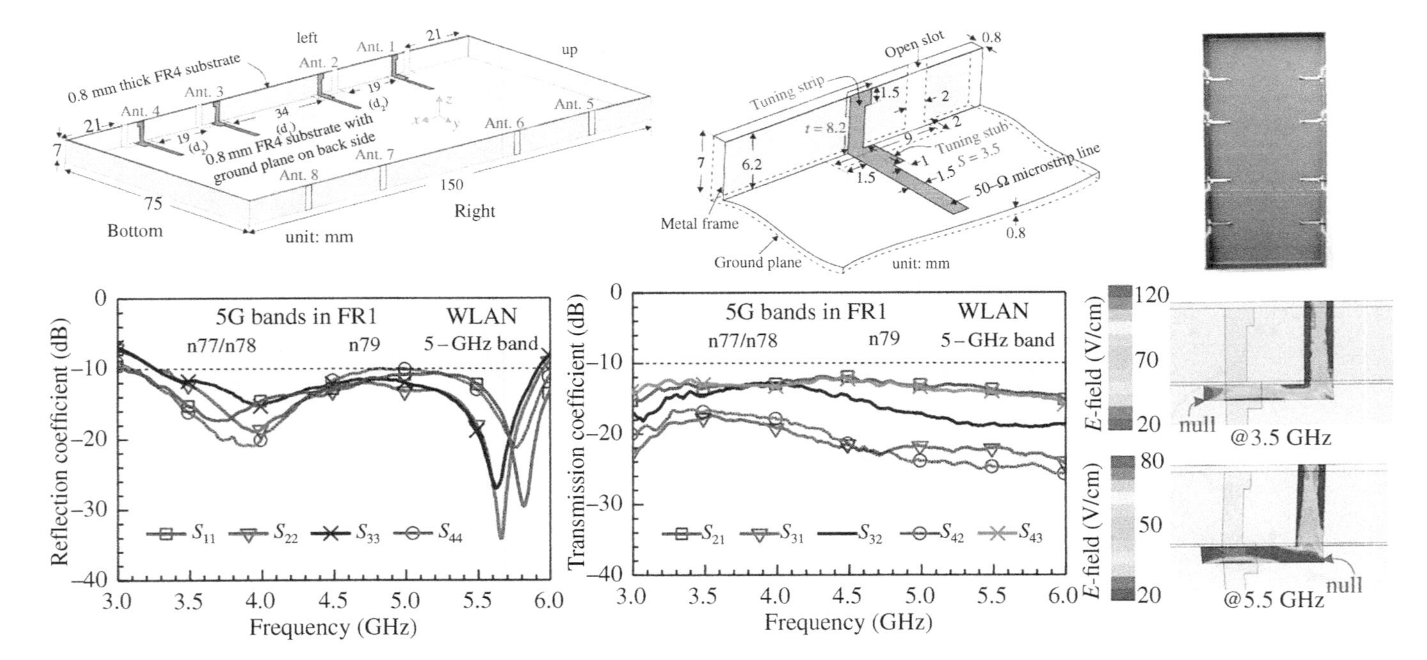

Figure 4.48 The layout of the eight-antenna array element and the detailed configuration of the open slot radiator in [54]. Its corresponding measured *S*-parameters and simulated *E*-field distributions are also included. *Source:* From [54]. ©2020 IEEE. Reproduced with permission.

open slot radiator, as shown in Figure 4.49. The total size of this integrated open slot radiator is 17 mm × 5.7 mm, and it is fed by an L-shaped feeding line, which, in turn, yielded four different resonance modes for achieving wide stacked 6-dB impedance bandwidth of 3300–6000 MHz. To further enhance the isolation from 13 to 18 dB between any two adjacent open slot radiators, it has applied the same decoupling technique as in [54], but the decoupling slot structure in this case is an H-shaped resonator type for trapping the surface current. The measured antenna efficiency was larger than 40% across the bands of interest.

Compared with [53–55], the work reported in [56] has for the first time successfully constructed a wideband antenna array for 5G smartphone applications that nearly satisfied the 5G NR band n77/n78/n79 and 5G NR-U band n46/n96. As shown in Figure 4.50, the L-shaped open slot radiator applied to the metallic chassis, in this case, is analogous to [54], which is the integration of a narrow horizontal slot (12.4 mm × 1.5 mm) and a narrow vertical slot (7 mm × 2.4 mm). To achieve good impedance matching, a tuning stub and an impedance transformer are

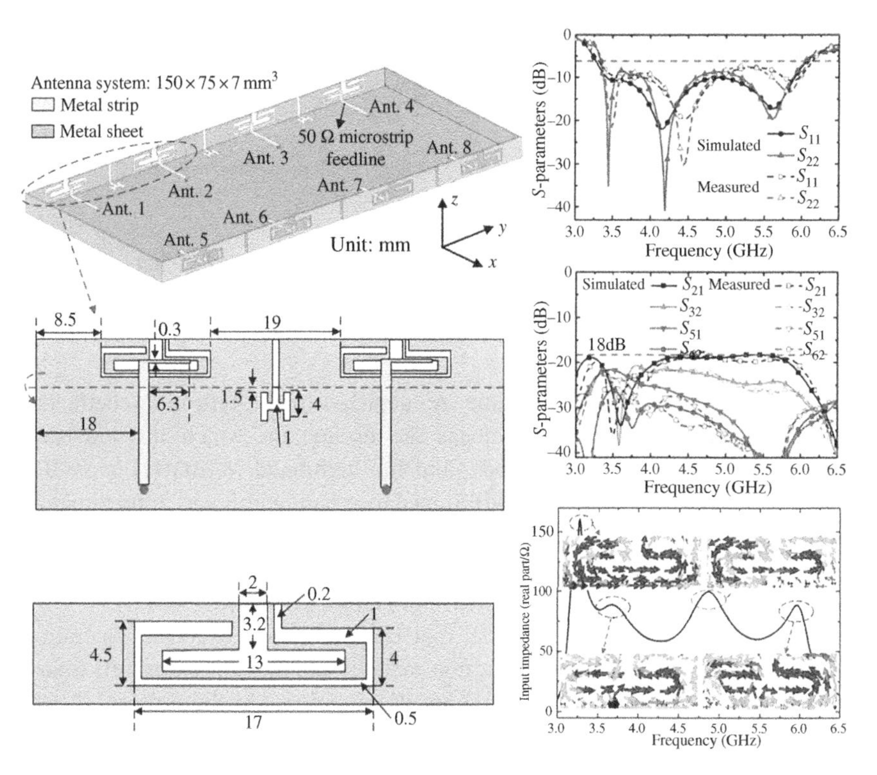

Figure 4.49 The layout of the eight-antenna array element and the detailed configuration of the open slot radiator in [55]. Its corresponding measured S-parameters and simulated current distributions are included. *Source:* From [55]. Under a Creative Commons License.

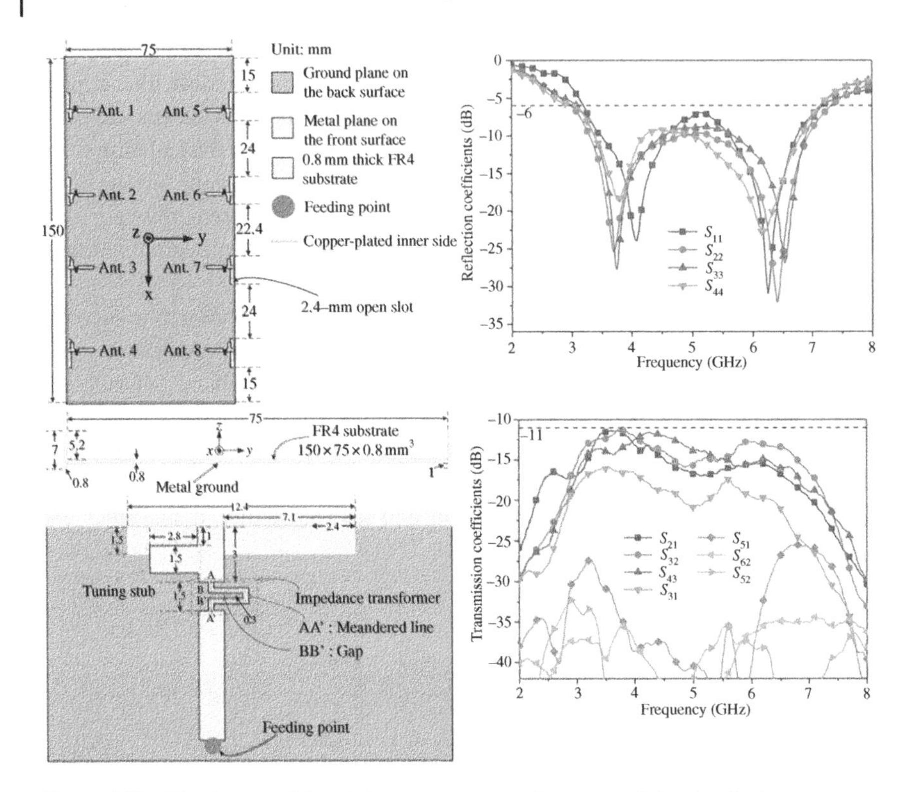

Figure 4.50 The layout of the eight-antenna array element and the detailed configuration of the open slot radiator in [56]. Its corresponding measured *S*-parameters and simulated current distributions are included. *Source:* From [56]. Under a Creative Commons License.

embedded into the 50Ω feeding line. As depicted in the reflection coefficient diagram, two major resonance modes are excited, in which the low-band (4.22 GHz) resonance is a slot mode, and the high-band (6.76 GHz) is an IFA mode. As the stacked measured 6-dB impedance bandwidth was approximately 3300–7100 MHz, it almost covered the bands of interest, including 5G NR-U band n96 (5925–7125 MHz). Lastly, the isolation levels and total efficiency measured across the bands of interest were better than 11 dB and 47%, respectively.

Besides the earlier-reported works that can yield very wideband operation beyond 5G NR band n77/n78/n79 for non-metallic and metallic chassis 5G smartphone applications, another two works are also discovered by the authors [57, 58]. However, the feeding mechanism of these two works is rather unusual as all the SMA connectors are soldered from the direction devised for placing the smartphone chassis, and thus the SMA ground size would have a serious effect on the antenna performances. For brevity, we have not included these works.

4.5 Conclusion

In this chapter, we covered many different areas in the design of the 5G FR1 antenna elements with multi-band and wideband operations for smartphone applications. We begin by considering the different types of planar antenna design structures (loop, monopole, slot, and IFA) that can only cover partial 5G NR band n77/n78 and band n79; followed by reviewing the various hybrid antenna design methods (such as the separated, integrated, or tightly arranged) that usually combine two different types of antenna structures. The co-existence of 3G/4G antenna and 5G FR2 mmWave antenna array in relation to the 5G FR1 antenna array are also introduced, and many examples are shown, especially the new techniques (AiPiA) that may influence the future integrated 5G FR1 and FR2 antenna array design. Finally, a very detailed description of all the recent reported antenna array designs that can yield wideband operation covering the 5G NR band n77/n78/n79, as well as those that can also cover the 5G NR-U band n46 and 5G NR-U band n96 are summarized. These wideband operational antenna designs are also categorized into smartphone applications that are non-metallic chassis type and metallic chassis type.

References

1 (Online) Sporton Lab, 5G NR Conformance and 3GPP Status Update, https://files.keysightevent.com/files/2020KWT/(A2)%205G%20NR%20conformance%20and%203GPP%20status%20update_Sporton%20Hendry%20Hsu_Keysight.pdf.

2 W. Jiang, B. Liu, Y. Cui, and W. Hu, "High-isolation eight-element MIMO array for 5G smartphone applications," *IEEE Access*, vol. 7, pp. 34104–34112, Mar. 2019.

3 W. Hu, L. Qian, S. Gao, L. H. Wen, Q. Luo, H. Xu, X. Liu, Y. Liu, and W. Wang, "Dual-band eight-element MIMO array using multi-slot decoupling technique for 5G terminals," *IEEE Access*, vol. 7, pp. 153910–153920, 2019.

4 (Online) https://www.ieice.org/cs/isap/ISAP_Archives/2018/pdf/FrE1-2.pdf.

5 Z. Ren, A. Zhao, and S. Wu, "MIMO antenna with compact decoupled antenna pairs for 5G mobile terminals," *IEEE Antennas Wireless Propag. Lett.*, vol. 18, no. 7, pp. 1367–1371, May 2019.

6 Z. Ren, and A. Zhao, "Dual-band MIMO antenna with compact self-decoupled antenna pairs for 5G mobile applications," *IEEE Access*, vol. 7, pp. 82288–82296, Jun. 2019

7 A. Zhao and Z. Ren, "Wideband MIMO antenna systems based on coupled-loop antenna for 5G n77/n78/n79 applications in mobile terminals," *IEEE Access*, vol. 7, pp. 93761–93771, May 2019.

8 Y. L. Ban, C. Li, C. Y. D. Sim, G. Wang, and K. L. Wong, "4G/5G multiple antennas for future multi-mode smartphone applications," *IEEE Access*, vol. 4., pp. 2981–2988, Jun. 2016.

9 Y. Li, C. Y. D. Sim, Y. Luo, and G. Yang, "12-port 5G massive MIMO antenna array in sub-6 GHz mobile handset for LTE bands 42/43/46 applications," *IEEE Access*, vol. 6, pp. 344–354, Oct. 2017.

10 M. Peng, H. Zou, Y. Li, M. Wang, and G. Yang, "An eight-port 5G/WLAN MIMO antenna array with hexa-band operation for mobile handsets," *in 2018 IEEE International Symposium on Antennas and Propagation & USNC/URSI National Radio Science Meeting*, pp. 39–40, Boston, MA, USA, Jul. 2018.

11 I. Dioum, I. Diop, L. Sane, M. Khouma, and K. Diallo, "Dual band printed MIMO antennas for 5G handsets," *in 2017 International Conference on Wireless Technologies, Embedded and Intelligent Systems (WITS)*, Fez, Morocco, pp. 1–4, Apr. 2017.

12 D. Serghiou, M. Khalily, V. Singh, A. Araghi, and R. Tafazolli, "Sub-6 GHz dual-band 8 × 8 MIMO antenna for 5G smartphones," *IEEE Antennas Wireless Propag. Lett.*, vol. 19, no. 9, pp. 1546–1550, Sept. 2020.

13 L. Cui, J. Guo, Y. Liu, and C. Y. D. Sim, "An 8-element dual-band MIMO antenna with decoupling stub for 5G smartphone applications," *IEEE Antennas Wireless Propag. Lett.*, vol. 18, no. 10, pp. 2095–2099, Oct. 2019.

14 Y. Li, C.Y.D. Sim, Y. Luo, and G. Yang, "Multiband 10-antenna array for sub-6 GHz MIMO applications in 5-G smartphones," *IEEE Access*, vol. 6, pp. 28041–28253, Jun. 2018.

15 L. Sun, Y. Li, and Z. Zhang, "Wideband decoupling of integrated slot antenna pairs for 5G smartphones," *IEEE Trans. Antennas Propag.*, vol. 69, no. 4, pp. 2386–2391, Apr. 2021.

16 X. Zhao, S. P. Yeo, and L. C. Ong, "Decoupling of inverted-F antennas with high-order modes of ground plane for 5G mobile MIMO platform," *IEEE Trans. Antennas Propag.*, vol. 66, no. 9, pp. 4485–4495, Jun. 2018.

17 Z. Qin, W. Geyi, M. Zhang, and J. Wang, "Printed eight-element MIMO system for compact and thin 5G mobile handest," *Electron. Lett.*, vol. 52, no. 6, pp. 416–418, Mar. 2016.

18 Z. Tian, R. Chen, and C. Li, "Dual-band inverted F-shaped antenna array for Sub-6 GHz smartphones," *in 2019 IEEE 89th Vehicular Technology Conference (VTC2019-Spring)*, Kuala Lumpur, Malaysia, 2019.

19 Y. J. Deng, J. Yao, D. Q. Sun, and L. X. Guo, "Ten-element MIMO antenna for 5G terminals," *Microwave Opt. Technol. Lett.*, vol. 60, no. 12, pp. 3045–3049, Dec. 2018.

20 X. Shi, M. Zhang, S. Xu, D. Liu, H. Wen, and J. Wang, "Dual-band 8-element MIMO antenna with short neutral line for 5G mobile handset," *in Proc. 11th Eur. Conf. Antennas Propag. (EUCAP)*, pp. 3140–3142, Paris, France, Mar. 2017.

21 W. J. Zhang, Z. B. Weng, and L. Wang, "Design of a dual-band MIMO antenna for 5G smartphone application," *in 2018 Int. Workshop Antenna Technol. (iWAT)*, Nanjing, China, Jun. 2018.

22 M. Y. Li, Y. L. Ban, Z. Q. Xu, C. Y. D. Sim, K. Kang, and Z. F. Yu, "Eight-port orthogonally dual-polarized antenna array for 5G smartphone applications," *IEEE Trans. Antennas Propag.*, vol. 64., no. 9, pp. 3820–3830, Sept. 2016.

23 H. Zou, Y. Li, C. Y. D. Sim, and G. Yang, "Design of 8 × 8 dual-band MIMO antenna array for 5G smartphone applications," *Int. J. RF Microwave Comput. Aided Eng.*, e21420, vol. 28, no. 9, Nov. 2018.

24 L. Sun, Y. Li, Z. Zhang, and Z. Feng, "Wideband 5G MIMO antenna with integrated orthogonal-mode dual-antenna pairs for metal-rimmed smartphones," *IEEE Trans. Antennas Propag.*, vol. 68, no. 4, pp. 2494–2503, Apr. 2020.

25 L. Sun, H. Feng, Y. Li, and Z. Zhang, "Compact 5G MIMO mobile phone antennas with tightly arranged orthogonal-mode pairs," *IEEE Trans. Antennas Propag.*, vol. 66, no. 11, pp. 6364–6369, Nov. 2018.

26 A. Zhao and Z. Ren, "Size reduction of self-isolated MIMO antenna system for 5G mobile phone applications," *IEEE Antennas Wireless Propag. Lett.*, vol. 18, no. 1, pp. 152–156, Jan. 2019.

27 A. Zhao and Z. Ren, "5G MIMO antenna system for mobile terminals," *2019 IEEE International Symposium on Antennas and Propagation and USNC-URSI Radio Science Meeting*, Atlanta, GA, USA, pp. 427–428, July, 2018.

28 A. Zhao, Z. Ren, and S. Wu, "Broadband MIMO antenna system for 5G operations in mobile phones," *Int. J. RF Microw. Comput. Aided Eng.*, e21857, vol. 29, no. 10, pp. 1–10, Oct. 2019.

29 C. Z. Han, L. Xiao, Z. Chen, and T. Yuan, "Co-located self-neutralized handset antenna pairs with complementary radiation patterns for 5G MIMO applications," *IEEE Access*, vol. 8, pp. 73151–73163, Apr. 2020.

30 H. Piao, Y. Jin, and L. Qu, "A compact and straightforward self-decoupled MIMO antenna system for 5G applications," *IEEE Access*, vol. 8, pp. 129236–129245, Jul. 2020.

31 K. L. Wong, B. W. Lin, and W. Y. Li, "Dual-band dual inverted-F/loop antennas as a compact decoupled building block for forming eight 3.5/5.8-GHz MIMO antennas in the future smartphone," *Microwave Opt. Technol. Lett.*, vol. 59, no. 11, pp. 2715–2721, Aug. 2017.

32 W. Li, W. Chung, and K. L. Wong, "Compact quad-offset loop/IFA hybrid antenna array for forming eight 3.5/5.8 GHz MIMO antennas in the future smartphone," *in International Symposium on Antenna and Propagation*, Busan, Korea (South) 2018.

33 C. You, D. Jung, M. Song, and K. L. Wong, "Advanced coupled-fed MIMO antennas for next generation 5G smartphones," *in International Symposium on Antenna and Propagation*, Busan Korea (South), 2018.

34 H. Wang, R. Zhang, Y. Luo and G. Yang, "Compact eight-element antenna array for triple-band MIMO operation in 5G mobile terminals," *IEEE Access*, vol. 8, pp. 19433–19449, Jan. 2020.

35 M. Y. Li, C. Li, Y. L. Ban, and K. Kang, "Multiple antennas for future 4G/5G smartphone applications," *in 2016 IEEE MTT-S International Microwave*

Workshop Series on Advanced Materials and Processes for RF and THz Applications (IMWS-AMP), Chengdu, China, Jul. 2016.

36 Y. L. Ban, Z. X. Chen, Z. Chen, K. Kang, and J. L. W. Li, "Decoupled closely spaced heptaband antenna array for WWAN/LTE smartphone applications," *IEEE Antennas Wireless Propag. Lett.*, vol. 13, pp. 31–34, 2014.

37 Q. Chen, H. Lin, J. Wang, L. Ge, Y. Li, T. Pei, and C.Y.D. Sim, "Single ring slot based antennas for metal-rimmed 4G/5G smartphones," *IEEE Trans. Antennas Propag.*, vol. 67, no. 3, pp. 1476–1487, Mar. 2019.

38 I. Syrytsin, S. Zhang, G. F. Pedersen, and A. S. Morris, "User-shadowing suppression for 5G mm-wave mobile terminal antennas," *IEEE Trans. Antennas Propag.*, vol. 67, no. 6, pp. 4162–4172, Jun. 2019.

39 R. Hussain, A. T. Alreshaid, S. K. Podilchak, and M. S. Sharawi, "Compact 4G MIMO antenna integrated with a 5G array for current and future mobile handsets," *IET Microwave Antennas Propag.*, vol. 11, no. 2, pp. 271–279, Feb. 2017.

40 M. M. Samadi Taheri, A. Abdipour, S. Zhang, and G. F. Pedersen, "Integrated millimeter-wave wideband end-fire 5G beam steerable array and low-frequency 4G LTE antenna in mobile terminals," *IEEE Trans. Veh. Technol.*, vol. 68, no. 4, pp. 4042–4046, Apr. 2019.

41 J. Kurvinen, H. Kähkönen, A. Lehtovuori, J. Ala-Laurinaho, and V. Viikari, "Co-designed mm-Wave and LTE handset antennas," *IEEE Trans. Antennas Propag.*, vol. 67, no. 3, pp. 1545–1553, Mar. 2019.

42 H. Wang, M. Wang, F. Nian. Y. Luo, and G. Yang, "Small-size four-element antenna system for 2×2 LTE LB and 4×4 LTE M/HB MIMO operations in 5G mobile terminals," *Int. J. RF Microwave Comput. Aided Eng.*, e22328, vol. 30, no. 9 Sept. 2020.

43 G. Yang, "Design and analysis of mobile phone antenna system with integration of LTE 4G, Sub-6G and millimeter wave 5G technologies," *in 2019 International Conference on Microwave and Millimeter Wave Technology (ICMMT)*, Guangzhou, China, May 2019.

44 H. C. Huang and J. Lu, "Evolution of innovative 5G millimeter-wave antenna designs integrating non-millineter-wave antenna functions based on antenna-in-package (AiP) solution to cellular phones," *IEEE Access*, vol. 9, pp. 72516–72523, May 2021.

45 H. C. Huang, Z. Qi, D. Gao, J. Liu, Y. Zhou, J. Li, and H. Lin, "5G miniaturized module of wideband dual-polarized mm-wave antennas-in-package integrating non-mm-wave antennas (AiPiA) for cell phones," *in Proc. IEEE Asia–Pacific Microw. Conf. (APMC)*, pp. 63–65, Dec. 2020.

46 H. C. Huang, Z. Qi, D. Gao, J. Liu, Y. Zhou, J. Li, and H. Lin, "Miniaturized 5G module of wideband dual-polarized mm-wave antennas-in-package integrating non-mm-wave antennas (AiPiA) for MIMO in cellular phones," *in* Proc. 15th Eur. Conf. Antennas Propag. (EuCAP), Mar. 2021.

47 H. C. Huang, Z. Qi, D. Gao, J. Liu, Y. Zhou, J. Li, and H. Lin, "Miniaturized 5G module of wideband dual-polarized mm-wave antennas-in-package as non-mm-wave antennas (AiPaA) for cell phones," *in Asia–Pacific Microw. Conf. (APMC)*, Nov. 2021.

48 H. C. Huang, Z. Qi, D. Gao, J. Liu, Y. Zhou, J. Li, and H. Lin, "Miniaturized 5G module of wideband dual-polarized mm-wave antennas-in-package as non-mm-wave antennas (AiPaA) for MIMO functions in cell phones," *in* Proc. Int. Conf. Microw. Millimeterw. Technol. (ICMMT), May 2021.

49 K. L. Wong, Y. H. Chen, and W. Y. Li, "Decoupled compact ultra-wideband MIMO antennas covering 3300~6000 MHz for the fifth-generation mobile and 5GHz-WLAN operations in the future smartphone," *Microwave Opt. Technol. Lett.*, vol. 60, no. 10, pp. 2345–2351, Sept. 2018.

50 C. Y. D. Sim, H. Y. Liu, and C. J. Huang, "Wideband MIMO antenna array design for future mobile devices operating in the 5G NR frequency bands n77/n78/n79 and LTE band 46," *IEEE Antennas Wireless Propag. Lett.*, vol. 19, no. 1, pp. 74–78, Jan. 2020.

51 H. Wang, R. Zhang, Y. Luo, G. Yang, "Design of MIMO antenna system operating in wideband of 3300 to 6400 MHz for future 5G mobile terminal applications," *Int. J. RF Microwave Comput. Aided Eng.*, e22426, vol. 30, no. 12, Sept. 2020.

52 X. T. Yuan, Z. Chen, T. Gu, and T. Yuan, "A wideband PIFA-pair-based MIMO antenna for 5G smartphones," *IEEE Antennas Wireless Propag. Lett.*, vol. 20, no. 3, pp. 371–375, Mar. 2021.

53 X. Zhang, Y. Li, W. Wang, and W. Shen, "Ultra-wideband 8-port MIMO antenna array for 5G metal-frame smartphones," *IEEE Access*, vol. 7, pp. 72273–72282, Jun. 2019.

54 H. D. Chen, Y. C. Tsai, C. Y. D. Sim and C. Kuo, "Broadband 8-antenna array design for future sub-6GHz 5G NR bands metal-frame smartphone applications," *IEEE Antennas Wireless Propag. Lett.*, vol. 19, no. 7, pp. 1078–1082, Jul. 2020.

55 X. T. Yuan, W. He, K. D. Hong, C. Z. Han, Z. Chen, and T. Yuan, "Ultra-wideband MIMO antenna system with high element-isolation for 5G smartphone application," *IEEE Access*, vol. 8, pp. 56281–56289, Mar. 2020.

56 Q. Cai, Y. Li, X. Zhang, and W. Shen, "Wideband MIMO antenna array covering 3.3–7.1 GHz for 5G metal-rimmed smartphone applications," *IEEE Access*, vol. 7, pp. 142070–142084, Sept. 2019.

57 N. O. Parchin, Y. I. A. Al-Yasir, A. M. Abdulkhaleq, H. J. Basherlou, A. Ullah, and R. A. Abd-Alhameed, "A new broadband MIMO antenna system for sub 6 GHz 5G cellular communications," *in Proc.14th European Conference on Antennas and Propag.*, Copenhagen, Denmark, pp. 1–4, Mar. 2020.

58 A. Singh and C. E. Saavedra, "Wide-bandwidth inverted-F stub fed hybrid loop antenna for 5G sub-6 GHz massive MIMO enabled handsets," *IET Microwave Antennas Propag.*, vol. 14, no. 7, pp. 677–683, Jun. 2020.

5

MIMO-Based 5G FR1 Band Mobile Antenna

In Chapters 3 and 4, we have covered many antenna designs for fifth-generation Frequency Range 1 (5G FR1) band smartphone applications, from single band to multi-band, followed by wideband antenna type that can cover the 5G (New Radio) NR band n77/n78/n79 and 5G NR-U band n46/n96. As most of these designs are mostly focused on the single antenna element/radiator design and their respective antenna performances, their corresponding (Multiple-Input Multiple-Output) MIMO antenna array design topologies and their related performances such as Envelope Coefficient Correlation (ECC), Mean Effective Gain (MEG), and channel capacity, are not illustrated. Furthermore, because of the finite space in a smartphone, it has become an important matter to comprehend the design of MIMO antenna array beginning from one single antenna element, let alone the lack of summarizing the various feasible methods that can enhance the isolation level between two adjacent antenna elements, as they are very much closely spaced in a 5G smartphone. Therefore, 5G antenna engineers must possess sufficient knowledge to achieve maximum MIMO performances for antenna arrays collocated in a smartphone via the different methods that have been reported so far.

The MIMO antenna technology for wireless communications is realized by applying multiple antennas at both the source (transmitter) and the end-user (receiver). In the 5G scenario, it can be seen as the 5G base station (or CPE – Customer Premise Equipment) and the 5G User Equipment (UE) or smartphone. Here, the 5G antenna array at both ends is combined to optimize data transmission speed, yield a lower latency rate, and enhance the channel capacity (or throughput) by allowing the transmission data to travel over many signal paths simultaneously. By creating multiple traveling paths of the same signal, it will yield more chances for the transmission data to arrive at the receiving antenna

Microwave and Millimeter-Wave Antenna Design for 5G Smartphone Applications,
First Edition. Wonbin Hong and Chow-Yen-Desmond Sim.

Published 2023 by John Wiley & Sons, Inc.

array even when it is being affected by various fading effects (such as multipath fading) and free-space path loss, which, in turn, increases the signal to noise ratio (SNR) and error rate. By amplifying the channel capacity of the wireless communication systems, the MIMO technology can therefore produce a much more stable wireless communication network.

According to the 3rd Generation Partnership Project (3GPP), it has already added the MIMO technology to the Mobile Broadband (MB) Standard. Presently, the MIMO technology is used in WiFi networks, Fourth-Generation (4G) Long-Term Evolution (LTE) cellular, and Fifth-Generation (5G) MB technology and wireless network. Furthermore, the MIMO technology has supported all wireless products with Wireless Local Area Network (WLAN) 802.11n/ac, not to mention the WLAN 802.11ax, also known as WiFi 6. As the wireless industry has been working toward accommodating more antennas, networks, and end-user devices (or UE), thus, the 5G MIMO technology continues to reform and expand through its use in massive new applications, which is now known as the 5G massive MIMO (mMIMO). As the name “massive” has indicated, the application of massive MIMO technology involves a much larger scale for greater network coverage and capacity, which signify that the massive MIMO will apply large number of transmitting and receiving antenna elements to send and receive more data simultaneously. The advantage of massive MIMO is to further enhance the transmission gain, which results in many more end-users or UEs connecting to the network simultaneously while maintaining high throughput.

Unlike 4G MIMO, which applies the Frequency Division Duplex (FDD) system for supporting multiple UEs, the 5G massive MIMO has involved a different setup known as the Time Division Duplex (TDD) that offers several advantages over the FDD. Even though the FDD and TDD techniques are both applied in the mobile communication networks, in the FDD mode, the uplink and downlink of data transmission can be performed at the same time at different spectrum frequencies; however, it requires a reserved spectrum (known as the guard band), an unused part of the spectrum that separates the uplink and downlink spectrum frequencies, so as to prevent interference. In the TDD mode, both the uplink and downlink of data transmission use the same spectrum frequencies but over different times, thus allowing wider bandwidth and more users. That is the main reason for all 5G NR bands such as n77/n78/n79 and those in the 5G NR-U band n46/n49 are using the TDD mode.

5.1 Motivation and Requirements

Before MIMO, there were other types of antenna technologies with different configurations, such as the conventional Single-Input Single-Output (SISO), Single-Input Multiple-Output (SIMO), and Multiple-Input Single-Output (MISO).

Therefore, it can be considered that the MIMO is built on these technologies, let alone the massive MIMO that is supposed to use many more antennas than the number of UE in the cell.

As there is no specific minimum number of antennas required for the application of MIMO system, the commonly accepted threshold for a system is more than eight transmitter (Tx) and eight receiving (Rx) antennas, and the number of Tx/Rx antennas can be much higher, extending to massive MIMO systems with tens or even hundreds of Tx/Rx antennas.

Figure 5.1 shows the channel capacity (bps/Hz) of a MIMO system with an increasing number of Tx (referred to as the number of antenna elements in the transmitting end, such as the base station) and Rx (referred to as the number of antenna elements in the receiving end, such as the smartphone). As can be seen, if the Rx users end (smartphone) has only one antenna element, even if increasing the Tx to an eight-antenna array system, the channel capacity has merely increased from 5.88 to 6.56 bps/Hz for a 1×8 MIMO. However, if the Rx is an eight-antenna array type, such as those antenna array designs introduced in Chapters 3 and 4, by slowly increasing the Tx from a single antenna element to an eight-antenna array, the differences in channel capacities between the 8×1 MIMO and 8×8 MIMO is near 4.6 times, from 9.55 to 43.97 bps/Hz. For a 200 MHz operational bandwidth, said from 3400 to 3600 MHz, disregarding all the other factors such as signal fading and losses, the maximum channel capacity, in this case, can be reached up to 8.794 Gbps ($43.97 \times 200 \times 10^6$). This is the main motivation as to why the MIMO technology has prevailed and been selected over many others of its predecessors.

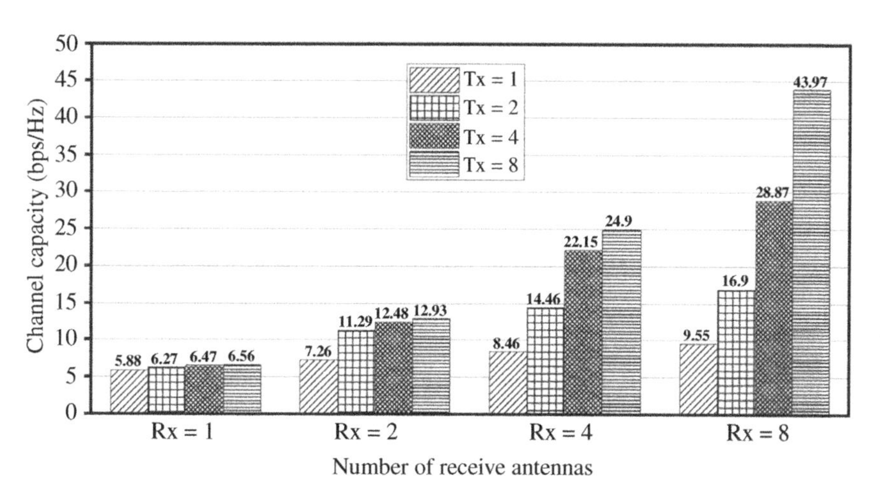

Figure 5.1 The channel capacity of a MIMO system with different number of receiving antennas (Rx) and transmitting antennas (Tx).

Obviously, like many other new technologies, implementing the massive MIMO technology into the 5G network comes with several hefty prices. One is the additional antenna techniques known as the beam steering (or beamforming), which is one of the key technologies for delivering such high channel capacity (or throughput) demanded by the 5G MIMO system due to the high number of Tx/Rx antennas that would be applied, followed by the high manufacturing cost during the process of realizing the MIMO system and its beamforming technique, as high precision beamforming requires high-cost Radio Frequency (RF) front-end circuit for the phased-array antenna system to focus the signal in a chosen direction, normally toward a specific receiving UE. This results in an improved signal between the transmitting end and the UE, with less interference between individual UE signals. As this chapter focuses on the 5G FR1 antenna design technologies, we would not further discuss the beamforming techniques for brevity.

To justify the potency of any MIMO antenna, some important diversity performances such as the channel capacity, Envelope Correlation Coefficient (ECC), Diversity Gain (DG), MEG, Channel Capacity Loss (CCL), and Total Active Reflection Coefficient (TARC), are essential, and they must be calculated and verified through simulation as well as measurement. As for the isolation level between adjacent antenna elements, it is a very vital parameter as well, and the methods to achieve very good isolation for the 5G FR1 antenna array will be explicitly introduced in Section 5.2.

5.1.1 Channel Capacity

The channel capacity is used to compute the multiplexing performance of the MIMO antenna. Based on the channel conditions, the MIMO system will yield higher channel capacity according to the modified Shannon equation given by [1]

$$C = MB\log_2\left(1 + N/M \times \mathrm{SNR}\right) \tag{5.1}$$

where C is the channel capacity (bps), M is the number of transmission antennas, N is the number of receiving antennas, B is the operational bandwidth (Hz), and SNR is the Signal-to-Noise Ratio. A more accurate modified channel capacity equation for M and N, with no channel state information, Gaussian distributed signals, and identity covariance matrix is given by [2]

$$C = B\log_2\left(\det\left(\mathbf{I}_N + P_\mathrm{T}/\left(\sigma^2 M\right) \times \mathbf{H}\mathbf{H}^\mathrm{H}\right)\right) \tag{5.2}$$

where min (N, M) is the minimum number of independent channels in the wireless environment, P_T is the equally distributed input power among the elements, σ^2 is the noise power, $\mathbf{I}_N$ is the $N \times N$ identity matrix, and $\mathbf{H}$ is the complex channel matrix.

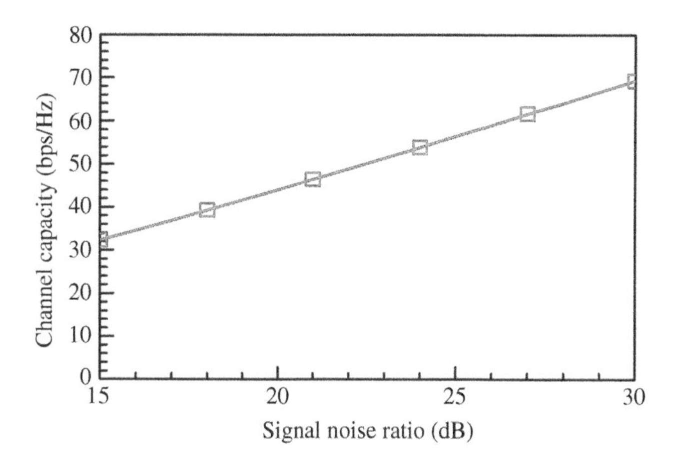

Figure 5.2 Calculated peak channel capacity with respect to SNR for an 8×8 MIMO system in [3].

By observing and comparing Figure 5.1 to Eqs. (5.1) and (5.2), one can see that a linear increase in the number of antennas at both transmitting and receiving ends will ensure higher channel capacity, especially when it reaches 8×8 MIMO, but it is based on the assumption of an ideal environment, in which the multipath channels are not correlated. As for the SNR, it is very vital to maintain a high level of SNR, as it would highly affect the peak channel capacity yielded by the MIMO system. One example is the wideband eight-antenna array design reported in [3] that can cover the 5G NR band n77/n78/n79 and 5G NR-U n46 for metallic chassis smartphone applications. Here, for an 8×8 MIMO system, Figure 5.2 shows the peak channel capacities of the eight-antenna array with respect to different SNR in the receiver, and they were calculated by averaging 100 000 independent and identically distributed (i.i.d.) Rayleigh fading channel realizations [4]. In this figure, the calculated peak channel capacity (at 15 dB SNR) was 32.28 bps/Hz. When the SNR was increased to 20 dB SNR, the calculated peak channel capacity was improved to 43.93 bps/Hz, and when the SNR was further increased to 30 dB, the calculated peak channel capacity had reached a much higher level of 69.25 bps/Hz. Therefore, besides implementing more antenna elements for the MIMO system (in both transmitting and receiving ends), maintaining a high level of SNR is also essential for seeking higher channel capacity.

5.1.2 Envelope Correlation Coefficient

The ideal of ECC is to learn how independently two adjacent antenna elements can radiate throughout the operational bandwidth, and thus it is a very essential parameter to be investigated. Ideally, if one antenna is completely Horizontally

Polarized (HP), and the other one is completely Vertically Polarized (VP), the ECC value should be equal to zero, which indicates that the antenna elements radiate independently by producing uncorrelated radiations. However, in a rich fading environment, the ECC value is not equivalent to zero.

In [2], it has been reported that many researchers have provided separate and unconnected ground for different antenna elements in a MIMO antenna system, which is clearly infeasible, as, in a real system, the signals should have a common reference ground plane, so that all signal levels within the system can be interpreted properly based on that reference level. Thus, it is vital to avoid multiple separated (unconnected) ground planes for the MIMO antenna system. To identify the desirable ECC value, the far-field radiation patterns of the MIMO antenna system should be acquired, as they directly affect the channel capacity between the transmitting and receiving end, and the ECC that takes into account the antenna's radiation pattern shape, polarization, and relative phase between two adjacent antenna elements is given by [2]

$$|\rho_{12}|^2 = \rho_e(\text{ECC}) = \frac{\left|\iint_{4\pi}\left[\vec{F_1}(\theta,\varnothing)\times\vec{F_2}(\theta,\varnothing)\right]\mathrm{d}\Omega\right|}{\iint_{4\pi}\left|\vec{F_1}(\theta,\varnothing)\right|^2\mathrm{d}\Omega\iint_{4\pi}\left|\vec{F_2}(\theta,\varnothing)\right|^2\mathrm{d}\Omega} \tag{5.3}$$

where $\vec{F_n}$ $(\theta, \varnothing)$ is the Three-Dimensional (3-D) far-field pattern for antenna n, and Ω is the solid angle. Notably, Eq. (5.3) is based on the assumption of having an isotropic wireless environment. As indicated in [2], for many years, a more commonly misused equation for calculating the ECC based on the port parameters (S-parameters) is given in

$$\text{ECC} = \rho_e = \frac{\left|S_{11}{}^{*}S_{12} + S_{21}{}^{*}S_{22}\right|^2}{\left(1-|S_{11}|^2-|S_{21}|^2\right)\left(1-|S_{22}|^2-|S_{12}|^2\right)} \tag{5.4}$$

However, the condition to be noted is that the two antenna elements must have antenna efficiency of >90% or >−1 dB (lossless antenna) so that a much more accurate ECC can be determined. As most antennas are lossy elements, unless the efficiency of the MIMO antenna is very high, then Eq. (5.4) should not be applied because it is not valid. Thus, it is always better to apply Eq. (5.3) via the far-field radiation patterns to determine the ECC, as the calculation of ECC via this equation is more accurate. One point to be noted is the acceptable ECC value in a MIMO antenna system. For mobile antenna system, the commonly acceptable ECC threshold value is 0.5, thus as long as the ECC value of a MIMO antenna system is smaller than 0.5, the antenna elements are uncorrelated from each other. However, such ECC value has been challenged recently in many reported wideband 5G smartphone antenna papers [3], and it has since been reduced to 0.3.

5.1.3 Diversity Gain

As aforementioned, for diversity and MIMO application, the correlation between signals received at the same side of a wireless link by the involved antenna element is an important figure of merit for the whole system. Thus, the ECC is usually applied to evaluate the diversity capability of the MIMO antenna system, and possessing a very low ECC value is also an indication of achieving high isolation and large DG between adjacent antenna elements. The DG is a metric of interest, as it measures the effect of diversity on the communication system. The DG is defined as the difference between the combined signal from all the antennas of the diversity system and the signal from a single antenna element. Therefore, it is always applied to evaluate the diversity performance of MIMO antenna [5]. To calculate the DG, one must obtain Eq. (5.3) and apply it to this formula:

$$\mathrm{DG} = 10\sqrt{1 - |\rho_\mathrm{e}|^2} \tag{5.5}$$

If the DG value is high, the improvement in diversity performance is better. One point to note is the desirable DG value should be close to 10 dB, and under such conditions, the antenna elements can be regarded as strongly uncorrelated with each other.

5.1.4 Mean Effective Gain

The MEG is a vital parameter for the MIMO antenna system, and it is usually applied to determine the performance of antenna elements in a rich multipath fading practical environment. The MEG is defined as the ratio of mean received power to the mean incident power at the antenna element, and it is calculated using the S-parameters of the MIMO antenna system given in

$$\mathrm{MEG}_i = 0.5\eta_{i,\mathrm{rad}} = 0.5\left(1 - \sum_{j=1}^{M} |S_{ij}|^2\right) \tag{5.6}$$

where i is the exciting antenna, $\eta_{i,\mathrm{rad}}$ is the radiation efficiency of the ith antenna, M is the number of antenna elements in the MIMO system. If only two adjacent antenna elements (Ant. 1 and Ant. 2) are considered, Eq. (5.6) can be further expanded, in which the MEG of each antenna element can be calculated by

$$\mathrm{MEG}_1 = 0.5\left(1 - |S_{11}|^2 - |S_{12}|^2\right) \tag{5.7}$$

$$\mathrm{MEG}_2 = 0.5\left(1 - |S_{21}|^2 - |S_{22}|^2\right) \tag{5.8}$$

where MEG_1 and MEG_2 are referring to the MEG of Ant. 1 and Ant. 2, respectively. Here, it is noteworthy that the calculated MEGs of the two antenna elements

Table 5.1 Calculated MEGS of the eight-antenna array in [6].

Freq. (GHz)	MEG1	MEG2	MEG3	MEG4	MRG5	MEG6	MEG7	MEG8
2.550	−5.747	−5.642	−5.904	−5.611	−5.467	−5.768	−5.365	−5.328
2.600	−5.460	−5.628	−5.666	−5.606	−5.025	−5.256	−4.791	−4.862
2.650	−5.454	−5.787	−5.650	−5.782	−5.069	−5.185	−4.656	−4.819

Source: From [6]. ©2016 IEEE. Reproduced with permission.

should have a difference of less than 1 dB across the entire operational band, or their corresponding ratio (MEG_1/MEG_2) is near to 1, which validates good diversity performance from the proposed MIMO antenna under a very rich multipath fading environment. One good example is the one reported in [6] for an eight-antenna array for smartphone applications, as shown in Table 5.1. In this table, the calculated MEGs of all the eight antenna elements (from MEG1 to MEG8) have exhibited MEG differences of less than 1 dB in all three operating frequencies and their corresponding MEG ratios are closer to 1 as well.

5.1.5 Total Active Reflection Coefficient

The TARC was initially proposed to evaluate the performance of a multiple port radiator [7], and it was only at a later stage that it was used as a MIMO metric for antenna systems. As the TARC has considered the inter-port coupling, port matching, and the effect of the random phases of the incoming signals into each antenna element, it can therefore yield valuable insights about its behavior when encountering real communication channels [8]. As reported in [7], for an N-port lossless antenna with a scattering matrix $[S]$, the TARC is defined as the ratio of the square root of total reflected power to the square root of total incident power as given by [8]

$$\Gamma_{\mathrm{a}}^{\mathrm{t}} = \sqrt{\sum_{i=1}^{N} \left|b_i\right|^2} \Bigg/ \sqrt{\sum_{i=1}^{N} \left|\mathrm{a}_{\mathrm{i}}\right|^2} \tag{5.9}$$

$$\left[b\right] = \left[S\right]\left[a\right] \tag{5.10}$$

where vector $[a]$ is the summation of the incident power at all excitation ports, a_i is the ith element of the vector $[a]$, the power reflected to the source due to antenna port mismatching (or inter-port transferred power) is the vector $[b]$, and b_i is the ith element of the vector $[b]$. As the TARC is a value between 0 and 1, the value 0 refers to all power delivered to the port being radiated, whereas TARC = 1 refers to all power being reflected to the ports (or transferred to other ports).

The TARC equation for a two-antenna port MIMO system using S-parameters is given in [9]

$$\Gamma = \sqrt{\left(\left|S_{11} + S_{12}e^{j\theta}\right|^2\right) + \left(\left|S_{21} + S_{22}e^{j\theta}\right|^2\right)} \Big/ \sqrt{2} \tag{5.11}$$

where θ is the input random phase angle varied between 0 and 2π, S_{11} and S_{22} are the reflection coefficients (dB) of the two antenna elements (Ant. 1 and Ant. 2), respectively, and S_{12} and S_{21} are the transmission coefficients of the two antenna elements. Therefore, it is easy to see that the TARC accounts for both coupling and random signals combining. One point to take note of is that a TARC value of less than 0 dB is acceptable for the MIMO antenna system.

5.1.6 Channel Capacity Loss

The CCL is an essential metric to feature the diversity performance of a MIMO antenna system, as it describes the maximum attainable limit of information transmission rate up to which the signal can be easily transferred with a significant loss [10]. In another word, the CCL helps in defining the loss of transmission (bps/Hz) in a high data rate transmission. Notably, to ensure high data rate transmission, the minimum acceptable limit of CCL over which the high data transmission is feasible is defined as 0.4 bps/Hz. Thus, the CCL values should be kept below 0.4 bps/Hz across the operational band. The CCL can be calculated using the following set of equations by [10, 11]

$$\text{CCL} = -\log_2 \det(\alpha) \tag{5.12}$$

where α is the correlation matrix of the receiving antenna and it is expressed in

$$\alpha = \begin{bmatrix} \rho_{11} & \cdots & \rho_{18} \\ \vdots & \ddots & \vdots \\ \rho_{81} & \cdots & \rho_{88} \end{bmatrix} \tag{5.13}$$

and

$$\rho_{ii} = 1 - \left(\sum_{j=1}^{M} \left|S_{ij}\right|^2\right), \rho_{ij} = -\left|S_{ii}^* S_{ij} + S_{ji}^* S_{jj}\right| \tag{5.14}$$

5.2 Antenna Isolation Techniques

As described in the previous section, attaining a high isolation level between adjacent antenna elements, especially for the eight-antenna array applied in the 5G smartphone, is a vital feature for improving the diversity performances of the

MIMO antenna system. Therefore, this section summarizes the various isolation techniques commonly applied in eight-antenna arrays for 5G FR1 smartphone applications. Furthermore, we will also explicitly describe the in-depth topologies of achieving good isolation between adjacent antenna elements. However, before discussing the various isolation or decoupling methods, we should comprehend the main factors that cause poor port isolation and high coupling between two adjacent (closely arranged) printed antennas in a MIMO antenna system that shared the same ground plane.

Figure 5.3 shows a conceptual diagram of a two-antenna array (Ant. 1 and Ant. 2) closely arranged on the same substrate with a common reference ground plane. Here, when only Port 1 is excited, the main factors that usually result in high coupling or poor port isolation between the two antennas are as follows: (i) the Over-The-Air (OTA) radiation patterns, which include the surface wave and reactive Near-Field (NF) experienced by the Ant. 2 from Ant. 1, and (ii) the ground current generated due to port 1 excitation that flows from Port 1 to Port 2. The OTA surface wave usually happens to the printed antenna as the antenna elements have shared the same substrate, and the propagation of signals (or electromagnetic [EM] waves) will flow to the adjacent antenna. Notably, this phenomenon can also happen to be a non-printed antenna, as two closely arranged antennas (with a gap distance $< 0.25\lambda_0$) will incur high mutual coupling as their radiation patterns are pointing at each other [12].

The reactive NF phenomenon is not widely studied in many reports, and most antenna engineers would usually ignore its existence because if the two antennas are placed much farther away from each other (for example, a gap distance $> 0.5\lambda_0$), this phenomenon will cease to exist. Figure 5.4 shows a good example of the simulated reactive NF effects from Ant. 1 to Ant. 2 when only Ant. 1 is excited at different frequencies from 2.1 to 3.5 GHz. Here, the two antennas are very closely arranged at a gap distance of $D = 9$ mm. As can be seen, when the exciting frequency is at 2.1 GHz, due to its longer electrical wavelength, the reactive NF

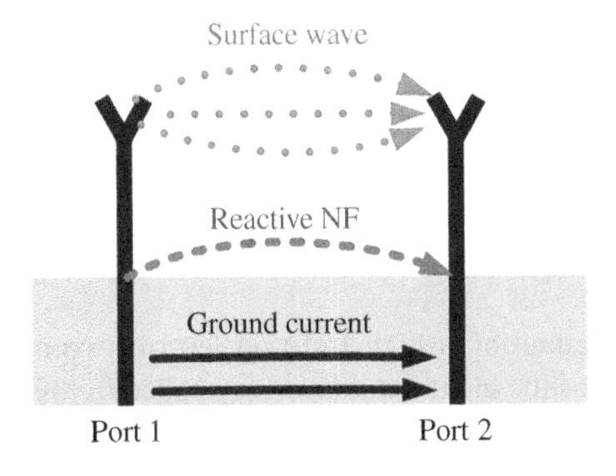

Figure 5.3 The main factors that result in high coupling or poor port isolation between two adjacent antennas that shared a common reference ground plane. In this case, only port 1 is excited.

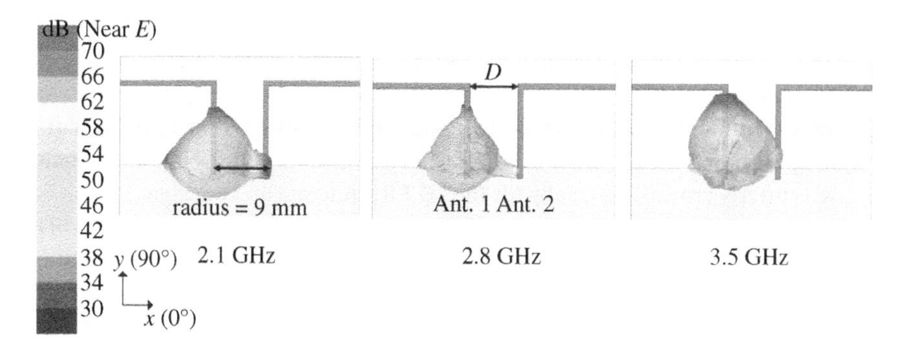

Figure 5.4 Simulated reactive near-field radiation patterns at 2.1, 2.8, and 3.5 GHz. *Source:* From [12]. ©2018 IEEE. Reproduced with permission.

strength experienced by Ant. 2 (from Ant. 1) is very much stronger than those working at a higher frequency. At a higher frequency (especially at 3.5 GHz), the reactive NF strength from Ant. 1 to Ant. 2 has significantly reduced because a higher frequency has a shorter electrical wavelength, which results in a much longer electrical distance between the two antennas. Therefore, it is always a good practice for antenna engineers to observe the reactive NF of two very closely spaced antennas.

As mentioned earlier by [2], a feasible MIMO antenna system should have a common reference ground plane for sending the signal to its respective antenna element within the antenna array. Therefore, the two adjacent antennas that appeared in Figure 5.3 (sharing the same common reference ground) will generally excite surface current on the ground plane that will be flowing in the transverse direction from Port 1 to Port 2 (as only Port 1 is excited). This, in turn, will yield undesirable ground current that "leak" into Port 2 (generated by Port 1), causing poor port isolation.

From the previous two chapters, it is learned that the 5G smartphone has a very limited finite space for placing the desired 5G antenna array (e.g. eight-antenna type) along the two longer side edges of the chassis. Therefore, the distance between two adjacent antenna elements of such kind would usually be less than half-wavelength (@3300 MHz, $0.5\lambda_0$ ~45 mm), or even lesser than quarter-wavelength (@3300 MHz, $0.25\lambda_0$ ~ 23 mm) because the usual system ground plane size of a common non-metallic chassis 5G smartphone is approximately between $136 \times 68\,\text{mm}^2$ and $150 \times 80\,\text{mm}^2$, which includes the clearance for main/diversity 4G LTE antennas. To collocate four antenna elements on each longer side edge of the frame (or system ground), one can calculate that each antenna element has only approximately 34–37.5 mm (in length) to spare. Thus, the length of the antenna element has become a crucial factor because the longer the antenna element, the smaller the gap distance will be between the two adjacent antenna elements. As explained earlier, two closely spaced antenna elements (with a very small gap distance) will incur poor port isolation and high coupling effects due to the aforementioned OTA effects (surface wave and reactive NF) and ground current distribution. Therefore, the knowledge to improve the port isolation for such a case is utterly critical, especially for the 5G FR1 antenna array design.

To accommodate the 5G FR1 antenna elements in a finite compact size 5G smartphone, as well as achieve a desirable isolation level, from the open literature, the most commonly applied isolation (or decoupling) techniques for 5G FR1 smartphone applications are mainly categorized as follows:

1) Spatial and polarization diversity
2) Neutralization line
3) Slotted element
4) Wave trap element
5) Self-isolated element
6) Others

In the following sub-sections, we would briefly describe the various methods applied to achieve good port isolation between two adjacent 5G FR1 antenna elements with examples, and some of their in-depth topologies are also explicitly introduced.

5.2.1 Spatial and Polarization Diversity

As indicated in [13], the two most commonly applied methods for achieving good port isolation and mutual decoupling for two adjacent antennas sharing the same common reference ground are spatial diversity and polarization diversity. The idea of the spatial diversity method is to separate the two antennas farther away from each other with a desirable gap distance of $>0.5\lambda_0$. By applying this technique, no additional decoupling structure and alteration to the antenna array structure are required. Even though such a technique can yield good port isolation of larger than 20 dB, the required gap distance between two adjacent antennas working in the 5G n77/n78 band, theoretically, must be at least 45 mm ($0.5\lambda_0$ at 3300 MHz), which is not a feasible solution for FR1 5G smartphone applications. Thus, to avoid such a dilemma, the polarization diversity technique is introduced, as it only requires two antennas to be displaced in an orthogonal fashion so that their respective excited electric-field (E-field) or polarization senses are orthogonal as well, which can yield good port isolation of at least 15 dB [13]. One good example is from work reported in [14], as shown in Figure 5.5, in which Ant. 1 to Ant. 5 are allocated along the shorter edges of the ground system, while (Ant. 2 to Ant. 4) and (Ant. 6 to Ant. 8) are collocated along the two long side edges. Here, one can see that the

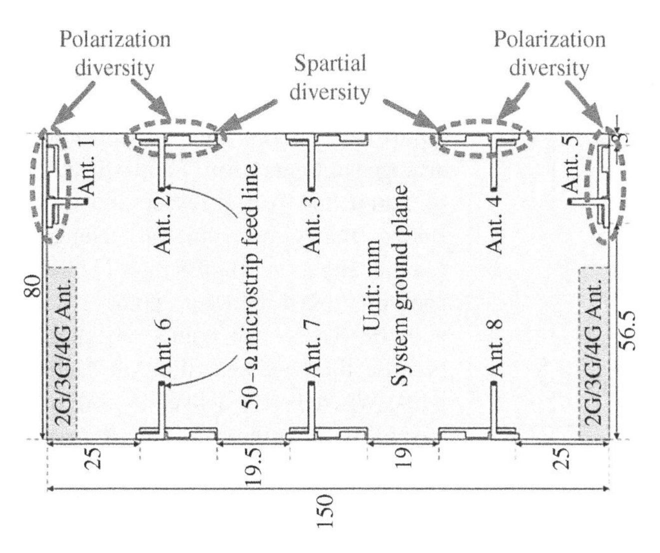

Figure 5.5 The antenna layout of [14] that includes polarization diversity (Ant. 1 and Ant. 2) and spatial diversity (Ant. 2 and Ant. 4). *Source:* From [14]. ©2019 IEEE. Reproduced with permission.

polarization diversity technique was applied to (Ant. 1 and Ant. 2) and (Ant. 4 and Ant. 5), as the two adjacent slot radiators are in orthogonal positions. On the other hand, the spatial diversity technique was applied to Ant. 2 and Ant. 4 (or Ant. 1 and Ant. 5) as they have a gap distance of more than $0.5\lambda_0$. To further realize other features of the polarization diversity method for antenna arrays working in the 5G FR1 band smartphone applications, we recommend the progressive works reported by the same group of authors in [15–17]. Notably, from these recommended works, another name, "orthogonal mode," was given to the spatial diversity technique when applied to the 5G FR1 antenna array for smartphone applications. An antenna pair was usually involved and had produced two different excitation modes with orthogonal polarization (see Figure 4.24). Other interesting designs that have shared the same idea of introducing different excitation modes for improving the isolation level to more than 17 dB have also been reported in [18, 19].

5.2.2 Neutralization Line

The NL technique has been a topic of interest for over a decade. It provides a feasible method to improve the port isolation between two closely arranged antennas without resorting to complicated decoupling structures or lump elements. But before we introduce the various works that have applied the NL technique, we

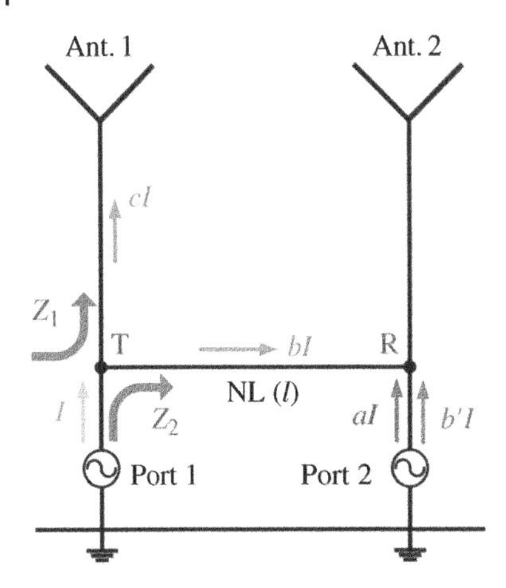

Figure 5.6 The conceptual equivalent circuit diagram of a closely arranged two-antenna system that shared the same common reference ground, and has a neutralization line (NL) for improving the port isolation.

should comprehend the circuit model of a two-antenna system that has involved the NL technique, as depicted in Figure 5.6. In this figure, two closely arranged antennas, Ant. 1 and Ant. 2, fed by Port 1 and Port 2, respectively, have shared the same common reference ground. Here, an NL of length l is horizontally linked between point T and point R. These two points are chosen because they are near the two feeding ports. When Port 1 delivers a current I toward Ant. 1, at the same time, a current aI is coupled to Ant. 2. Because of the NL connected at the two points, the current I (from Port 1) is split into bI and cI, where bI runs along with the NL and cI flows toward Ant. 1. Notably, due to the current flowing on the NL, another new current $b'I$ is coupled to Ant. 2, and the summation of these two coupled currents in an ideal circuit environment should be equals to zero, as in

$$aI + b'I = 0 \tag{5.15}$$

and one can see that the main solution to perfect decoupling is to select an appropriate coupling coefficient b'. To further explain the decoupling relationship between the NL and the two adjacent antennas, two input impedances are defined, in which the input impedance observing from point T to Ant. 1 is Z_1, and the input impedance looking from point T toward the NL is Z_2. By applying the equivalent circuit theory, the coefficients b and c can be given as

$$b = Z_1/(Z_1 + Z_2), c = Z_2/(Z_1 + Z_2) \tag{5.16}$$

As the current wave bI is propagating on the NL (with a total length l) from point T to point R, the relationship between the coefficient b and the coupling coefficient b' can be approximated by considering the transmission line theory as

$$b' = be^{-j\beta l} \tag{5.17}$$

where β is the phase shift coefficient.

From this analysis, a design guideline can be concluded as follows:

1) The coupling coefficients can be affected/varied by applying different operating frequencies, tuning the position of the two connecting points (Point T and Point R), and varying the length (l) and width of the NL.
2) The position of the two connecting points and the width of the NL determine the magnitude, whereas the operating frequency and the length (l) of the NL mainly affect the phase shift.

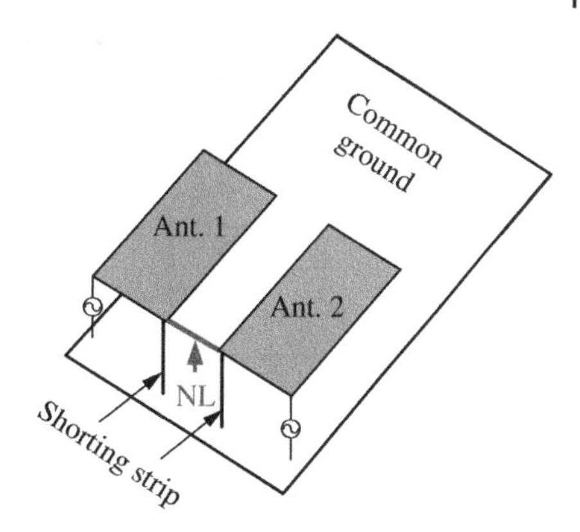

Figure 5.7 The conceptual diagram of two inverted-F antennas closely arranged on a common reference ground, and has implemented a neutralization line for improving the port isolation.

Section 9.1.2 has also provided another perspective in describing the mutual coupling effects between two adjacent antennas.

Some of the earliest works that have applied the NL technique to a mobile terminal have been reported in [20, 21]. In these reports, two inverted-F antennas (IFAs) were analyzed, as shown in Figure 5.7, and it has been concluded that

1) The NL must be connected to the points that exhibit the smallest impedance.
2) Increasing the port isolation level up to 10 dB is possible by carefully positioning the IFAs on the top edge of the common reference ground with their shorting strips facing each other.
3) The NL can be connected between the two feeding lines or the two shorting strips of the IFAs. The NL applied in both cases has been shown to act like neutralization devices withdrawing an amount of the signal on one IFA and bringing it back to the other so that the mutual coupling is reduced.

Interestingly, the above findings have also been verified in [22, 23], when it has physically connected an NL between the shorting points or feeding points of an IFA pair. However, a careful reminder was also stated that the length and width of the NL must be carefully tuned to create an additional coupling path for canceling out the existing mutual coupling; if not, an additional mismatch will incur. According to the authors' personal experience, finding the exact dimension (length and width) of the NL and the positions of the connecting points have proven to be a very tedious process.

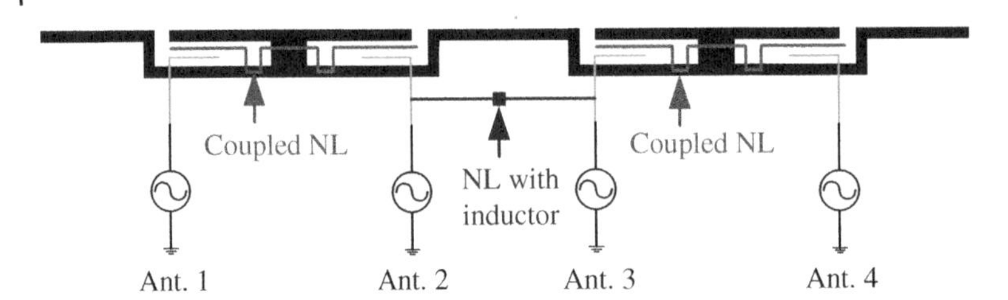

Figure 5.8 The conceptual diagram of a two-antenna slotted pairs design that has applied the coupled neutralization line and loaded a lumped element into the neutralization line for improving the port isolation [24].

Besides the usual method of linking the NL physically between the two adjacent antennas for a mobile device, the idea of applying coupled NL and loading a lumped element (chip inductor of 8.2 nH) into the NL was introduced in [24], and its conceptual antenna design diagram is as shown in Figure 5.8. In this case, as each slotted antenna pair (Ant. 1 and Ant. 2) and (Ant. 3 and Ant. 4) is only 19 mm × 3 mm, the NL must be devised as a meandering shape. Because the gap distance between the two antenna pairs is approximately 12 mm, which is much shorter than the quarter-wavelength of the operating C-band frequency at 3.5 GHz, an additional inductor (8.2 nH) was loaded into the NL for achieving better mutual decoupling. Notably, this lump-loading technique has also been applied to a slotted ground-coupled loop-type decoupling structure for improving the port isolation of a slotted antenna pair, but in this case, a chip capacitor of 0.79 pF was applied [25].

From this analysis, it has been proven that the loading of NL (working as the decoupling structure) can successfully improve the port isolation of two adjacent antennas. By further observing [26], as long as the integrated antenna (single-fed but with two different radiators) has a separate radiator for each different operating band, as shown in Figure 5.9, the loading of the NL between an antenna pair (of the same radiator, in this case, is the gap-coupled loop branch) will have no effects on the other radiator. More detail of this dual band 5G FR1 eight-antenna array design can be seen in Figure 3.21. As shown in Figure 5.9, without loading the NL, a high coupling effect is experienced by the gap-coupled loop branch of Ant. 3 when the one in Ant. 2 is excited. After loading the NL, the surface current distribution in Ant. 3 has significantly reduced because the intense current on the NL has resulted in the strong phase-reversal coupling, which offsets the original mutual coupling. Notably, this figure has also shown the simulated effects on the isolation (S_{32}) between the two adjacent antennas by tuning the position (d) of the NL from 0 mm (without NL) to 2.9 mm. Here, the isolation between Ant. 2 and Ant. 3 has increased by approximately 3 dB in the low-band when $d = 2.9$ mm. In contrast, tuning parameter d has shown very little effect on the S_{32} across the

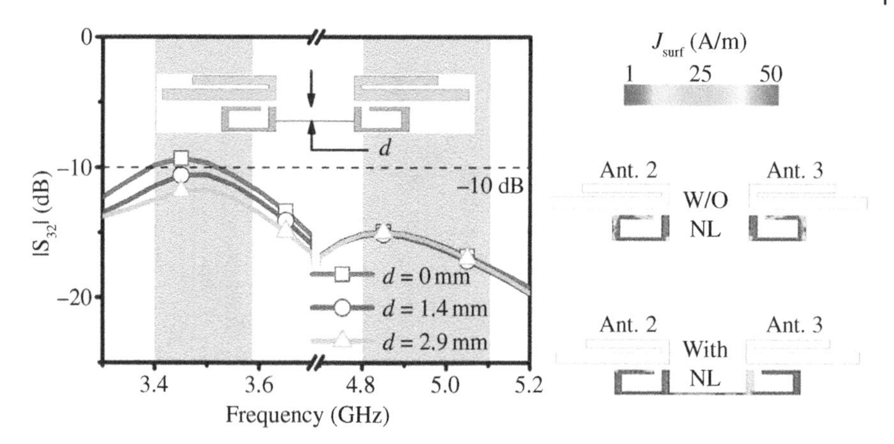

Figure 5.9 The simulated surface current distribution diagrams between Ant. 2 and Ant. 3 with and without loading the neutralization line between the two gap-coupled loop branch radiators, and its corresponding simulated isolations across the two operating bands between Ant. 2 and Ant. 3 when tuning the NL position *d*. *Source:* From [26]. ©2018 IEEE. Reproduced with permission.

high-band, which signifies that the loaded NL can only improve the low-band isolation between Ant. 2 and Ant. 3, while the high-band has remained undisturbed. For multi-band applications, this independent characteristic could be very useful.

Even though this NL technique has proved very successful in reducing the mutual coupling between two adjacent antennas for 5G FR1 smartphone applications, however, the above works [24, 26] can only be applied to a narrow bandwidth antenna element (3400–3600 MHz). Therefore, to allow the NL to be applied to antenna elements with dual band and wideband characteristics, [27] has introduced a Wideband NL (WNL) structure. As shown in Figure 5.10, a WNL is connected between antenna pair elements for metallic chassis 5G FR1 smartphone applications. The structure of this WNL is very different from the conventional type, as the width of the NL (along path QR) is a unit-step type with its middle section wider than the two connecting sides, which is analogous to an inverted-E-shaped structure. On the other hand, the conventional NL usually has the same line width. By further observing this figure, the two apparent transmission peaks (−6.8 and −8.3 dB) are significantly reduced after the loading of this WNL. However, as indicated in [28] that has applied a similar WNL structure, this wider middle section of the WNL may improve the port isolation of the low-band but increase the mutual coupling in the high-band between the antenna pair elements. Furthermore, a similar phenomenon was also observed in their corresponding return losses (impedance matching) across the low-band and high-band.

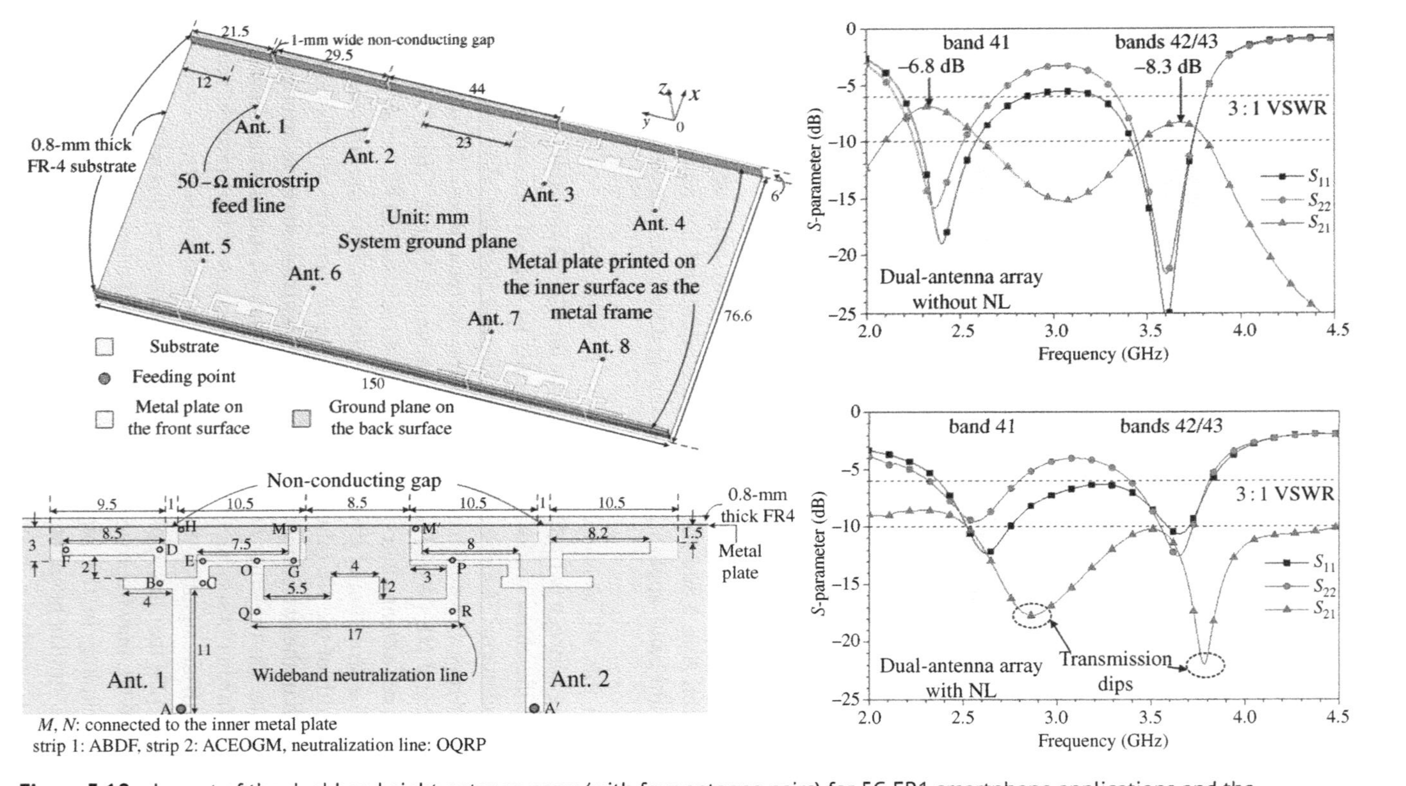

Figure 5.10 Layout of the dual band eight-antenna array (with four antenna pairs) for 5G FR1 smartphone applications and the configuration of the antenna pair loaded with the wideband neutralization line (across path OQRP). The simulated transmission coefficients (or isolations) between the antenna pair elements with and without loading the wideband neutralization line. *Source:* From [27]. ©2019 Wiley. Reprinted with permission.

Therefore, if this WNL structure is to be implemented, one must prepare to yield a good trade-off between isolations as well as return losses in the low-band and high-band. Nevertheless, the WNL or the conventional NL techniques are very difficult to apply to an ultra-wideband 5G FR1 antenna working in the 5G NR band n77/n78/n79 and 5G NR-U band n46/n96.

5.2.3 Slotted Element

The slotted element is the simplest decoupling structure for achieving good port isolation between two adjacent antenna elements that share the same common reference ground. It is also suitable to be applied for ultra-wideband antenna elements, which is an added advantage, as the NL or WNL technique is not suitable for wideband operation. As indicated earlier in Figure 5.3, the ground current generated by the excited antenna element will have a detrimental effect on the port isolation of its adjacent antenna element. To successfully reduce/block the ground current from flowing into its adjacent antenna elements, a slotted element can be loaded into the ground plane between the two closely arranged antenna elements. For the case when the metallic chassis is involved in the antenna design for smartphone applications, as shown in Figure 5.11, one easy way to improve the isolation is to extend the slotted element from the ground plane into the metallic chassis [29, 30]. Even though such measures can easily yield better port isolation for these wideband designs working across the 5G NR band n77/n78/n79 and 5G NR-U n46 (from 3300 to 5925 MHz), the metallic chassis must be cut into several separate sections, which, in turn, may result in weakening the overall sturdiness of the metallic chassis. Therefore, applying such a method to the metallic chassis of the smartphone must be seriously considered. Notably, there is no design guideline in choosing the appropriate length or width of the slotted element, but selecting a quarter-wavelength long-slotted element will be a good starting point.

5.2.4 Wave Trap Element

From the earlier investigation, we have learned various techniques to reduce the mutual coupling between two closely arranged antennas. Besides the two diversity methods applied to increase the far-field distance or minimize the reactive NF strength, the NL technique creates an additional coupling path for canceling out the existing mutual coupling. In contrast, the slotted element applied to the antenna array with metallic chassis is mainly used to block the ground current traveling from the excited antenna to its adjacent one. Therefore, to further mitigate the OTA surface wave and reactive NF between two closely arranged antennas depicted in Figure 5.3, a well-defined decoupling structure has recently been

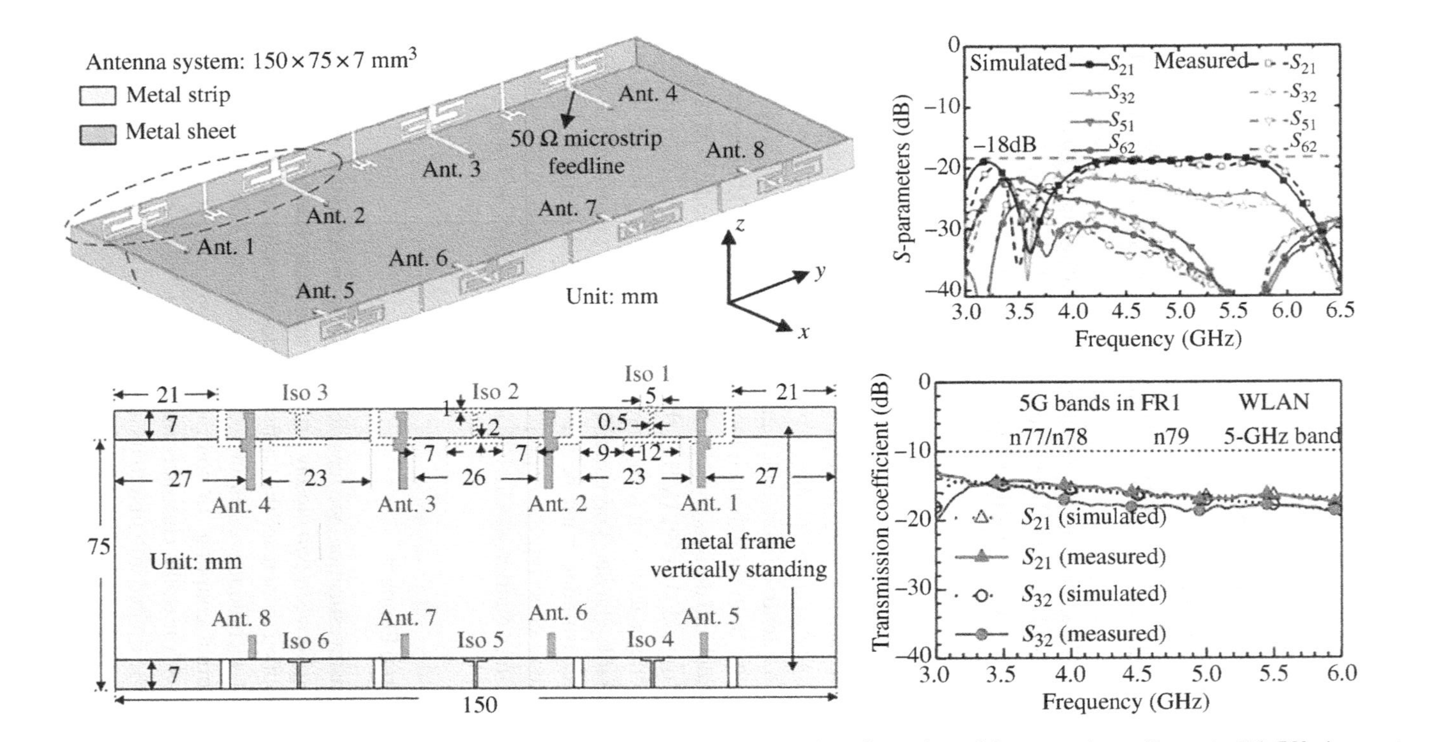

Figure 5.11 The layout of the eight-antenna array elements and the detailed configuration of the open slot radiators in [29, 30], shown at the top and bottom, respectively. Their corresponding measured *S*-parameters are included. *Source:* From [29]. Under a Creative Commons License; From [30]. ©2020 IEEE. Reproduced with permission.

unveiled for 5G FR1 antenna array, especially when a wideband operation is required, which is known as the wave trap element (sometimes known as the protruded element).

From past literature, two different types of wave trap elements have been commonly applied for mobile devices; the first type is usually a quarter-wavelength or half-wavelength narrow strip/stub that protrudes from the ground between two antenna elements [31, 32]. It acts like a resonator that can function as an effective wave trap element to either trap or block the surface wave or reactive NF radiation from one antenna to the other and vice versa. On the other hand, the second type of wave trap element is usually a much wider element that also protrudes from the ground, but it has a single or multiple slots (open slot or closed slot) loaded into the protruded element [33]. Owing to this wide protruded element, it can effectively block the OTA surface wave and reactive NF radiation, and the open slot is also acting as a resonator that can trap the exited surface currents flowing along the common reference ground between two adjacent antennas, which is analogous to the slotted element introduced in Section 5.2.3.

A good example to show the working mechanism of the first type of wave trap element for wideband 5G FR1 antenna array smartphone applications is reported in [34], and the layout of the eight-antenna array as well as the two ground-connected wave trap elements (denoted as the T-shaped decoupling stub, TSDS) can be seen in Figure 4.6. Here, two different types of TSDS are applied between the antenna elements, but they shared the same working principles. As depicted in Figure 5.12, a high mutual coupling can be observed between the two adjacent antenna elements when the TSDS is absent, but when it is protruded from the

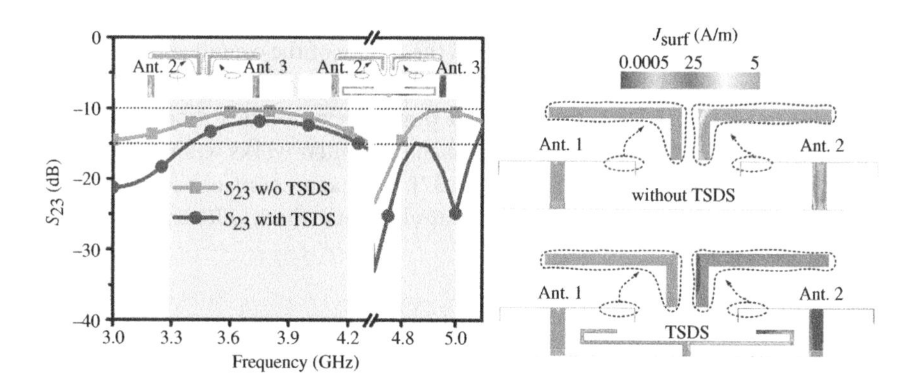

Figure 5.12 Simulated transmission coefficients of Ant. 2 and Ant. 3 with/without loading the protruded element (TSDS), and their corresponding current distributions at high-band 4.9 GHz. The current distributions at low-band 3.6 GHz, with/without loading the protruded element (TSDS) for Ant. 1 and Ant. 2 are also plotted on the right side. *Source:* From [34]. ©2019 IEEE. Reproduced with permission.

ground, the mutual coupling effect is significantly mitigated. However, one needs to notice the current distributions on the left section of the TSDS, as a high surface current intensity was trapped in there. Thus, it further verified that the TSDS is also acting as a wave-trapping decoupling structure. By observing the isolation level between Ant. 2 and Ant. 3, the isolations across the low-band (5G NR-band n77/n78) and high-band (partial 5G NR band n79) are improved by approximately 2 and 5 dB, respectively.

Figure 5.13 shows the work reported in [35] that has applied the second type of wave trap element for wideband 5G FR1 antenna array smartphone applications. Here, one can see that the wave trap element protruding from the ground between two adjacent antenna elements (dual-loop antennas) is very much different from the first type in [34]. The size of this wave trap element protruded from the ground is 13 mm × 7 mm, and it is loaded by three open slots (also known as decoupling slots) with identical widths of 0.4 mm. From the simulated transmission coefficients, the wave trap element alone (without the three decoupling slots) cannot improve the isolations across the two wide 5G FR1 operational bands. However, after loading the three decoupling slots, the isolation levels (S_{12}, S_{23}, and S_{34}) between the four antenna elements collocated along the same long edge of the system ground are significantly improved to approximately 16 dB. To verify its potential to achieve good wave trapping across the two operational bands (3400–3800 MHz and 4800–5000 MHz), from the simulated current distributions, one can see that the low-band (3.6 GHz) radiation was trapped by Slot 1 and Slot 2, while the high-band (4.9 GHz) radiation was trapped by Slot 2. Even though this work has successfully performed good mutual decoupling between the dual-loop antenna elements, however, the dual-loop antenna element has a length of 14.8 mm and a profile of only 4.3 mm, whereas this wave trap element (or decoupling structure) has undesirably increased the profile of the overall antenna array to 7 mm. Therefore, one should consider applying the wave trap element (TSDS) proposed in [34], as the height of the wave trap element is much shorter than the main radiators, which is rather unusual as many similar works will implement a wave trap element with higher profile [36, 37]. On that account, a second thought should be given when applying this wave trap element to the 5G FR1 antenna array.

5.2.5 Self-Isolated Element

As shown in Figure 5.3, besides the OTA (surface wave and reactive NF) issues that result in poor port isolation, one of the main problems is the flowing of the ground current between two closely arranged antennas that shared the same common reference ground. As we have investigated in the earlier sub-sections, the NL technique is unable to resolve this problem, while the slotted element or the wave trap element is an additional decoupling structure that has to be meticulously

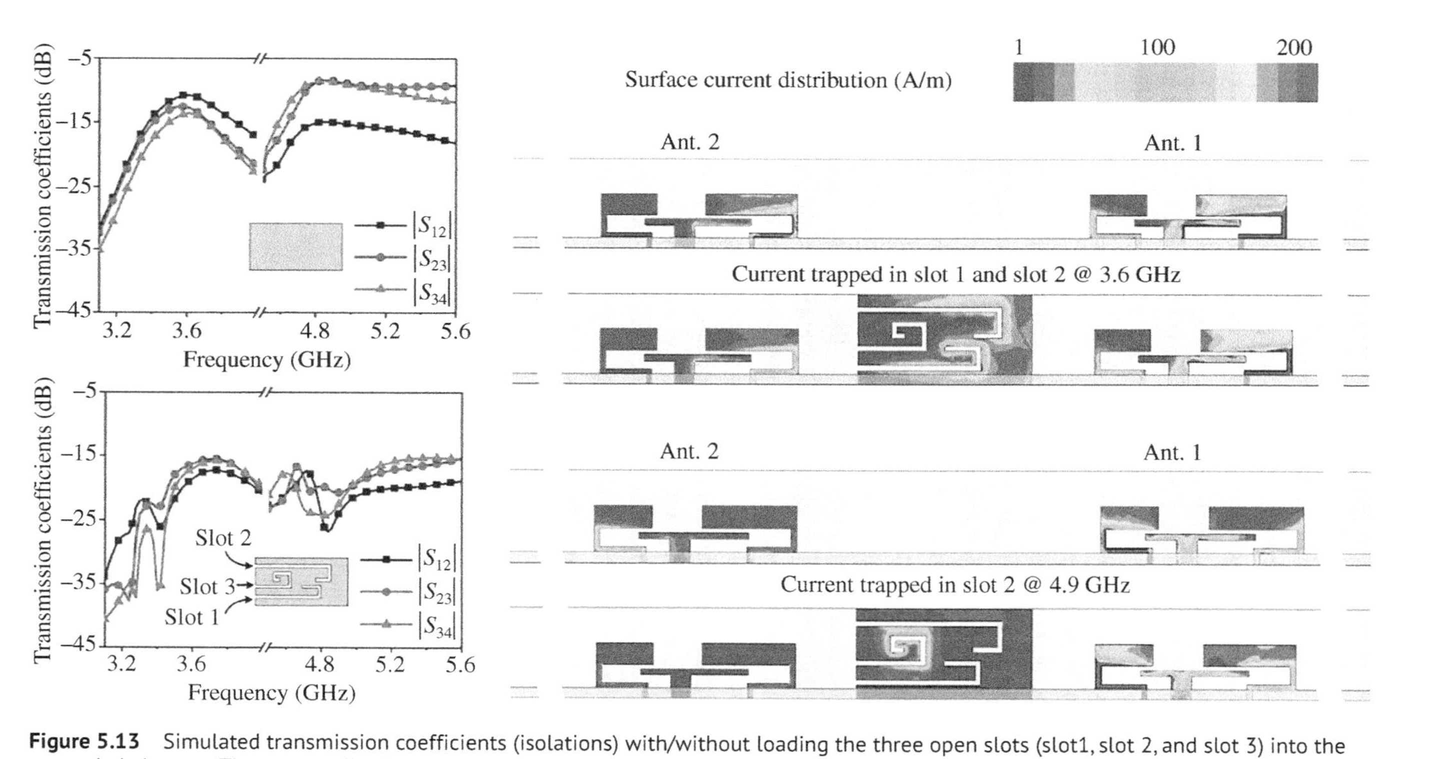

Figure 5.13 Simulated transmission coefficients (isolations) with/without loading the three open slots (slot1, slot 2, and slot 3) into the protruded element. The current distributions at low-band 3.6 GHz and high-band 4.9 GHz. *Source:* From [35]. Under a Creative Commons License.

settled between the collocated antenna arrays without disturbing the other typical performances of the antenna array. However, as we have already pointed out, the slotted element technique will weaken the overall sturdiness of the metallic chassis, as it has to be cut into several separate sections, not to mention that these slots are also extended to the system ground. The wave trap element, on the other hand, if inappropriately applied, may increase the overall profile of the antenna array. Therefore, it is important to seek another isolation technique for 5G FR1 antenna element that can avoid any alteration (slot-loading or cutting) to the metallic chassis and ground system, as well as refraining from loading additional decoupling structures into the 5G FR1 antenna array.

One of the earliest works that we have discovered over the years that can isolate (or trap) the ground current generated below the antenna is reported in [38]. The idea of this work is to generate a weakly ground-coupled current below the excited antenna element so that most of the current induced on the ground plane (across multiple frequencies) was at a maximum in the area beneath the antenna, whereas a relatively weak current distribution appeared on the rest of the ground plane. The name given to this technique is known as the "balance antenna design," in which the antenna structure must be balanced (or symmetrical), meaning that the exciting current distribution along the antenna structure at certain desired frequencies must be symmetrical. Another good example is the work reported in [39], which shared the same design methodology with [38] using the balance-folded monopole/dipole/loop antenna structure. As shown in Figure 5.14, one can see that at the different resonance frequencies (also known as balance modes), the current distributions along the entire antenna structure are symmetrical, and their respective ground currents are mostly isolated/concentrated beneath the main antenna structure, whereas the rest of the ground plane has shown weak current distribution [39].

Based on these findings, a high isolation eight-antenna array for 5G FR1 smartphone applications is reported in [14]. In this case, the antenna element is an open slot antenna type that has a balanced slotted structure, as shown in Figure 5.15. By further observing its simulated electric field (E-field), equivalent magnetic current distributions, and surface electric current distributions at slot mode 3.5 GHz, a distinct E-field null point is shown in the middle section of the balance slot, which is also symmetrically distributed along with the slotted structure. By exciting one of the balance slot antenna elements (Ant. 3), one can see that the maximum current is concentrated around Ant. 3, whereas the two adjacent antennas, Ant. 2 and Ant. 4, that are located on the left and right side, respectively, have shown weak current distributions around the slotted structures. Notably, this work has achieved a good isolation level of better than 17.5 dB without the need to apply any additional decoupling structures or NL. The eight-antenna array layout of this work can also be seen in Figure 3.12.

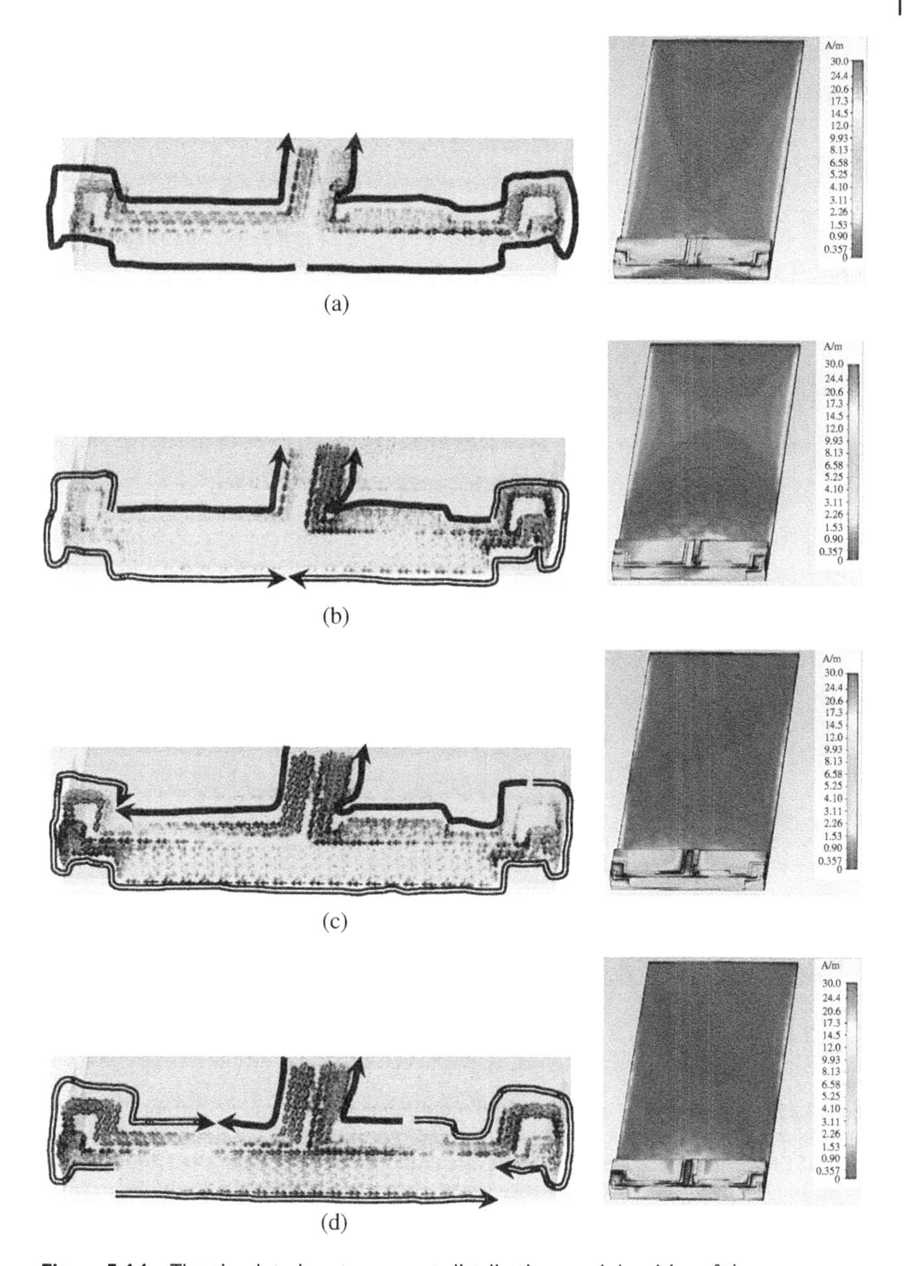

Figure 5.14 The simulated vector current distributions and densities of the balanced antenna at different frequencies. (a) $f = 900$ MHz, $2 \times 1/4\lambda$ folded monopol, (b) $f = 1700$ MHz, $2 \times 3/4\lambda$ folded monopol, (c) $f = 1900$ MHz, $2 \times 1/2\lambda$ folded monopol, (d) $f = 2300$ MHz, $4 \times 1/2\lambda$ folded monopol. *Source:* From [39]. ©2012 IEEE. Reproduced with permission.

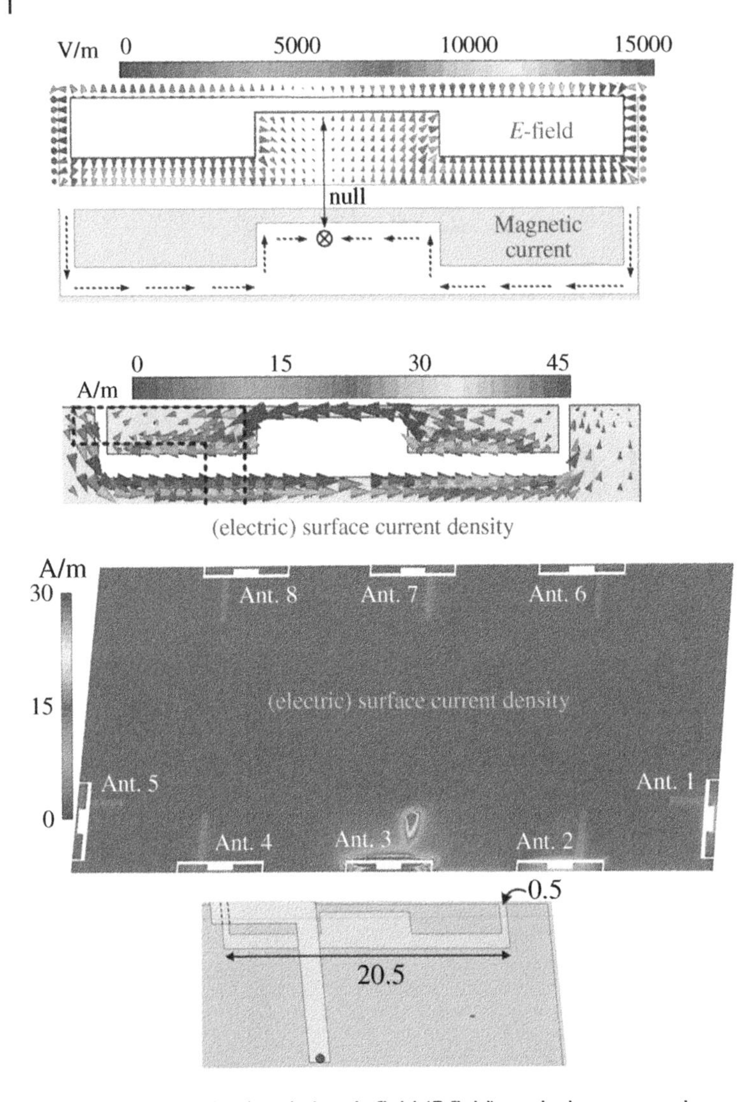

Figure 5.15 The simulated electric field (*E*-field), equivalent magnetic current distributions, and surface electric current distributions of the balance slot antenna at slot mode 3.5 GHz. The surface electric current distributions on the ground when Ant. 3 is excited and the dimension of this balance slotted structure are included. *Source:* From [14]. ©2019 IEEE. Reproduced with permission.

As the antenna array element of [14] is a balanced slot antenna structure loaded into the ground plane, it is not a suitable design for the 5G FR1 antenna array that is required to be collocated along with the chassis (frame) of the smartphone. Thus, a progressive report by the same authors in [40–42] has also introduced the same idea of applying a coupled-fed loop antenna structure that can excite a balance mode for achieving good port isolation between adjacent antenna elements. For brevity, only [41] will be discussed, and its corresponding coupled-fed loop antenna structure, and eight-antenna array layout has been depicted in Figures 3.18 and 3.19, respectively. Here, another name is given to this coupled-fed antenna loop structure, which is the "self-isolated" antenna, meaning that the induced current distributing around the coupled-fed loop structure is isolated (or concentrated) within close proximity of the excited coupled-fed loop structure, and it can be observed in Figure 5.16. In this case, a symmetrical current distribution is observed around the coupled loop structure, signifying the excitation of a balanced mode excited by a balanced antenna structure. Furthermore, one can also see that the excited antenna will have very minimum mutual coupling effects on its adjacent antenna elements. One good advantage of applying such a method is the high isolation level of better than 19.1 dB across the partial 5G FR1 n77/n78 band.

Another progressive work [43] that is stemmed from [40–42] has also been reported by the same group of authors, and it has applied a similar coupled-fed balance loop structure, but the two identical antennas are grounded using the

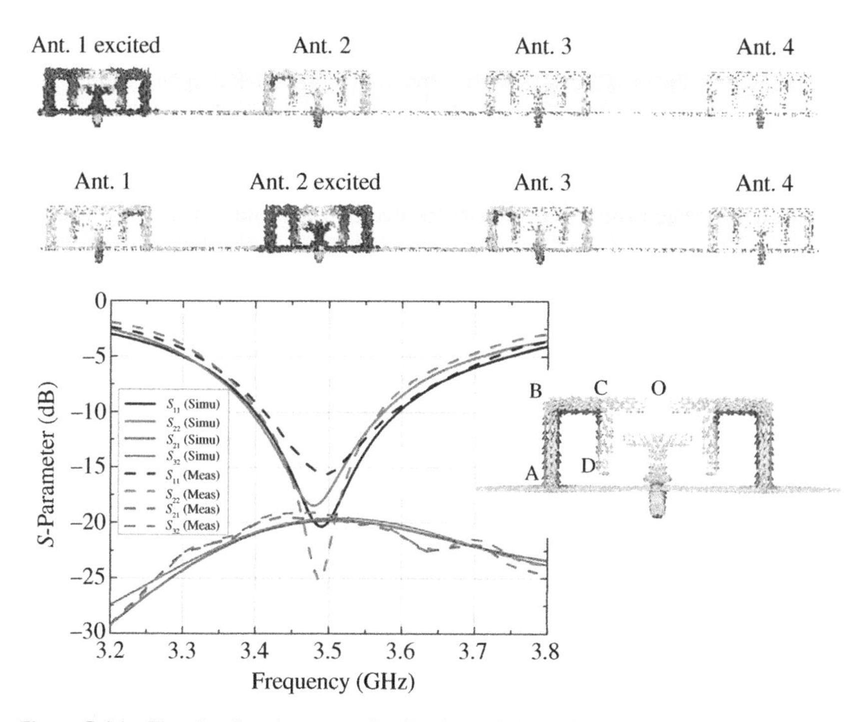

Figure 5.16 The simulated current distributions of the balance loop antenna at loop mode 3.5 GHz, and the current distributions across the four-antenna array when only one antenna is excited. The measured and simulated *S*-parameters are included. *Source:* From [41]. ©2019 IEEE. Reproduced with permission.

same coupled ground strip loaded between them. In this compact self-decoupled antenna pair design, the gap distance between the two adjacent coupled-fed loop antennas is only 1.2 mm (or 0.014 wavelength @3.5 GHz), and good isolation of better than 17 dB was achieved. To further achieve dual band operation (3400–3600 MHz and 4800–5000 MHz), [43] is further modified as a dual band type with asymmetrical dual self-decoupled antenna pairs, and the isolation level is also better than 17 dB [44].

5.2.6 Others

Even though we have categorized some of the major isolation (or decoupling) techniques applied for the 5G FR1 antenna array smartphone applications, many other unique design methods are still available and interesting to unveil in this chapter. For brevity, we can only briefly describe them as follows:

1) Hybrid decoupling structures [45, 46]:
   In [45] (see Figure 3.20), two decoupling structures are simultaneously applied to the eight-antenna array that has two different types of coupled-fed loop antenna elements. Here, an NL is connected between Ant. 2 and Ant. 3 located in the middle section, whereas an inverted I-shaped ground slot structure (wave trap element) is applied between Ant. 1 and Ant. 2 (also Ant. 3 and Ant. 4). As for [46], it is a 10-antenna array design, and each antenna element is a dual-mode IFA. Here, a wave trap element (narrow protruded stub) is loaded between the antenna pairs located on both ends, and an NL is connecting the two antenna elements located in the middle section. Even though [45, 46] have applied the hybrid decoupling structures, they can only improve the isolation to >12 and >15 dB, respectively.
2) Decoupled building block [47–49]:
   The idea of realizing a building block for an antenna array in the smartphone is to achieve a compact size for an antenna pair that has two identical antennas having an asymmetrical mirrored structure with respect to the system ground in the smartphone. Figure 5.17 shows three progressive works reported by the same authors' group that have demonstrated the aforementioned characteristics, as the total size of the antenna pair can be as small as 10 mm × 7 mm, and the size of a single gap-coupled (or coupled-fed) antenna element for all cases is only 10 mm × 3.1 mm [47–49]. Here, [47] is mainly exciting a loop mode for the 3.5 GHz band, while [48, 49] have excited the IFA mode and loop mode simultaneously for dual band 3.5/5.8 GHz operation. One can also see Figure 4.32 for more details on the excitation of the hybrid IFA/loop mode. Even though this decoupling building block design can successfully achieve a very compact size for the two mirrored antenna pair structures,

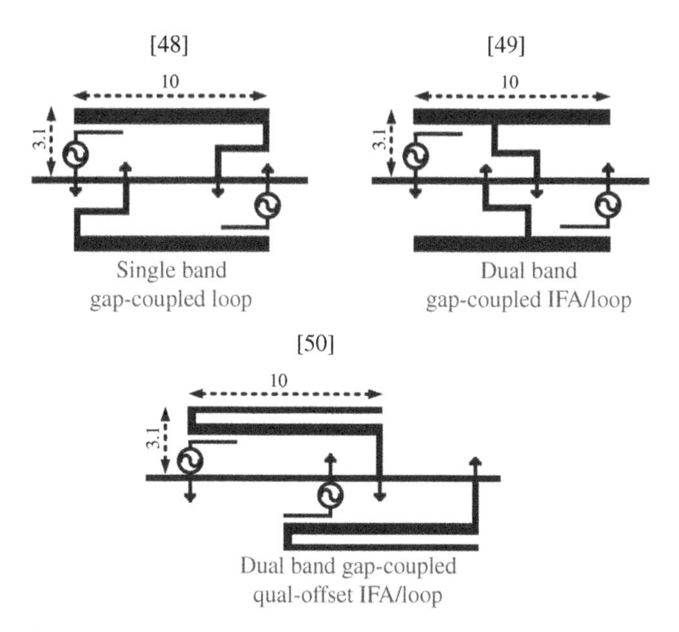

Figure 5.17 Conceptual diagrams of the three decoupled building blocks reported in [47–49] that have applied a gap-coupled (or coupled-fed) loop structure, as well as hybrid IFA/loop structure.

due to the closely arranged position between the antenna pair, the isolation level across the low-band operation (3400–3600 MHz) is approximately 10 dB.

3) Metallic chassis with orthogonal mode [17, 19, 50]:

Except for the slotted element technique mentioned in Section 5.2.3, most of the isolation techniques introduced are unsuitable for smartphone applications with metallic chassis (sometimes known as a metallic frame, bezel, or rimmed). The main reason is that the nature of the metallic chassis is a component that will block the EM wave, and therefore, the applied antenna array must be integrated into the chassis. Even though the antenna array in [29, 30] can yield very wide operational bandwidth, they cannot excite an orthogonal mode, which is now a research topic of interest.

To excite an orthogonal mode, [19, 50] have proposed loading an inverted T-shaped slot. The vertical slot section is loaded into the metallic chassis, and the horizontal slot section is embedded into the ground plane. By exciting this T-shaped slot using two different feeding ports (of different feeding mechanisms), an orthogonal mode (slot mode and loop mode) can be induced. However, the above two works can only be applied for narrow 5G FR1 band operation (3400–3600 MHz). Therefore, the work reported in [17] has introduced a wideband tightly arranged hybrid slot and monopole radiator that forms an antenna pair for metallic chassis 5G FR1 smartphone applications. This antenna pair can cover the 5G NR band n77/n78/n79 (3300–5000 MHz). In Figures 4.22 and 4.23, one can see that it has applied two T-shaped slots (analogous to [19, 50]) as the antenna pair. As the excited orthogonal modes have been defined in Figure 4.24, we would not discuss it any further for brevity. Nevertheless, one should note that one of the feeding ports applied in [17] is a differential feeding type. By observing the work reported in [51], the differential feeding technique can yield good pattern diversity when in-phase and out-of-phase signals are utilized, with isolation of better than 21 dB achieved between the antenna pair.

4) Stable current nulls [52, 53]:

The technique of selecting the stable current/*E*-field null-amplitude points for allocating the remaining antenna elements with reference to the first antenna element placed within the finite ground plane of a smartphone is introduced in [52]. The characteristic mode analysis (CMA) was applied in this work to study the fundamental and high-order mode of the E-field distribution along the edges of the ground plane. It has been concluded that when a single antenna element (e.g. planar inverted-F antennas [PIFA]) is placed on a finite ground plane analogous to the one used in the smartphone when the antenna element is excited, two types of traveling waves (*E*-field) will incur along the edges of the ground plane, namely the traveling wave and non-traveling wave. For the traveling waves, the null-amplitude points of the *E*-field distributed along the edges of the ground plane will move with different phase ranges (e.g. from 0° to 180°). In contrast, the null-amplitude points of the *E*-field will remain undisturbed with different phase ranges, and the name "stable null point" was given to it. Therefore, the idea of this stable current null method is to allocate the remaining IFAs into the positions of the stable null points, and according to [52], isolation improvements of between 20 and 40 dB can be achieved when the positions and orientations of the two IFAs are optimized. However, one must take note that this method is only good for narrow 5G FR1 band operation (3400–3600 MHz) because when it was applied to a wideband loop antenna array operating in the 5G NR band n77/n78/n79 (3300–5000 MHz), as depicted in Figure 5.18, in which the remaining wideband loop antennas are allocated at Null points #1 to #3, the isolation level was improved by only 3 dB. Nevertheless, such improvement is considered an incremental approach as neither additional decoupling structures nor alterations to the loop antenna structure are required.

Our previous statement that concludes the stable current nulls method is only good for narrow 5G FR1 band operation (3400–3600 MHz) has also been verified by the report in [53], as very high isolation was achieved between a two-antenna building block that is composed of a gap-coupled loop antenna

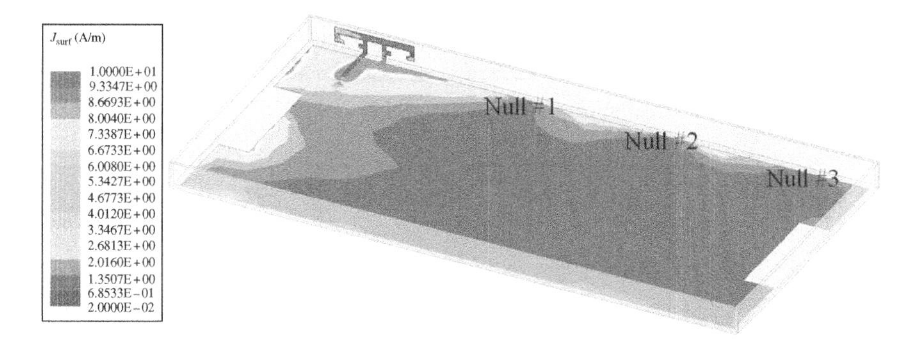

Figure 5.18 Simulated current distribution diagram of an excited wideband loop antenna and its corresponding current null point positions distributed on the edge of the finite ground system of a smartphone.

(Ant. 1) and a loop antenna (Ant. 2). Here, Ant. 1 was meticulously designed to excite a standing wave region (null amplitude point) by sending two opposite traveling waves onto the structure of Ant. 2, as depicted in Figure 5.19, and high isolation of better than 26 dB can be achieved. By observing the simulated surface current distributions of the building block with different phase ranges (from 0° to 135°) at 3.5 GHz, a stable current null point can be seen throughout the entire phase range. Notably, the extended four-antenna and eight-antenna MIMO arrays by applying this two-antenna building block can yield desirable measured isolation of better than 23 and 17.9 dB, respectively, across the narrow 5G FR1 band (3400–3600 MHz).

5) Self-curing [54, 55]:

The basic concept of this decoupling technique is to reduce the mutual coupling between two closely spaced IFAs by simply putting a capacitive load into the shorting arm of the two IFAs [54]. It is called the self-curing decoupling technique because no additional circuitry or structure is connecting the two coupled antennas, directly or indirectly. By further observing its progressive work in [55], which changes the position of the loaded capacitor (from direct contact with the shorting arms to inlay it into a reentrant opening on the ground plane near but not connected to the shorting arm of each IFA), as well as proving the concepts on the other two conventional antenna types (monopole and loop), it has claimed that the isolation between two monopole antennas, two loop antennas, and two IFAs, can be improved by at least 12 dB. However, one must take note that this technique is only good for low-frequency applications, and one capacitive load only affects the antenna characteristics in a designated frequency band with narrow band operation. As the selection procedure of attaining the accurate value of the capacitor load on

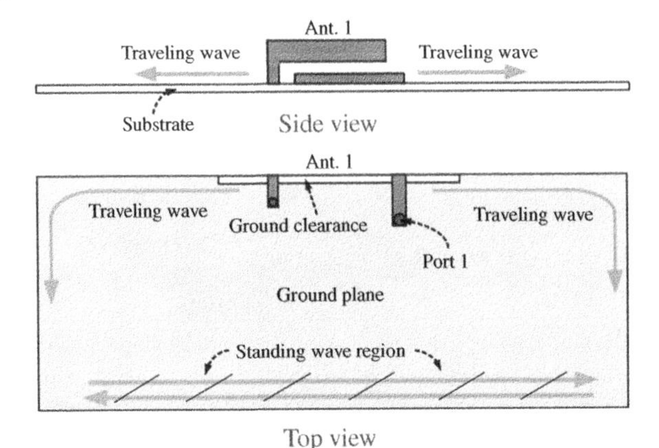

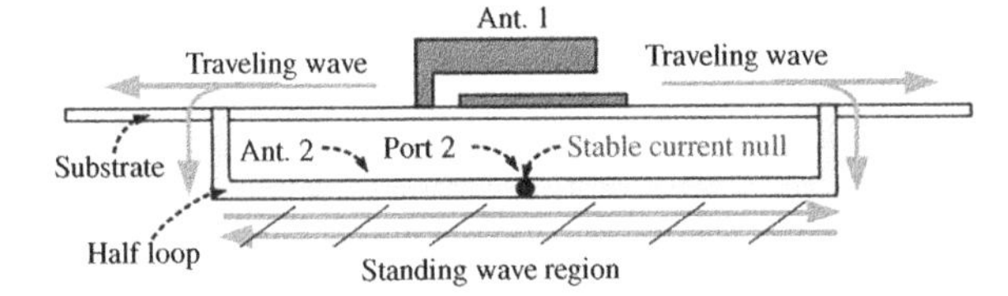

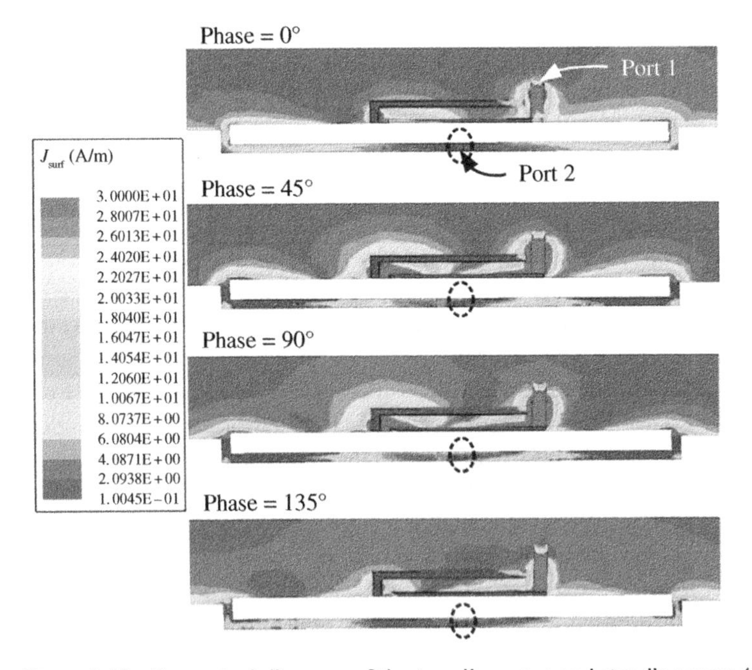

Figure 5.19 Conceptual diagrams of the traveling wave and standing wave (null amplitude point) region of the two-antenna decoupling building block. The simulated surface current distributions of the building block with different phases at 3.5 GHz, when Ant. 1 is excited. *Source:* From [53]. Under a Creative Commons License.

each antenna type is not described, this self-curing decoupling technique remains an enigma to the antenna society.

6) Double grounded stubs for full metallic chassis [56]:

On several occasions, the authors were challenged by antenna designers regarding the isolation of the 5G antennas built on a full metallic chassis. One practical method is to apply the double grounded stubs [56], as shown in Figure 5.20. As depicted in Figure 5.20a, if Ant. 2 is excited, currents may transmit along with the slot, and the isolation level (S_{23}) can be seriously degraded. To resolve this problem, inserting properly grounded stubs can suppress the currents. As shown in Figure 5.20b, when the length of slot$_{2\text{-}3}$ is $\lambda_g/2$, the grounded stub of Ant. 3 is located at the position where the current has reached a maximum. Thus, the effect of the current suppression is poor. As shown in Figure 5.20c, when the length of slot$_{2\text{-}3}$ is $\lambda_g/4$, the grounded stub of Ant. 3 will be shifted to the position where the current has reached a minimum, achieving maximum current suppression. Therefore, to enhance the isolation between Ant. 2 and Ant. 3 (S_{23}), the best solution is to select the length of slot$_{2\text{-}3}$ (L_{12}) to be $\lambda_g/4$. The same design methodology is applied to the lengths of slot$_{3\text{-}5}$ and slot$_{2\text{-}4}$. Other figures of this antenna can be seen in Figures. 4.35–4.38.

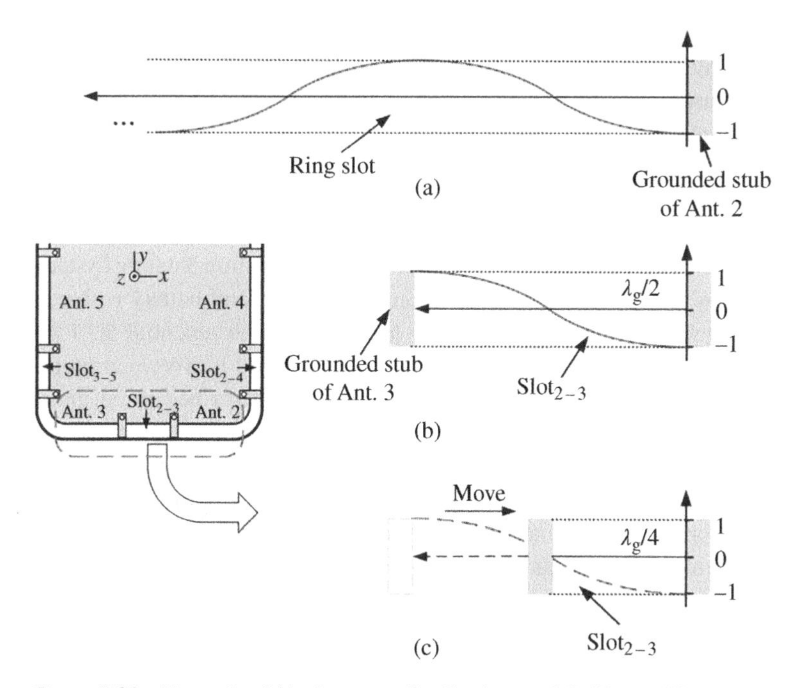

Figure 5.20 Normalized ideal current distributions at 3.5 GHz in different statuses. (a) Only grounded stubs of Ant 2 on the ring slot. (b) When slot2-3 is λg/2 long. (c) When slot2-3 is λg/4 long. *Source:* From [56]. ©2019 IEEE. Reproduced with permission.

5.3 Practical Considerations and Challenges

As the smartphones, tablets, and other mobile devices continue their advancement to satisfy the requirement of higher data rates, negligible latency, and superior reliability, their corresponding RF system design has once again become a barrier to the cellular devices or networks that aim to deliver more data to the end-users in many more demanding cases. Even though the 3rd Generation Partnership Project (3GPP) and the International Telecommunication Union (ITU) have released new specifications to address the needs and requirements for the mobile/cellular industry to push deeper into the design of the next-generation 5G UE products, its corresponding 5G antenna design is still subjected to the end device form factor and Original Equipment Manufacturer's (OEMs) preferences.

From Chapters 3 to 5, we have addressed many different types of MIMO antenna array designs for the functioning of the 5G NR band n77/n79/n79 and 5G NR-U band n46/n96. However, it is still crucial to comprehend the practical consideration and challenges ahead, especially those that may influence the 5G FR1 antenna array performances. As reported in [57], the three potential features impacting the antennas in 5G smartphones are, namely, high screen-to-body ratios, large battery sizes, and wireless charging function. As the present flagship smartphone has raced to high screen-to-body ratios of up to 90%, followed by leading to higher power consumption that requires a significant increase in the battery capacities (sizes), and the requirement to use a large recharging coil antenna for the wireless charging functions, the placement locations for the 4G/5G antennas, effective volumes of the antennas, the gap distances between MIMO antenna array, etc., will be further restricted. As a result, they will incur higher risks to antenna performances.

As depicted in Figures 3.1 and 3.5, by further considering the existing 4G LTE antennas, WiFi antennas, Bluetooth antennas, Global Navigation Satellite System (GNSS) antennas, Near-Field Communication (NFC) coils, and battery recharging coil, very limited volume spaces will be available for the potential 5G FR1 band (or sub-6 GHz bands) antenna array and 5G FR2 band mmWave antenna array, as shown in Figure 3.2. Therefore, based on the described three main potential features in the 5G smartphones, the overall environment is practically unfriendly for the 5G antenna performances [57]. Nevertheless, the difficulties and challenges for 5G FR1 antenna array smartphone applications must be resolved by exploring new approaches and better techniques with practical novelties. The following practical considerations (based on 5G FR1 antenna array designs) are some of the concerns raised by the cellular industry that awaits the antenna community to deliver better solutions:

1) Antenna array for metallic body
2) Compact size (building block) antenna
3) Wideband antenna array (5G NR band n77/n78/n78 and 5G NR band n46/n96)

4) Antenna pair with dual polarization (orthogonal mode)
5) Antennas with very high isolation
6) Tunable antennas
7) Low-profile antennas with wideband characteristics
8) Integrated 4G LTE and 5G FR1/FR2 antennas

As many of these concerns have been discussed in Chapters 3–5, we will not be elaborating on each concern for brevity. Besides the above-stated concerns, other research topics of interest that may shed new light on the cellular industry were also appealed to by Dr. Hanyang Wang (*IEEE Fellow*) during our many encounters:

a) Body-centric antennas for wearable devices
b) Phase antennas for 5G mobile terminals
c) MIMO throughput measurement system
d) 5G mmWave AiP solutions
e) High Impedance Surface (HIS) and Artificial Magnetic Conductor (AMC) for mobile applications
f) Characteristics Mode Analysis (CMA)

Based on the personal experiences of the authors, a few existing problems on 5G FR1 antennas have persisted over the years as follows:

i) 10-dB impedance bandwidth and antennas' profile:
   The impedance bandwidths in many reported works are usually of 6-dB threshold, and very few works [3, 30] can exhibit wideband characteristics with 10-dB impedance bandwidth across the 5G NR band n77/n78/n79 (3300–5000 MHz) and 5G NR-U band n46 (5150–5925). Notably, [3, 30] have shared the same antenna profile (height) of 7 mm, which may not be a practical consideration for a modern smartphone. From personal experience, the dilemma is that if we would lower the antennas' profile, it becomes challenging to achieve wide 10-dB impedance bandwidths to cover the 5G NR bands n77/n78/n79, not to mention the extended 5G NR-U band n46/n96 (5150–7125 MHz). So, one has to find an acceptable tradeoff between achieving wide operating impedance bandwidths (especially those with a 10-dB threshold) and the antennas' profile. At the moment, [58] has reported a low-profile 5 mm antenna element (including the 0.8 mm ground substrate thickness) with a length of 16 mm. Even with a compact size, this reported antenna can cover a very wide 6-dB impedance bandwidth of 3300–6000 MHz. Nevertheless, one of the major challenges ahead is to lower the profile to below 5 mm (or even near 4 mm) and achieve wide operational bandwidth across 3300–7125 MHz.
ii) Integration and coexistence of 4G, 5G FR1, and 5G FR2 mmWave AiP:
   In Section 4.3, we have discussed the various designs that integrate the 4G LTE antenna and 5G FR1/5G FR2 antennas. Even so, it has come to our

attention that very little work has reported the integration of all these three antenna types with a detailed description of the selection of the 5G FR1 antenna arrays and the 5G FR2 mmWave AiP positions. Even though the concepts of 5G AiPiA may be a future solution to allow the coexistence of 5G FR1 antennas and 5G mmWave AiP [59], the idea of a 5G FR1 antenna array with a very small aperture may yield undesirable gain, total efficiency, and radiation patterns. Nonetheless, such challenges are much dependent on the packaging solution and integration with the front-end components, which are not the main focus of this book.

iii) High isolation mitigation for 5G FR1 eight-antenna arrays:

Even though many of the reported works in Section 5.2 have demonstrated very high isolation levels between the two antenna elements, dual-antenna pair, or building block, it has come to our attention that when these high isolation dual-antenna pairs are further applied as an eight-antenna array, the isolation level between adjacent antennas will reduce drastically. Two good examples are from the works reported in [17, 53]. In [17], the dual-antenna pair has applied the orthogonal-mode design scheme, and high isolation level of 21 dB was achieved. However, when it was applied as an eight-antenna array, the isolation level between any two adjacent antenna elements has reduced to 12 dB, across a wide bandwidth of 3300–5000 MHz. As for [53], the two-antenna decoupling building block that has applied the current null point method by exciting a standing wave region has demonstrated a very high isolation level of 26 dB, but when it was applied as a four-antenna array and an eight-antenna array, the isolation level was mitigated to 23 and 17.9 dB, respectively, across a narrow bandwidth of 3400–3600 MHz. By comparing these results and many others, the mitigation of isolation level can be a staggering 9 dB. Therefore, the future challenge for antenna engineers is to discover a better isolation solution that can avoid such drawbacks.

5.4 Conclusion

In this chapter, we covered the motivation for applying the MIMO technology to 5G antenna element and the MIMO antenna performances such as channel capacity, ECC, DG, MEG, TARC, and CCL are described explicitly. Next, complete descriptions of the typical antenna isolation/decoupling techniques are illustrated, and a brief outline was also given on some of the unique decoupling methods that may have great potential in future 5G FR1 antenna array designs. Finally, some of the practical considerations and concerns raised by the cellular industry regarding the new approaches to the 5G FR1 band antenna array are briefly illustrated. We have also outlined some other research topics of interest that may shed new light on the cellular industry, and a few existing problems based on the authors' experiences on 5G FR1 antenna array designs are illustrated for the readers to ponder.

References

1 M. S. Sharawi, *Printed MIMO Antenna Engineering*, Artech House, 2014.

2 M. S. Sharawi, "Current misuses and future prospects for printed multiple-input, multiple-output antenna systems," *IEEE Antennas Propag. Magn.*, vol. 59, no. 2, pp. 162–170, Apr. 2017.

3 C. Y. D. Sim, H. Y. Liu, and C. J. Huang, "Wideband MIMO antenna array design for future mobile devices operating in the 5G NR frequency bands n77/n78/n79 and LTE band 46," *IEEE Antennas Wireless Propag. Lett.*, vol. 19, no. 1, pp. 74–78, Jan. 2020.

4 R. Tian, B. K. Lau, and Z. Ying, "Multiplexing efficiency of MIMO antennas," *IEEE Antennas Wireless Propag. Lett.*, vol. 10, pp. 183–186, 2011.

5 S. S. Jehangir and M. S. Sharawi, "A single layer semi-ring slot Yagi-like MIMO antenna system with high front-to-back ratio," *IEEE Trans. Antennas Propag.*, vol. 65, no. 2, pp. 937–942, Feb. 2017.

6 M. Y. Li, Y. L. Ban, Z. Q. Xu, C. Y. D. Sim, K. Kang, and Z. F. Yu, "Eight-port orthogonally dual-polarized antenna array for 5G smartphone applications," *IEEE Trans. Antennas Propag.*, vol. 64., no. 9, pp. 3820–3830, Sept. 2016.

7 M. Manteghi and Y. Rahmat-Samii, "Multiport characteristics of a wide-band cavity backed annular patch antenna for multipolarization operations," *IEEE Trans. Antennas Propag.*, vol. 53, no. 1, pp. 466–474, Jan. 2005.

8 E. Fritz-Andrade, H. Jardon-Aguilar, and J. A. Tirado-Mendez, "The correct application of total active reflection coefficient to evaluate MIMO antenna systems and its generalization to N ports," *Int. J. RF Microwave Comput. Aided Eng.*, e22113, vol. 30, no. 4, Dec. 2019.

9 J. Kulkarni, A. Desai, and C. Y. D. Sim, "Two port CPW-fed MIMO antenna with wide bandwidth and high isolation for future wireless applications," *Int. J. RF Microwave Comput. Aided Eng.*, e22700, vol. 31, no. 8, Jul. 2021.

10 W. M. Abdulkawi, W. A. Malik, S. U. Rehman, A. Aziz, A. F. A. Sheta, and M. A. Alkanhal, "Design of a compact dual-band MIMO antenna system with high-diversity gain performance in both frequency bands," *Micromachines*, vol. 12, no. 4, Apr. 2021.

11 N. O. Parchin, Y. A. A. Al-Yasir, H. J. Basherlou, R. A. Abd-Alhameed, and J. M. Noras, "Orthogonally dual-polarized MIMO antenna array with pattern diversity for use in 5G smartphones," *IET Microwave Antennas Propag.*, vol. 14, no. 6, pp. 457–467, May 2020.

12 C. F. Ding, X. Y. Zhang, C. D. Xue, and C. Y. D. Sim, "Novel pattern-diversity-based decoupling method and its application to multielement MIMO antenna," *IEEE Trans. Antennas Propag.*, vol. 66, no. 10, pp. 4976–4985, Oct. 2018.
13 S. H. Chae, S. K. Oh, and S. O. Park, "Analysis of mutual coupling, correlations, and TARC in WiBro MIMO arrary antenna," *IEEE Antennas Wireless Propag. Lett.*, vol. 6, pp. 122–125, 2007.
14 Y. Li, C. Y. D. Sim, Y. Luo, and G. Yang, "High-isolation 3.5 GHz eight-antenna MIMO array using balanced open-slot antenna element for 5G smartphones," *IEEE Trans. Antennas Propag.*, vol. 67, no. 6, pp. 3820–3830, Jun. 2019.
15 L. Sun, H. Feng, Y. Li, and Z. Zhang, "Tightly arranged orthogonal mode antenna for 5G MIMO mobile terminal," *Microwave Opt. Technol. Lett.*, vol. 60, no. 7, pp. 1751–1756, Jul. 2018.
16 L. Sun, H. Feng, Y. Li, and Z. Zhang, "Compact 5G MIMO mobile phone antennas with tightly arranged orthogonal-mode pairs," *IEEE Trans. Antennas Propag.*, vol. 66, no. 11, pp. 6364–6369,Nov. 2018.
17 L. Sun, Y. Li, Z. Zhang, and Z. Feng, "Wideband 5G MIMO antenna with integrated orthogonal-mode dual-antenna pairs for metal-rimmed smartphones," *IEEE Trans. Antennas Propag.*, vol. 68, no. 4, pp. 2494–2503, Apr. 2020.
18 H. Zou, Y. Li, B. Xu, Y. Chen, H. Jin, and G. Yang, "Dual-functional MIMO antenna array with high isolation for 5G/WLAN applications in smartphones," *IEEE Access*, vol. 7, pp. 167470–167480, Nov. 2019.
19 A. Ren, Y. Liu, and C. Y. D. Sim, "A compact building block with two shared-aperture antennas for eight-antenna MIMO array in metal-rimmed smartphone," *IEEE Trans. Antennas Propag.*, vol. 67, no. 10, pp. 6430–6438, Oct. 2019.
20 A. Diallo, C. Luxey, P. L. Thuc, R. Staraj, and G. Kossiavas, "Study and reduction of the mutual coupling between two mobile phone PIFAs operating in the DCS1800 and UMTS bands," *IEEE Trans. Antennas Propag.*, vol. 54, no. 11, pp. 3063–3073, Nov. 2006.
21 A. Diallo, C. Luxey, P. L. Thuc, R. Staraj, and G. Kossiavas, "Enhanced two-antenna structures for universal mobile telecommunications system diversity terminals," *IET Microwave Antennas Propag.*, vol. 2, no. 1, pp. 93–101, Feb. 2008.
22 S. W. Su, C. T. Lee, and F. S. Chang "Printed MIMO-antenna system using neutralization-line technique for wireless USB-dongle applications," *IEEE Trans. Antennas Propag.*, vol. 60, no. 2, pp. 456–463, Feb. 2012.
23 S. W. Su and C. T. Lee, "Printed two monopole-antenna system with a decoupling neutralization line for 2.4-GHz MIMO applications," *Microwave Opt. Technol. Lett.*, vol. 53, no. 9, pp. 2037–2043, Sep. 2011.
24 K. L. Wong, J. Y. Lu, L. Y. Chen, W. Y. Li, and Y. L. Ban, "8-antenna and 16-antenna arrays using the quad-antenna linear array as a building block for the 3.5-GHz LTE MIMO operation in the smartphone," *Microwave Opt. Technol. Lett.*, vol. 58, no. 1, pp. 174–181, Jan. 2016.

25 L. Qu, R. Zhang, and H. Kim, "Decoupling between ground radiation antennas with ground-coupled loop-type isolator for WLAN applications," *IET Microwave Antennas Propag.*, vol. 10, no. 5, pp. 546–552, Apr. 2016.

26 J. Guo, L. Cui, C. Li, and B. Sun, "Side-edge frame printed 8 port dual-band antenna array for 5G smartphone applications," *IEEE Trans. Antennas Propag.*, vol. 66, no. 12, pp. 7412–7417, Dec. 2018.

27 Y. Li, C. Y. D. Sim, Y. Luo, and G. Yang, "Metal-frame-integrated eight-element multiple-input multiple-output antenna array in the long term evolution bands 41/42/43 for fifth generation smartphones," *Int. J. RF Microwave Comput. Aided Eng.*, e21495, vol. 29, no. 1, Jan. 2019.

28 H. Zou, Y. Li, B. Xu, Y. Luo, M. Wang, and G. Yang, "A dual-band eight-antenna multi-input multi-output array for 5G metal-framed smartphones," *Int. J. RF Microwave Comput. Aided Eng.*, e21745, vol. 29, no. 7, Jul. 2019.

29 X. T. Yuan, W. He, K. D. Hong, C. Z. Han, Z. Chen, and T. Yuan, "Ultra-wideband MIMO antenna system with high element-isolation for 5G smartphone application," *IEEE Access*, vol. 8, pp. 56281–56289, Mar. 2020.

30 H. D. Chen, Y. C. Tsai, C. Y. D. Sim, and C. Kuo, "Broadband 8-antenna array design for future sub-6 GHz 5G NR bands metal-frame smartphone applications," *IEEE Antennas Wireless Propag. Lett.*, vol. 19, no. 7, pp. 1078–1082, Jul. 2020.

31 T. W. Kang and K. L. Wong, "Isolation improvement of WLAN internal laptop computer antennas using dual-band strip resonator as a wave trap," in *Asia Pacific Microw. Conf.*, Sigapore, pp. 2478–2481, Dec. 2019.

32 J. H. Chou, H. J. Li, D. B. Lin, and C. Y. Wu, "A novel LTE MIMO antenna with decoupling element for mobile phone application," in *Int. Symp. Electromagnetic Compatibility*, Tokyo, pp. 697–700, May 2014.

33 H. J. Jiang, Y. C. Kao, and K. L. Wong, "High-osolation WLAN MIMO laptop computer antenna array," in *Asia Pacific Microw. Conf.*, Kaohsiung, Taiwan, pp. 319–321, Dec. 2012.

34 L. Cui, J. Guo, Y. Liu, and C. Y. D. Sim, "An 8-element dual-band MIMO antenna with decoupling stub for 5G smartphone applications," *IEEE Antennas Wireless Propag. Lett.*, vol. 18, no. 10, pp. 2095–2099, Oct. 2019.

35 W. Hu, L Qian, S. Gao, L. H. Wen, Q. Luo, H. Xu, X. Liu, Y. Liu, and W. Wang, "Dual-band eight-element MIMO array using multi-slot decoupling technique for 5G terminals," *IEEE Access*, vol. 7, pp. 153910–153920, Nov. 2019.

36 C. Yang, Y. Yao, J. Yu, and X. Chen, "Novel compact multiband MIMO antenna for mobile terminal," *Int. J. Antennas Propag.*, vol. 2012, 691681, pp. 1–9, 2012.

37 L. Wu, Y. Xia, X. Cao, and Z. Xu, "A miniaturized UWB-MIMO antenna with quadruple band-notched characteristics," *Int. J. Microwave Wireless Technol.*, vol. 10, no. 8, pp. 948–955, Oct. 2018.

38 D. Zhou, R. A. Abd-Alhameed, A. G. Alhaddad, C. H. See, J. M. Noras, P. S. Excell, and S. Gao, "Multi-band weakly ground-coupled balanced antenna design for portable devices," *IET Sci. Meas. Technol.*, vol. 6, no. 4, pp. 306–310, Jul. 2011.

39 M. Zheng, H. Wang, and Y Hao, "Internal hexa-band folded monopole/dipole/loop antenna with four resonances for mobile device," *IEEE Trans. Antennas Propag.*, vol. 60, no. 6, pp. 2880–2885, Jun. 2012.

40 A. Zhao and Z. Ren, "Multiple-input and multiple-output antenna system with self-isolated antenna element for fifth-generation mobile terminals," *Microwave Opt. Technol. Lett.*, vol. 61, no. 1, pp. 20–27, Jan. 2019.

41 A. Zhao and Z. Ren, "Size reduction of self-isolated MIMO antenna system for 5G mobile phone applications," *IEEE Antennas Wireless Propag. Lett.*, vol. 18, no. 1, pp. 152–156, Jan. 2019.

42 A. Zhao and Z. Ren, "5G MIMO antenna system for mobile terminals," *in 2019 IEEE International Symp. Antennas Propag. USNC-URSI Radio Science Meeting*, Atlanta, Georgia, USA, 2019.

43 Z. Ren, A. Zhao, and S. Wu, "MIMO antenna with compact decoupled antenna pairs for 5G mobile terminals," *IEEE Antennas Wireless Propag. Lett.*, vol. 18, no. 7, pp. 1367–1371, May 2019.

44 Z. Ren and A. Zhao, "Dual-band MIMO antenna with compact self-decoupled antenna pairs for 5G mobile applications," *IEEE Access*, vol. 7, pp. 82288–82296, Jun. 2019,

45 W. Jiang, B. Liu, Y. Cui, and W. Hu, "High-isolation eight-element MIMO array for 5G smartphone applications," *IEEE Access*, vol. 7, pp. 34104–34112, Mar. 2019.

46 W. Hu, X. Liu, S. Gao, L. H. Wen, L. Qian, T. Feng, R. Xu, P. Fei, and Y. Liu, "Dual-band ten-element MIMO array based on dual-mode IFAs for 5G terminal applications," *IEEE Access*, vol. 7, pp. 178476–178485, Dec. 2019.

47 K. L. Wong, C.Y. Tsai, and J. Y. Lu, "Two asymmetrically mirrored gap-coupled loop antennas as a compact building block for eight-antenna MIMO array in the future smartphone," *IEEE Trans. Antennas Propag.*, pp. 1765–1778, vol. 65, no. 4, Apr. 2017.

48 K. L. Wong, B. W. Lin, and W. Y. Li, "Dual-band dual inverted-F/loop antennas as a compact decoupled building block for forming eight 3.5/5.8-GHz MIMO antennas in the future smartphone," *Microwave Opt. Technol. Lett.*, vol. 59, no. 11, pp. 2715–2721, Aug. 2017.

49 W. Y. Li, W. Chung, and K. L. Wong, "Compact quad-offset loop/IFA hybrid antenna array for forming eight 3.5/5.8 GHz MIMO antennas in the future smartphone," *in Int. Symp. Antenna Propagation*, Busan, Korea (South), 2018.

50 L. Chang, Y. Yu, K. Wei, and H. Wang, "Polarization-orthogonal co-frequency dual antenna pair suitable for 5G MIMO smartphone with metallic bezels," *IEEE Trans. Antennas Propag.*, vol. 67, no. 8, pp. 5212–5220, Aug. 2019.

51 Z. Xu and C. Deng, "High-isolated MIMO antenna design based on pattern diversity for 5G mobile terminals," *IEEE Antennas Wireless Propag. Lett.*, vol. 19, no. 3, pp. 467–471, Mar. 2020.

52 X. Zhao, S. P. Yeo, and L. C. Ong, "Decoupling of inverted-F antennas with high-order modes of ground plane for 5G mobile MIMO platform," *IEEE Trans. Antennas Propag.*, vol. 66, no. 9, pp. 4485–4495, Sept. 2018,

53 A. Ren, Y. Liu, H. Yu, Y. Jia, C. Y. D. Sim, and Y. Xu, "A high-isolation building block using stable current nulls for 5G smartphone applications," *IEEE Access*, vol. 7, pp. 170419–170429, Dec. 2019.

54 J. Sui and K. L. Wu, "Self-curing decoupling technique for two inverted-F antennas with capacitive loads," *IEEE Trans. Antennas Propag.*, vol. 66, no. 3, pp. 1093–1101, Mar. 2018.

55 J. Sui, Y. Dou, X. Mei, and K. L. Wu, "Self-curing decoupling technique for MIMO antenna arrays in mobile terminals," *IEEE Trans. Antennas Propag.*, vol. 68, no. 2, pp. 838–849, Feb. 2020.

56 Q. Chen, H. Lin, J. Wang, L. Ge, Y. Li, T. Pei, and C.Y.D. Sim, "Single ring slot based antennas for metal-rimmed 4G/5G smartphones," *IEEE Trans. Antennas Propag.*, vol. 67, no. 3, pp. 1476–1487, Mar. 2019.

57 H. C. Huang, "Overview of antenna designs and considerations in 5G cellular phones," *in Int. Workshop Antenna Technol.*, Nanjing, China, Mar. 2018.

58 K. L. Wong, Y. H. Chen, and W. Y. Li, "Decoupled compact ultra-wideband MIMO antennas covering 3300–6000 MHz for the fifth-generation mobile and 5 GHz-WLAN operations in the future smartphone," *Microwave Opt. Technol. Lett.*, vol. 60, pp. 2345–2351, Oct. 2018.

59 H. C. Huang and J. Lu, "Evolution of innovative 5G millineter-wave antenna designs integrating non-millineter-wave antenna functions based on antenna-in-package (AiP) solution to cellular phones," *IEEE Access*, vol. 9, pp. 72516–72523, May 2021.

6

Millimeter-Wave 5G Antenna-in-Package (AiP) for Mobile Applications

From the perspective of carrier frequency in mobile communication systems, the mmWave band refers to the frequency range from 10 to 300 GHz with wavelengths in millimeters, as illustrated in Figure 6.1. Compared to the sub-6 GHz spectrum used for today's mainstream wireless networks, mmWave propagation channel inherently suffers from higher free-space path loss (FSPL) of more than 20 dB and additional propagation loss, such as penetration, diffraction, precipitation, and foliage losses [1]. To mitigate these loss factors and realize reliable wireless links at mmWave spectrum, real-time constructive interference of electromagnetic (EM) waves is essentially required in both base station (BS) and user equipment (UE), such as mobile devices and cellular handsets, as illustrated in Figure 6.2. Denoted as beamforming, this generates a highly directive antenna beam at a certain angle which is dictated by the unique amplitude and phase configuration of each antenna comprising the antenna array. This is in contrast to sub-6 GHz UE (see Figure 6.2) in which sub-6 GHz antennas are designed to formulate omnidirectional beams. This enables simultaneous spherical beam coverage to address unpredictable wireless channels caused by mobility.

Recently, it has been confirmed that mmWave beam-tracking and handover are practically possible while users are moving in various environments [2–4]. Nevertheless, the actual coverage area of a mmWave base station is inconsistent with a pre-defined cell boundary and is clearly dependent on the environment due to the real-time changes in penetration and blockage losses [4]. Therefore, it can be deduced that a dense deployment of base stations is essentially required for robust mm-Wave operation. On the UE side, it is confirmed that robust wireless links cannot be guaranteed even by using UE designs with beamforming coverage of 75% or more, as users' hands can lead to large spatial area blocking [5]. These observations suggest that subarray diversity in UE design is critical in overcoming

Microwave and Millimeter-Wave Antenna Design for 5G Smartphone Applications, First Edition. Wonbin Hong and Chow-Yen-Desmond Sim.

Published 2023 by John Wiley & Sons, Inc.

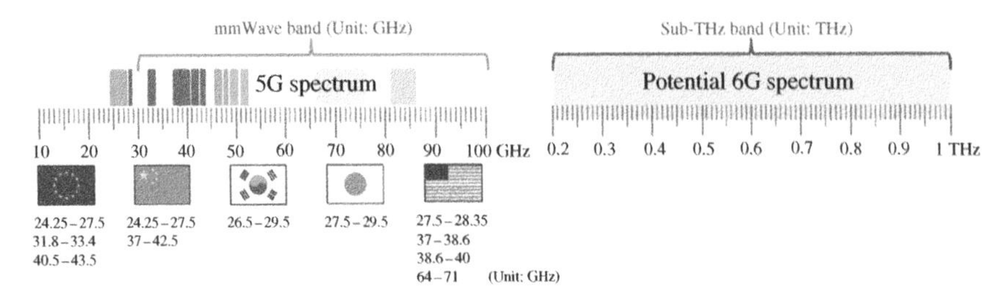

Figure 6.1 mmWave and sub-THz bands for next-generation wireless technology.

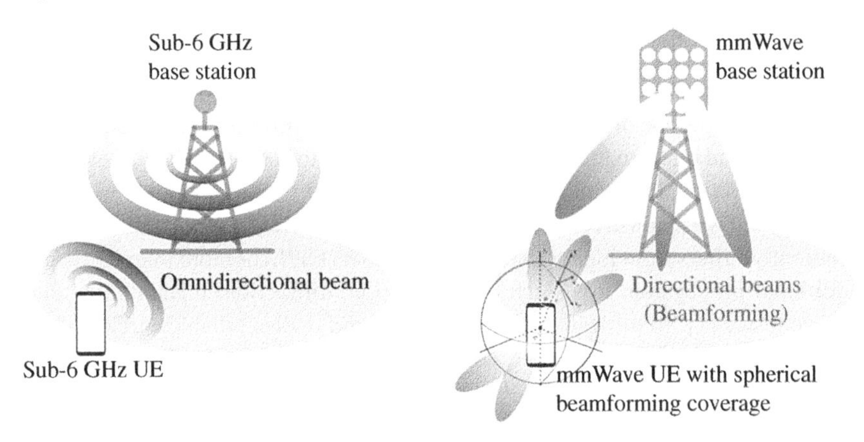

Figure 6.2 Conceptual illustration of the (a) sub-6 GHz and (b) mmWave mobile communication systems.

near-field obstructions as well as ensuring coverage at the UE side over the entire sphere [4]. Furthermore, multiple beams formed in mutually orthogonal directions in space can contribute to the realization of MIMO spatial multiplexing. In summary, it is imperative for mmWave UE to rapidly steer a highly directive beam across a spherical trajectory, introducing the need for highly sophisticated, spherical beamsteering phased–array antenna systems.

The antennas and RF front-ends for sub-6 GHz and mmWave cellular devices are compared as illustrated in Figure 6.3. In sub-6 GHz systems, discrete antennas are separately positioned outside the RF front-ends and connected to the radio frequency integrated circuits (RFICs) using interconnect structures, such as a micro coaxial cable due to the advantage of a relatively low interconnect loss at the corresponding frequency. Another key feature is that the metal rim of the cellular device is utilized as a main radiator, contributing to the transmission and reception of sub-6 GHz radio waves. Considering the omnidirectional radiation nature of sub-6 GHz antennas, the total radiated power (TRP) and total isotropic sensitivity (TIS) have been standardized as the key performance metrics of sub-6 GHz radio systems over the entire sphere of angles. In addition, the high isolation

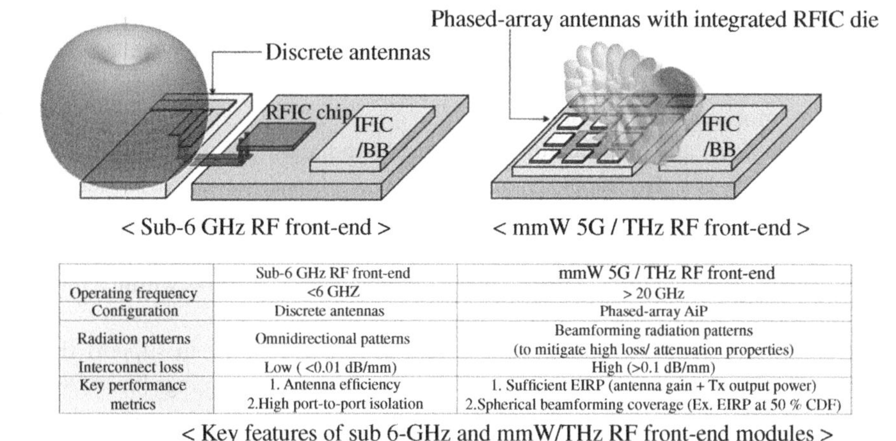

	Sub-6 GHz RF front-end	mmW 5G / THz RF front-end
Operating frequency	<6 GHZ	> 20 GHz
Configuration	Discrete antennas	Phased-array AiP
Radiation patterns	Omnidirectional patterns	Beamforming radiation patterns (to mitigate high loss/ attenuation properties)
Interconnect loss	Low (<0.01 dB/mm)	High (>0.1 dB/mm)
Key performance metrics	1. Antenna efficiency 2.High port-to-port isolation	1. Sufficient EIRP (antenna gain + Tx output power) 2.Spherical beamforming coverage (Ex. EIRP at 50 % CDF)

< Key features of sub 6-GHz and mmW/THz RF front-end modules >

Figure 6.3 Illustration and key features of the RF front-end modules for sub-6 GHz and mmWave cellular devices.

characteristics of multiple sub-6 GHz antennas have to be guaranteed to support multi-input multi-output (MIMO) antenna technology. The isolation level between sub-6 GHz antennas can be characterized by the envelope correlation coefficient (ECC) calculated from the complex data of far-field radiation patterns.

On the other hand, in mmWave systems, the implementation and efficient operation of beamforming antenna systems are essentially required to overcome the aforementioned attenuation factor of the mmWave propagation environments and to guarantee the reliability of the wireless link. Therefore, phased-array antennas and their beamforming architectures have become a dominant solution, especially for mmWave 5G cellular handsets. An antenna-in-Package (AiP) is the key technology that can solve the packaging issue since AiP can achieve a smaller form factor, and higher levels of integration for 5G RF front-end modules. The small physical size of millimeter-wave antennas allows the phased-array antennas to be easily embedded in the package and connected directly to an RFIC die as shown in Figure 6.3b [6]. Therefore, the AiP technology can minimize the interconnection loss between the RFIC and antennas, and reduce the overall height profile of modules.

6.1 Miniaturized Antenna-in-Package (AiP) Technology

6.1.1 Background and Challenges

For robust wireless channel conditions, mobile antenna systems require high gain, wide beamsteering range and polarization agility due to their mobility in Euler angles. Accordingly, an antenna system capable of varying polarizations and radiation patterns has been developed [7]. In addition, it is ascertained that phased arrays with dual-polarized beam switching configurations are effective in achieving a reliable quality of service (QoS) [8]. Therefore, broadside and endfire antenna arrays with orthogonal polarization are instrumental for realizing robust mmWave mobile antenna systems.

The mmWave antennas featuring broadside radiation patterns are widely reported in [9–17]. The topology of the mmWave broadside antenna include patch [11], slot [12, 13], aperture coupled patch [14] and substrate integrated waveguide (SIW) structure [15, 16]. Furthermore, the dual polarized broadside beam can be achieved using a planar patch and dual feed structures within a low-profile package [17].

The mmWave endfire antennas are preferred because they can support the desired line-of-sight (LOS) scenarios and achieve a wide beam coverage [8, 18–26]. A horizontally polarized (H-pol) endfire antenna can be implemented on the electrically thin planar structure, such as a dipole [18, 19], Yagi [20], and tapered slot [21]. In addition, their edge-positioned configuration is advantageous for space-efficient integration within modules and compact devices [22]. In contrast to the aforementioned antenna, realizing a vertically polarized (V-pol) endfire beam in thin substrates remains extremely challenging due to the required dominant electric field formed along the vertical axis. Recently, an open-ended SIW structure is adopted to implement the V-pol polarized endfire radiation [23, 24]. The V-pol endfire radiation is realized by implementing an electric dipole antenna using vertical vias at the PCB edge [25]. The half-wavelength vertical profile of the proposed antenna results in a few millimeter thickness, which will likely become problematic during the antenna implementation for the UE. A vertical and horizontal quasi-Yagi-Uda topology achieves the dual-polarized endfire radiation in [26]. Although the mesh-grid phased antenna featuring vertical polarization is implemented within a low-height profile module, electric field stored in the near field created between the planar patch results in low radiation efficiency [8].

6.1.2 Planar Folded Slot Antenna (PFSA) with Electrically Small and Low Profile

The planar folded slot antenna (PFSA) is formulated to realize a V-pol mode on a metallic surface as described in Figure 6.4. First, the planar slot is folded along the reference line (Figure 6.4b) to reduce the height of the entire structure. Afterward, a cavity (Figure 6.4c) is added behind the folded slot and can be utilized to maintain endfire. Finally, the proposed PFSA topology can achieve V-pol endfire radiation mode while keeping the longitudinal footprint to be minimal as illustrated in Figure 6.4d.

The devised topology of the mmWave V-pol endfire PFSA element is designed using low temperature co-fired ceramic (LTCC) process as illustrated in Figure 6.5 a.

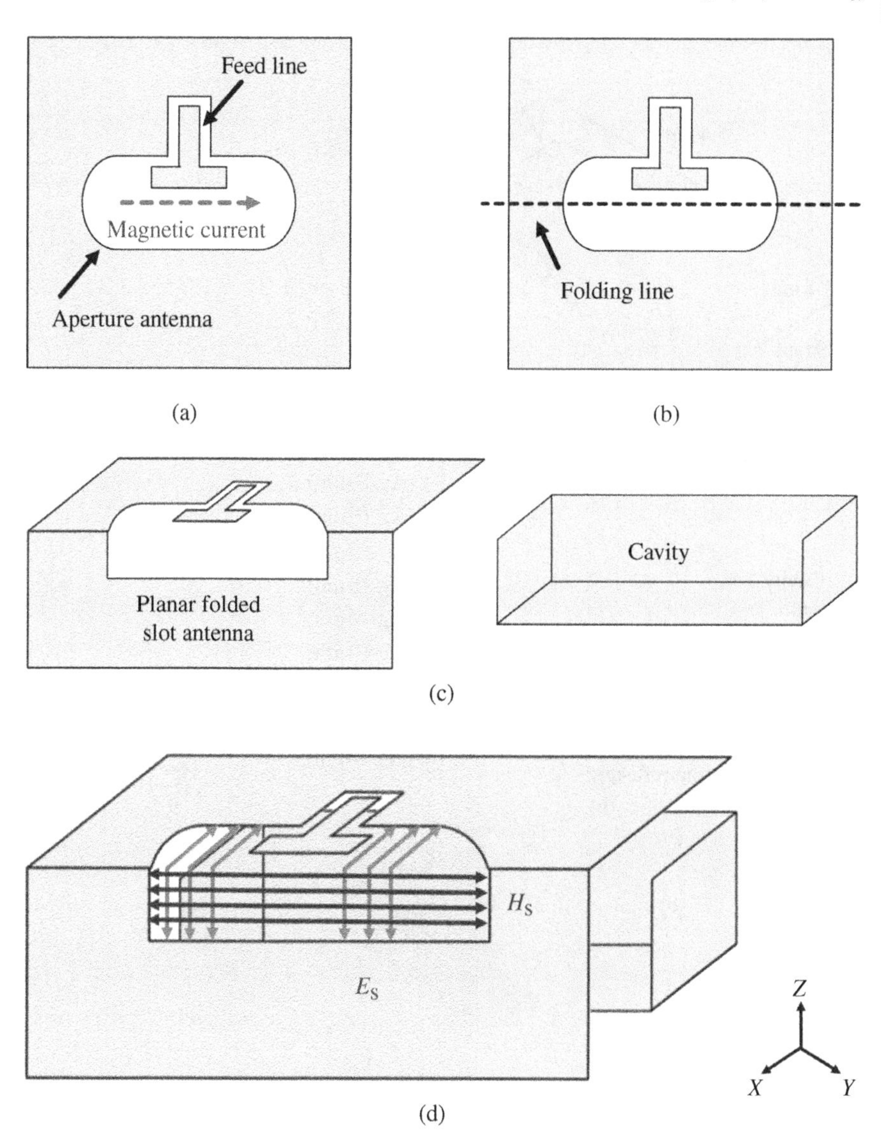

Figure 6.4 Conceptual illustration of devising a V-pol endfire PFSA topology. *Source:* From [6]. ©2019 IEEE. Reproduced with permission.

The overall size of the antenna element is 5.4 mm × 1.8 mm × 0.89 mm. The LTCC package consists of eight stacked layers, each having a thickness of 100 μm, relative permittivity $\varepsilon_r = 5.9$, and loss tangent $\tan\delta = 0.002$. Metal layers from L1 to L9 are implemented using 10 μm thick silver as shown in Figure 6.5b. The diameters of the vias and the capture pads are configured to be 100 and 150 μm, respectively,

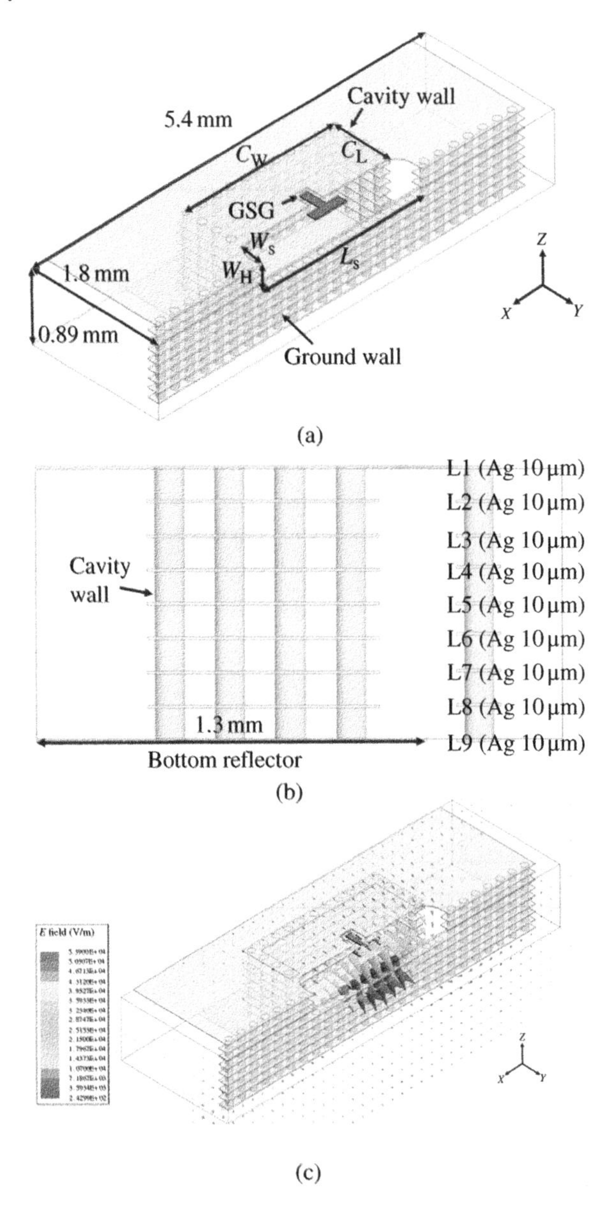

Figure 6.5 The proposed mmWave V-pol endfire PFSA element. (a) 3D view. (b) Cross-sectional view. (c) Electric-field distribution. *Source:* From [6]. ©2019 IEEE. Reproduced with permission.

throughout this paper. The folded slot aperture is situated on L1 ($W_S = 0.35$ mm) and is extended to the front-side face (z-x plane) of the antenna down to L4 ($W_H = 0.33$ mm). In contrast to the previous antenna designs, this folded implementation of the slot aperture allows the creation of vertical polarization while featuring a very electrically small antenna dimension. The resonant frequency of the antenna is predominantly realized by the length of the slot ($L_S = 2.2$ mm). The effect of the length of slot (L_S) is studied in Figure 6.6a, in which a resonance point is inversely proportional to the electrical length of the slot aperture. The quality factor Q can be further optimized by adjusting the width of the top slot (W_S) as studied in Figure 6.6b. The capacitive-type feed line is implemented on L1 for excitation and measurement using a 250-μm pitch ground-signal-ground (GSG) wafer probe. The width of the feed line and rectangular outer ground of the GSG port is configured to be 0.1 and 0.15 mm, respectively. Finally, a cavity wall is constructed and utilized to achieve wider bandwidth, maintain endfire radiation and suppress undesired backward radiation. The radiation characteristics, such as gain, efficiency, and front-to-back ratio (FTBR) are optimized by the cavity width ($C_W = 2.15$ mm) and length ($C_L = 0.77$ mm), which is approximately a quarter of the guided wavelength at 39 GHz. Figure 6.5c illustrates the electric field distribution of the proposed V-pol endfire PFSA element. It is noted that the V-pol electric fields are generated by the radiating magnetic currents.

The proposed V-pol endfire PFSA element is fabricated using LTCC process as shown in Figure 6.7. The 3-D EM simulator, ANSYS HFSS is used to simulate the antenna topology. The S-parameters are measured using N5247A KEYSIGHT PNA-X and the radiation patterns are measured using an anechoic far-field chamber at POSTECH, Pohang, South Korea. GGB picoprobe GSG wafer probe tips featuring 250 μm pitch are utilized to establish contact with the GSG pad on L1. The input reflection coefficient $|S_{11}|$ of the V-pol antenna element is illustrated in Figure 6.7 and is ascertained to respectively feature 4.02 GHz simulated and 3.76 GHz measured impedance bandwidths which are applicable to the spectrum at 37 GHz (37–38.6 GHz) and 39 GHz (38.6–40 GHz) bands assigned by the FCC. The normalized measured and simulated radiation patterns of the fabricated V-pol antennas element in the E-plane and H-plane at 39 GHz are illustrated in Figures 6.8a,b, respectively. Due to the limitation of the probe-based far-field chambers, the range of the measured plane of the proposed antenna is limited by blockage and scattering caused by adjacent metallic components, such as the wafer probe, probe stage, and microscope. The co- and cross-polarized radiation intensity difference is more than 20 dB in all directions of interest, which ascertains the desired V-pol radiation mode despite its extremely low height profile. The measured and simulated gain of the antenna element at 39 GHz is 2.43 and 2.39 dBi, respectively. The measured and simulated 3 dB half-power beamwidth in the E-plane are 190° and 221°, respectively and 117° and 100°,

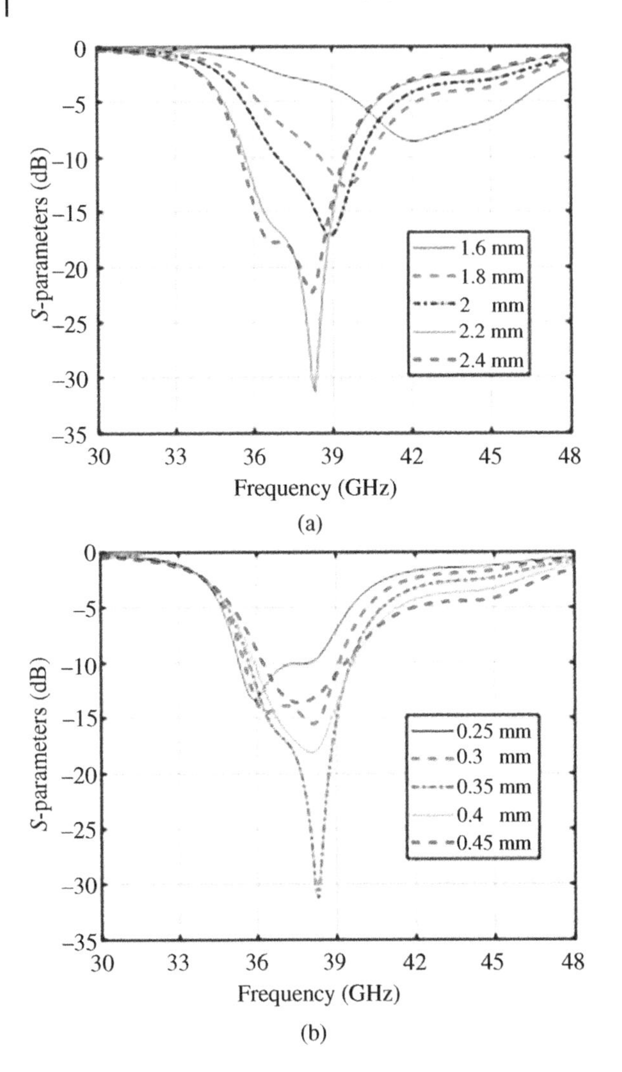

Figure 6.6 Simulated S_{11} of the V-pol endfire PFSA element. (a) S_{11} as a function of slot length L_S. (b) S_{11} as a function of top slot width W_S. *Source:* From [6]. ©2019 IEEE. Reproduced with permission.

respectively in the H-plane. The simulated antenna efficiency of the V-pol endfire PFSA element features more than 80% at 36–40.7 GHz as illustrated in Figure 6.9. Although the V-pol endfire PFSA is extremely compact in the axis of the dominant electric field (z-axis), the efficiency is comparable with that obtained in [26]. The calculated measured radiation efficiency of the PFSA element features 94.12% at 39 GHz.

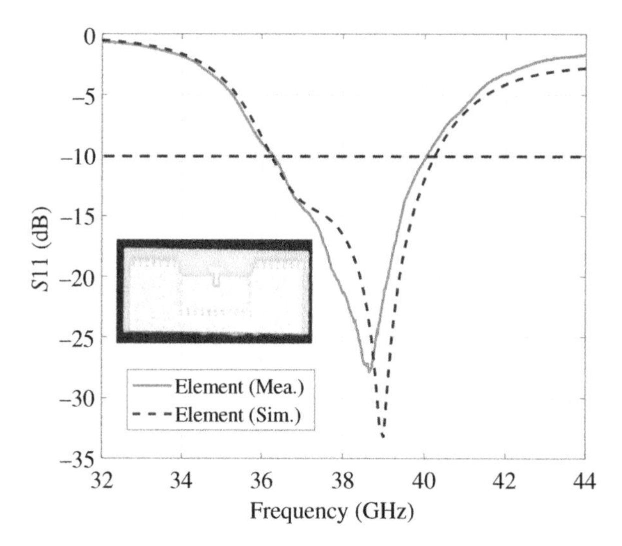

Figure 6.7 Measured and simulated input reflection coefficients of the V-pol endfire PFSA element. *Source:* From [6]. ©2019 IEEE. Reproduced with permission.

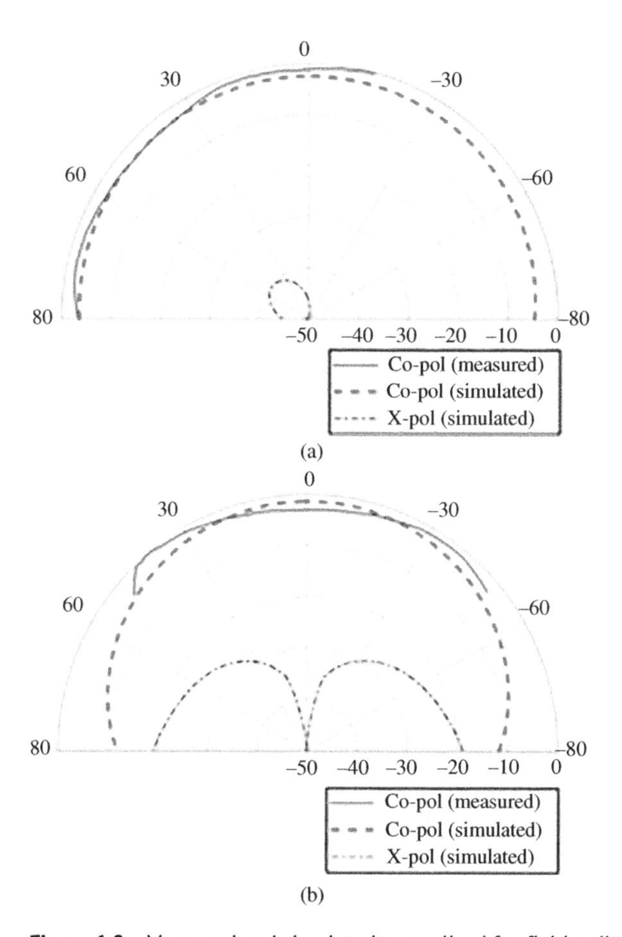

Figure 6.8 Measured and simulated normalized far-field radiation patterns of the V-pol endfire PFSA element. (a) *E*-plane (θ). (b) *H*-plane (φ) at 39 GHz. *Source:* From [6]. ©2019 IEEE. Reproduced with permission.

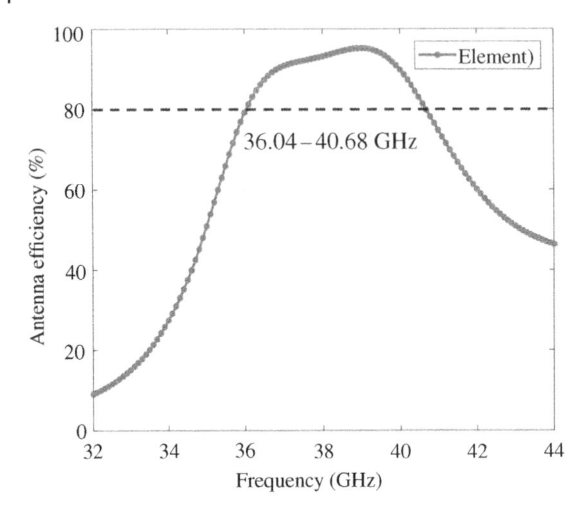

Figure 6.9 Simulated antenna efficiency of the V-pol endfire PFSA element. *Source:* From [6]. ©2019 IEEE. Reproduced with permission.

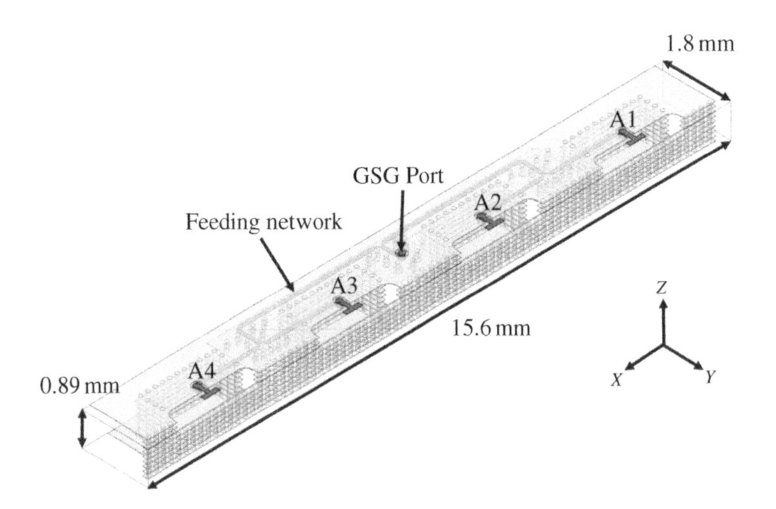

Figure 6.10 3D view of the proposed 1×4V-pol endfire PFSA array. *Source:* From [6]. ©2019 IEEE. Reproduced with permission.

The proposed PFSA element is extended to a 1×4 antenna array. Figure 6.10 illustrates the devised 1×4V-pol endfire PFSA array. The separation between adjacent antenna elements is 3.8 mm and the antenna array is connected to the antenna feeding network. A capacitive feed line located at L1 extends to L2 to establish an electrical connection between the antenna elements to the 1 : 4 strip-line power divider. The fabricated V-pol endfire PFSA element and array are photographed and presented in Figure 6.11.

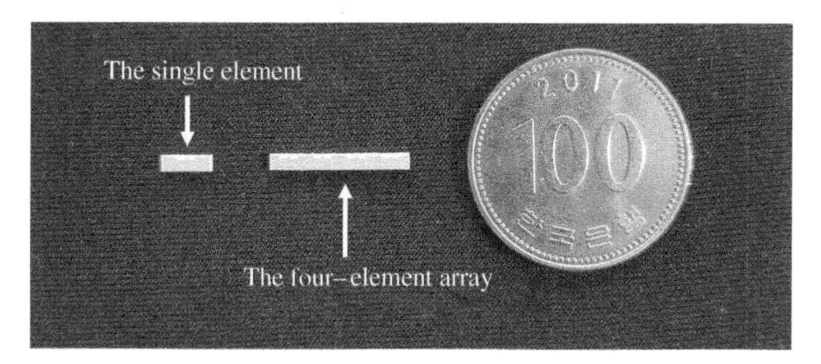

Figure 6.11 Photograph of the fabricated V-pol endfire PFSA element and array. *Source:* From [6]. ©2019 IEEE. Reproduced with permission.

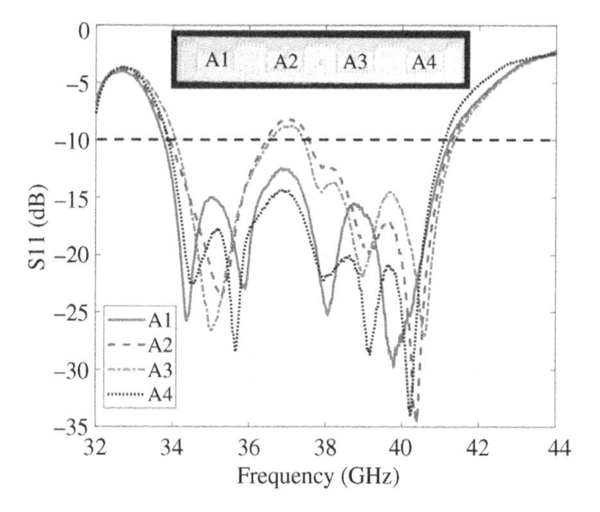

Figure 6.12 Measured input reflection coefficients of the 1×4V-pol endfire PFSA elements. *Source:* From [6]. ©2019 IEEE. Reproduced with permission.

The measured $|S_{11}|$ of the 1×4V-pol endfire PFSA elements is illustrated in Figure 6.12. The discrepancy between each element can be attributed to the contact loss of the wafer probes and manufacturing tolerances. The normalized measured and simulated radiation patterns of the fabricated 1×4V-pol endfire PFSA array in the *E*-plane and *H*-plane at 39 GHz are illustrated in Figures 6.13a,b, respectively. The measured and simulated gains are 7.7 and 8.09 dBi, respectively. Naturally, the *E*-plane characteristics remain similar to that of the single antenna element and the measured and simulated 3 dB half-power beamwidth in the H-plane are 34° and 30°, respectively.

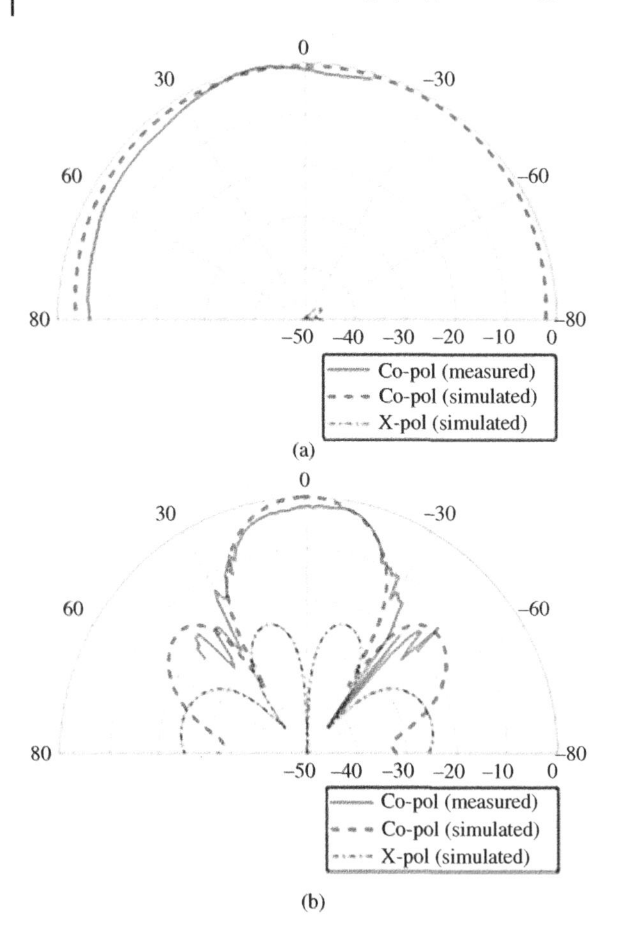

Figure 6.13 Measured and simulated normalized far-field radiation patterns of the 1×4V-pol endfire PFSA array. (a) *E*-plane (θ). (b) *H*-plane (φ) at 39 GHz. *Source:* From [6]. ©2019 IEEE. Reproduced with permission.

6.1.3 Low-Loss Interconnect Technology

This section presents an optimization methodology for vertical transitions that can improve the signal integrity of vertical interconnects in broadband with frequency-independent features. The interconnect structures play an important role in multi-layered mmWave antenna modules for circulating EM signals with extremely low loss. In particular, in order to fully adopt the 3D integration scheme and ensure the performance of the antenna system, low-loss vertical via transitions have to be realized within the package [27–30]. Vertical via transitions enable robust interconnection from RF components, such as monolithic microwave integrated circuits (MMICs) and filters, mounted on the top and bottom sides of packages, to the package-embedded transmission lines and antennas. Beyond the conventional configuration of package-to-board and package-to-chip, it will be more emphasized to optimize the vertical connection structure for the realization of advanced packaging systems, such as stacked packages of more than 1024 vertical I/Os. Poor optimization of the vertical transition can lead to increased insertion losses, reduced isolations, and undesired resonance, and fundamentally reduce system bandwidth. Depending on the type of package, the selection and optimization of a suitable vertical interconnect structure have become crucial.

Most of the vertical interconnect structures have an impedance lower than the characteristic impedance (Z_0) due to the large capacitance formed between the via and the reference plane, which is dominant over the inductance of the via itself [31]. Therefore, the reflected waves can be suppressed when the excess capacitive parasitic and the inductance of via are balanced by adjusting the anti-pad size. However, increasing the anti-pad size is restricted since another routing channel must also be accommodated. In this case, it can be a solution to add a patterned groove on the ground plane. For example, the proposed optimization method can be realized by adding the patterned grooves on the bottom reference plane of the stripline (L6) as shown in Figure 6.14. This method can reduce the capacitance between the via pad and ground plane and the fringing capacitance.

To verify the proposed optimization method of the vertical transition, a GCPW-to-stripline-to-GCPW vertical transition is designed using the LTCC process as shown in Figure 6.15. The design rule and material properties are the same as those used in the previous chapter. Both the GCPW and stripline have a characteristic impedance of 50 Ω. The GCPW center conductor and ground plane are placed on the top layer (L1) and one layer below (L2), respectively. The line width of the

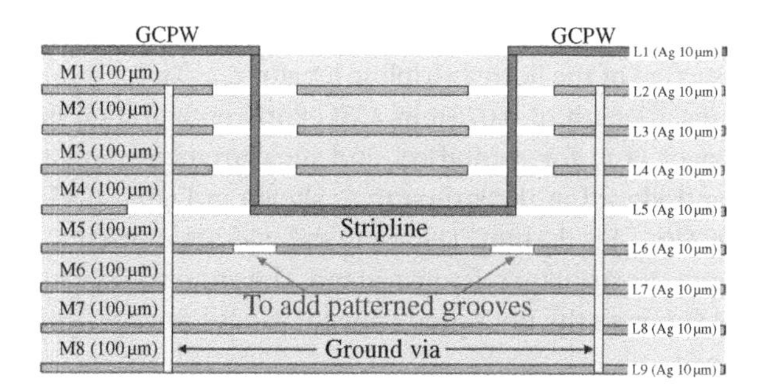

Figure 6.14 Optimization methodology to reduce the excess capacitive parasitic of the via to add patterned grooves on the ground plane. *Source:* From [6]. ©2019 IEEE. Reproduced with permission.

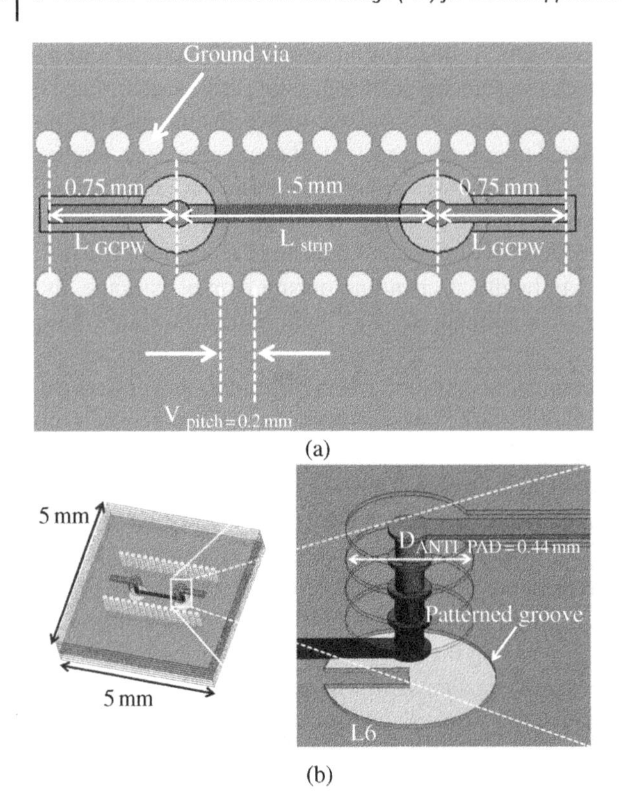

Figure 6.15 The designed GCPW-to-stripline-to-GCPW vertical transition structure. (a) Top view, (b) 3D and zoom-in view. *Source:* From [6]. ©2019 IEEE. Reproduced with permission.

GCPW is 100 μm and the distance between the ground on both sides and the GCPW signal line is 50 μm. The stripline ground is placed on L4 and L6 and the width of the stripline is 100 μm as shown in Figure 6.15a. The total length of the transition structures is 3 mm, the section of the buried stripline length (L_{strip}) is 1.5 mm, and each GCPW (L_{GCPW}) runs a length of 0.075 mm. GSG port configuration is implemented on the top layer (L1) for excitation and measurement. Some additional ground vias are introduced in the structure as shown in Figure 6.15. The pitch of the ground vias (V_{ptich}) is 0.2 mm. These ground vias are utilized to connect the GCPW ground and the stripline reference plane. The ground vias can also shield the EM field and suppress the undesired parasitic modes, such as the parallel plate wave-guide mode.

In order to reduce the excess capacitive parasitic of a coaxial-like via, the patterned grooves are added on the bottom reference plane of the stripline (L6) as shown in Figure 6.15b. A rectangular stub was added for guiding the signal of the

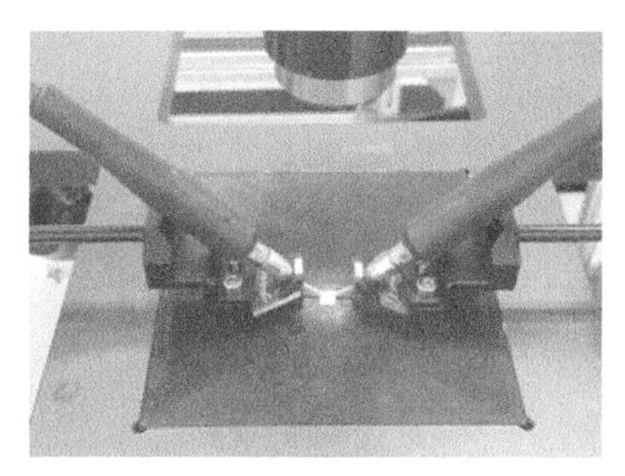

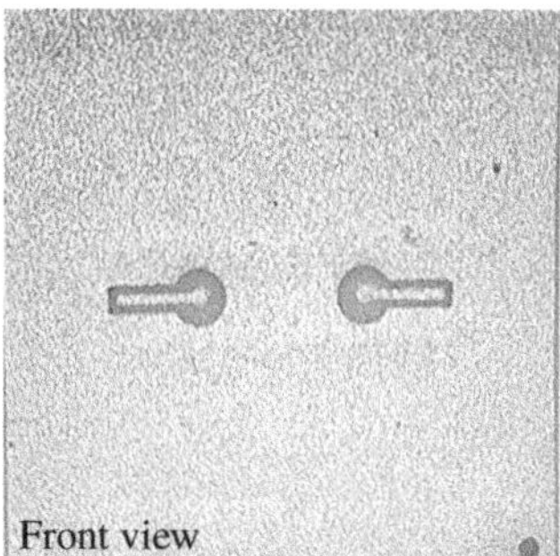

Figure 6.16 Photograph of the measurement setup and the fabricated prototype sample. *Source:* From [6]. ©2019 IEEE. Reproduced with permission.

stripline. The optimum sizes of the anti-pad and circular groove diameters are 440 and 600 μm, respectively.

The two prototypes are fabricated so as to verify the proposed approaches as shown in Figure 6.16. The KEYSIGHT PNA-X with probe station is used to measure the *S*-parameters of the two back-to-back transition prototypes in the frequency range from 100 MHz to 50 GHz. The fabricated prototype with patterned grooves can achieve insertion loss better than 0.5 dB over the entire frequency range except for around 32.5 GHz as illustrated in Figure 6.17a. This resonance is introduced by the distributed behavior caused due to the spacing between the signal and ground vias. The ground vias form an approximately rectangular cavity with a perfect electric conductor (PEC) boundary [32]. Figure 6.17b shows the reflection loss of the fabricated prototypes. The reflection loss is improved by 10 dB over 33 GHz when compared to the result without patterned grooves. The proposed optimization methodology improves signal integrity over the entire frequency spectrum.

In order to further clarify the proposed approach, the time-domain simulation is also conducted. For the time domain simulations, the transmitter and receiver are simply modeled as shown in Figure 6.18. Input signals are applied to use a pseudo-random bit sequence (PRBS) of 2^{10}-1 using an ADS channel simulation function and the input voltage swing is defined to 1.0 V. The target data rate of the transition channel is aimed 10 Gbps and the rising and falling times of the input signals are assumed as 10 ps. The simulation results in the time domain of the transition channel are plotted in Figure 6.19. The eye-opening voltages of the transition channel are 397 mV (79.4% of $V_{p\text{-}p}$) and the timing jitters of the channel are 1 ps. The eye diagram of the transition channel is widely open. As a result, it is shown that the signal integrity of the designed GCPW-to-stripline vertical transition channel is guaranteed.

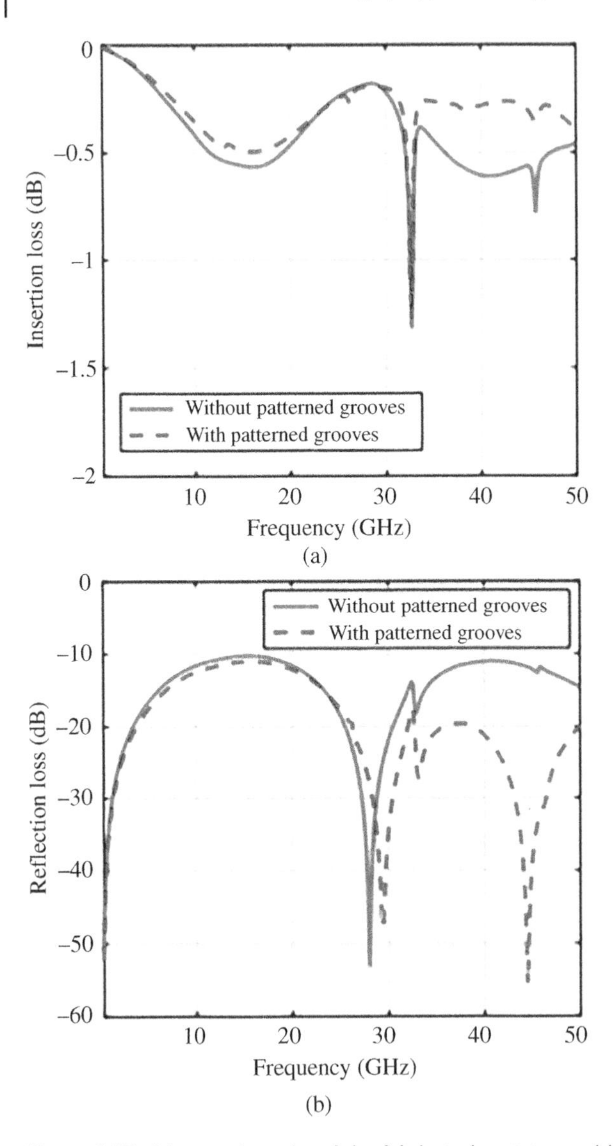

Figure 6.17 Measured results of the fabricated prototypes: (a) Insertion loss. (b) Reflection loss. *Source:* From [6]. ©2019 IEEE. Reproduced with permission.

6.1.4 Flip-Chip-Based Packaging of mmWave AiP

In this section, the mmWave beamforming module is designed and demonstrated using the proposed V-pol endfire PFSA (Section 6.1.2) and the optimization methodology (Section 6.1.3) for vertical transitions. At the time of this work, a mmWave beamformer IC operating at 26.5–29.5 GHz was the only available IC for

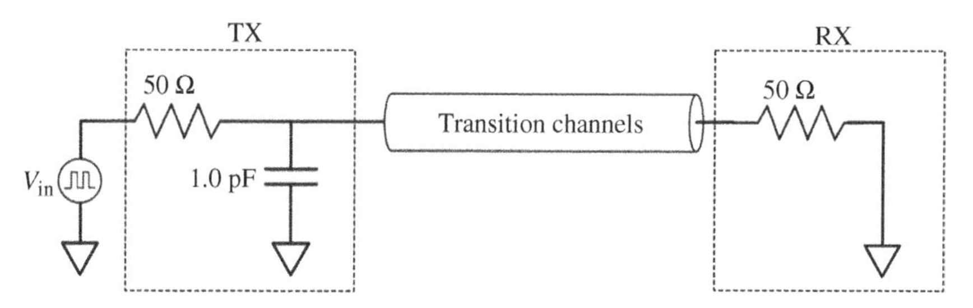

Figure 6.18 The simplified transmitter and receiver to simulate the eye diagram for the single-ended transition channel.

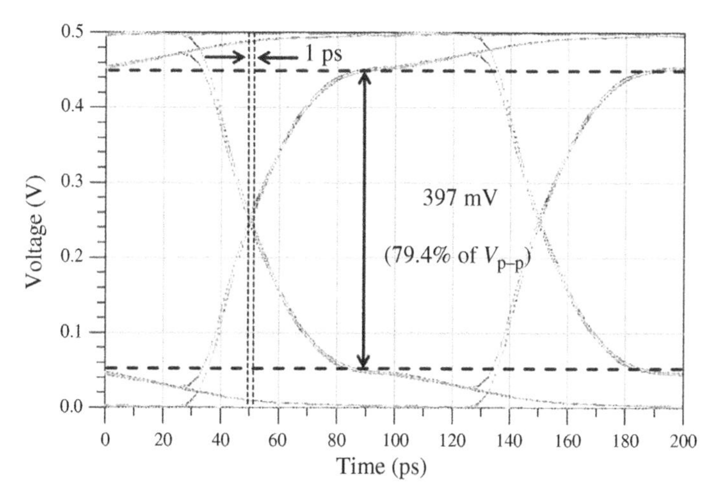

Figure 6.19 The eye diagram of the transition channel.

active verifications. Consequently, the V-pol endfire PFSA is rescaled to 28 GHz by adjusting the aperture and cavity size. The measured radiation patterns of the 28 GHz V-pol PFSA are almost identical to that of the 39 GHz V-pol PFSA. The fabricated 28 GHz V-pol PFSA element is confirmed to feature a gain of 2.95 dBi with a calculated radiation efficiency of 81.47%. Figure 6.20 illustrates the stack-up and conceptual diagram of the proposed mmWave beamforming module based on the V-pol endfire PFSA array and the mmWave beamformer IC. The V-pol endfire PFSA array is flip-chip mounted on the routing board consisting of eight stacked layers after aligning to the board edge. Afterward, the feed line of each PFSA element is electrically connected to the signal line of the routing board. Finally, the mmWave PFSA array is assembled into a carrier board with the beamformer IC.

The previously demonstrated optimization technology of the vertical interconnect structure is adopted for the design of the routing board as illustrated in Figure 6.21. The total length and width of the routing board are $L_m = 30\,\text{mm}$

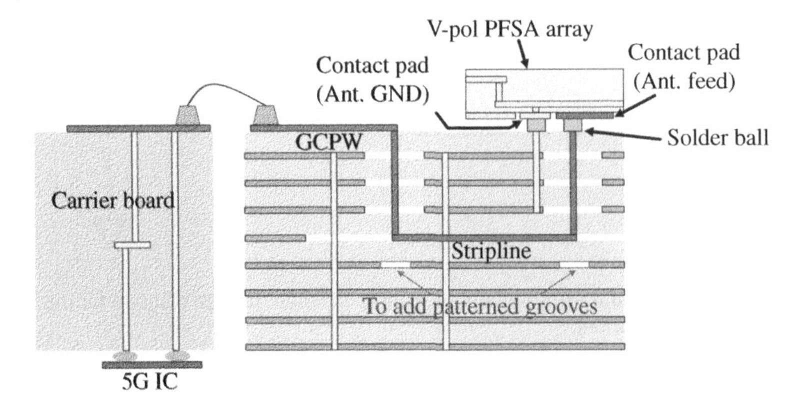

Figure 6.20 The stack-up and conceptual diagram of the proposed mmWave beamforming module using the V-pol endfire PFSA array and IC. *Source:* From [6]. ©2019 IEEE. Reproduced with permission.

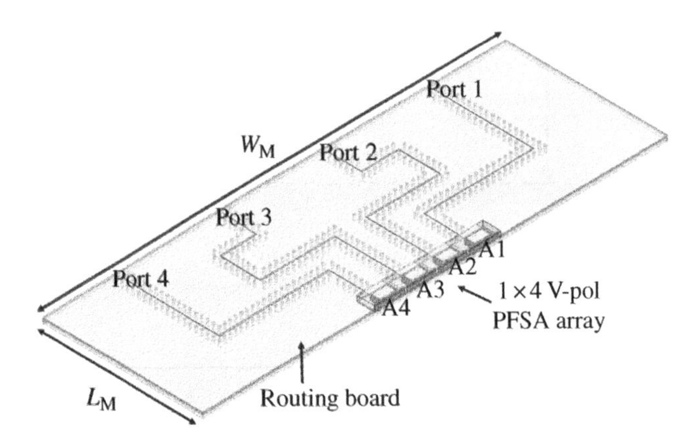

Figure 6.21 The 3-D view of the mmWave antenna array module consisting of the V-pol endfire PFSA array and routing board. *Source:* From [6]. ©2019 IEEE. Reproduced with permission.

and $W_m = 80\,\text{mm}$, respectively. To be assembled with mmWave beamforming module using RF connectors, each port of the PFSA element is electrically connected to signal lines of the routing board. The detailed interconnect structure between the V-pol PFSA array and routing board is illustrated in Figure 6.22. The flip-chip pad is implemented on the top layer for bonding alignment and the pitch between the signal and ground pad is 600 μm. The diameter and height of the ball are 150 and 100 μm, respectively. As described in Section 6.1.3,

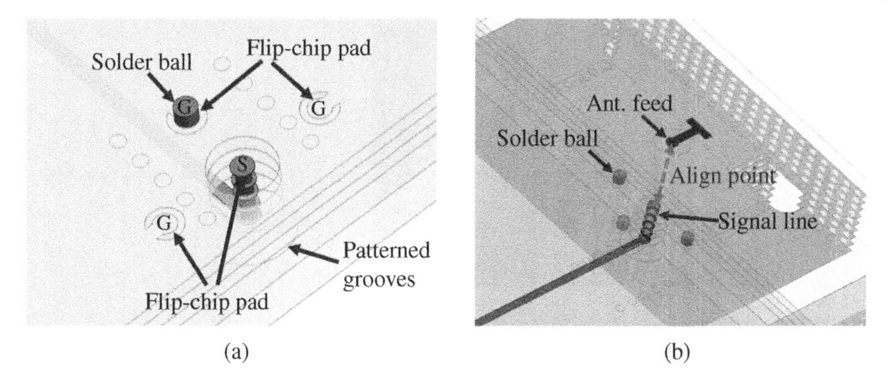

Figure 6.22 The zoom-in view of the vertical interconnection between the routing board and PFSA element. (a) The top layer of the routing board. (b) The vertical interconnection: the PFSA array is mounted on the top layer of the routing board using the alignment key. *Source:* From [6]. ©2019 IEEE. Reproduced with permission.

the coaxial-like via is transited to the stripline located on the L5 layer, and the patterned grooves are added on the bottom reference plane of the stripline (L6). Afterward, the PFSA array is flip-chip mounted on the top layer of the routing board using the alignment key as shown in Figure 6.22b. Finally, the designed antenna module is fabricated and connected to the beamforming module as illustrated in Figure 6.23.

To control the RFIC, one interposer board (USB–SPI interface) is connected to the beamforming board. Each channel is adjusted individually using PC control software and SPI interposer board. The beamforming properties are tested in the TX mode in the far field and measured using the anechoic far-field chamber at POSTECH. The proposed beamforming module using the PFSA array features a measured EIRP of 18.2 dBm and a scanning range of $\pm 50^{\circ}$ in azimuth (H-plane) at 28 GHz as illustrated in Figure 6.24a. It can be deduced that the proposed energy-efficient mmWave beamforming module with more than 4 RF chains can achieve spherical beamforming coverage. Gain control of one beam is exemplified in Figure 6.24b. In this experiment, identical power is fed to all ports. However, it can be deduced that the sidelobe level (SLL) can be further suppressed through tapering the power supplied to each port. Ultimately, to guarantee the stability of the beamforming, orthogonality between the phase and amplitude control is required in each RF channel.

Table 6.1 summarizes the performance comparison with recently reported V-pol endfire mobile antennas. It should be noted that the presented PFSA features the smallest height profile in comparison with previously reported V-pol endfire topologies.

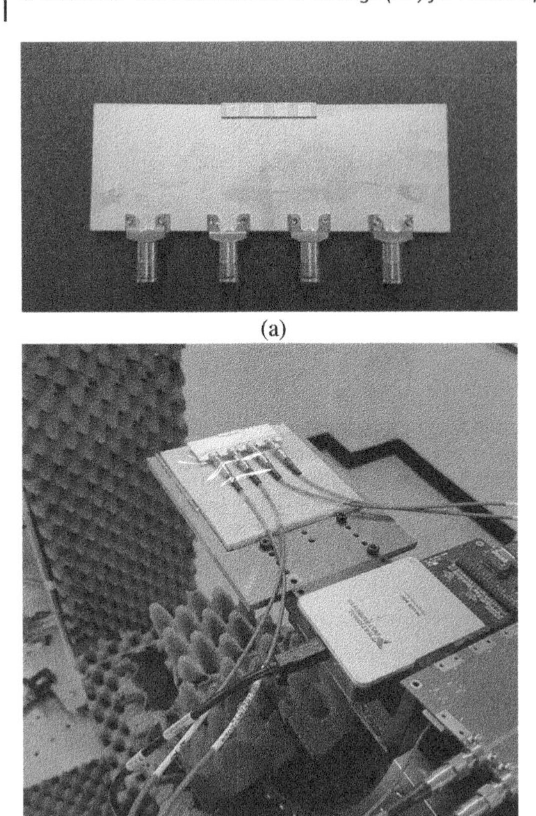

Figure 6.23 (a) Photograph of the fabricated mmWave antenna array module, and (b) measurement setup of the proposed mmWave beamforming module with the V-pol endfire PFSA array. *Source:* From [6]. ©2019 IEEE. Reproduced with permission.

6.2 Multi-Modal AiP Technology

From the UE's point of view, the most effective way to improve the QoS of the mmWave link without increasing the number of RF front-end modules is to develop a multi-modal AiP technology. One way for implementing the multi-modal AiP is to use a reconfigurable antenna (RA), which can simultaneously satisfy multiple design criteria using its versatile antenna modes [33–35]. RAs can alter their operation characteristics by rearranging the current distributions with the use of external passive and active components including PIN diodes, varactors, and MEMS. Despite the great potential of multi-functional capabilities, the adoption of RAs to phased arrays (see Figure 6.25a) poses great challenges due to

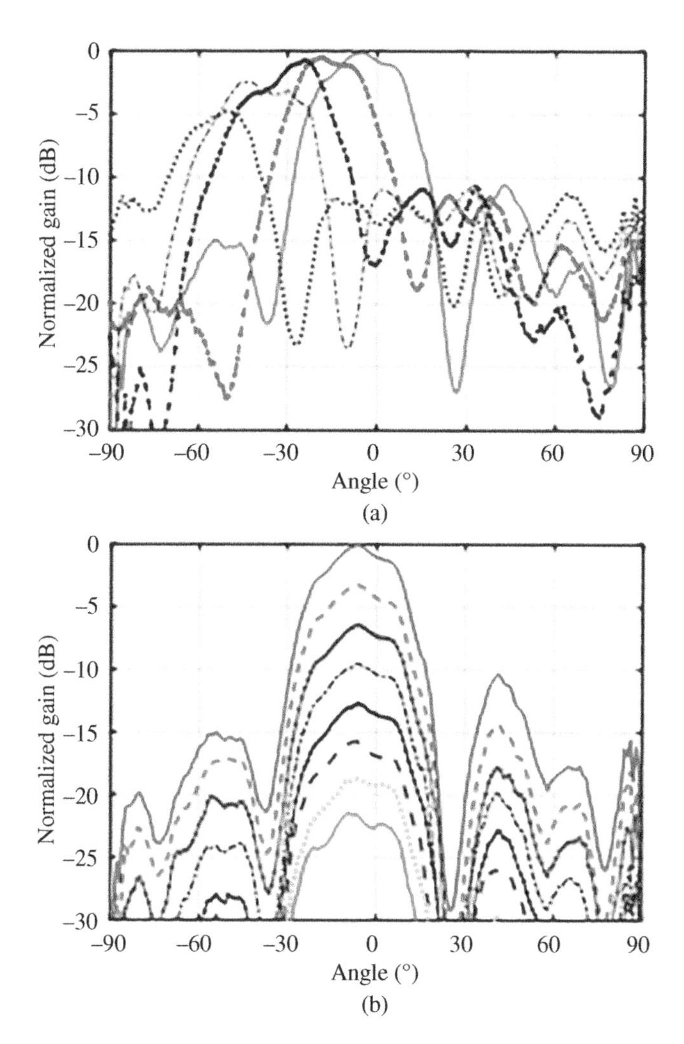

Figure 6.24 (a) Measured normalized far-field radiation patterns of the proposed mmWave beamforming module in the TX mode at 28 GHz, and (b) measured gain control radiation patterns of the proposed mmWave beamforming module at fixed phase configuration at 28 GHz. *Source:* From [6]. ©2019 IEEE. Reproduced with permission.

the following reasons: (i) The RF components and power supply circuitries added for the operation of the RA significantly increases the integration density of hardware. (ii) In order to be compatible with an adaptive beamforming system, the operation of RAs should be synchronized with other circuitries and controlled by the baseband blocks. Due to these packaging and operational issues, RA topologies have not been adopted in mmWave beamforming phased-array applications.

Table 6.1 Comparison with the state-of-the-art vertically polarized endfire mobile antennas.

	[23]	[24]	[25]	[26]		This work
Topology	**1 × 8 SIW array**	**SIW single element**	**1 × 4 electric dipole array**	**Dual-pol Yagi 1 × 4 array V-pol**	**Dual-pol Yagi 1 × 4 array H-pol**	**1 × 4 PFSA array**
Bandwidth of antenna element (GHz)	27.8	4.9	7.23	9	12.3	3.76
Gain (dBi)	12 (array)	11 (element)	12.61 (array)	10.8 (array)	11.8 (array)	7.7 (array)
3 dB beam scan range (°)	100	N/A	N/A	N/A	N/A	90
Vol (λ_0^3)	24.7 (4.53 × 11.6 × 0.47)	0.45 (1.28 × 1.74 × 0.2)	2.45 (2.8 × 1.9 × 0.46)	1.29 (3.65 × 1.54 × 0.23)		0.05 (1.98 × 0.23 × 0.11)
Height profile (λ_0)	0.47	0.2	0.46	0.23		0.11

Source: From [6]. ©2019 IEEE. Reproduced with permission.

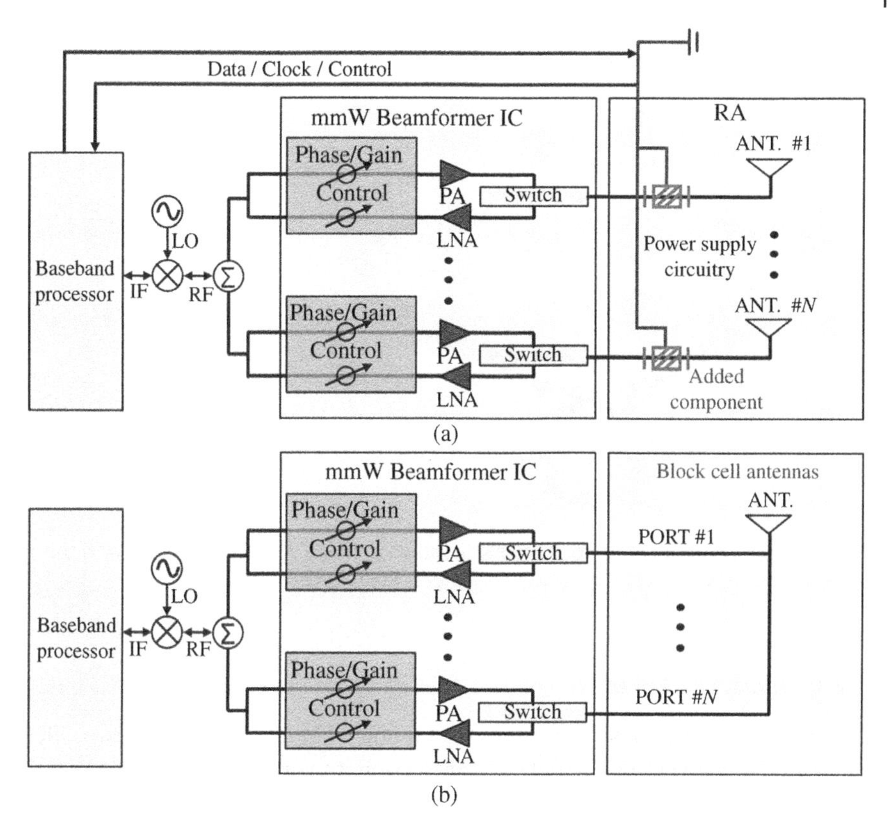

Figure 6.25 System block diagrams of mmWave phased array architectures. (a) The conventional architecture based on reconfigurable antenna arrays. (b) The proposed architecture comprised BCA. *Source:* From [34]. ©2019 IEEE. Reproduced with permission.

One feasible multi-modal AiP concept originates from a modular unit block cell antenna (BCA) structure that can be expanded along the horizontal axes in order to formulate an array [35]. The unit BCA consists of two ports, the weighting matrix is used to selectively excite the ports for operation modes. As the unit BCA is expanded two-dimensionally, an $N \times N$ BCA structure can be obtained. This two-dimensionally expanded structure can emulate various antenna characteristics including a wide range of directivity, polarization, and main beam direction in the form of distinct modes. Above all, the BCA concept only requires phase shifters and attenuators, while functioning as an RA. Therefore, the proposed BCA concept is fully compatible with currently existing mmWave 5G phased-array architectures, as illustrated in Figure 6.25b. In addition, the proposed BCA can be fully integrated with the mmWave 5G RFIC in the form of an AiP, as illustrated in Figure 6.26.

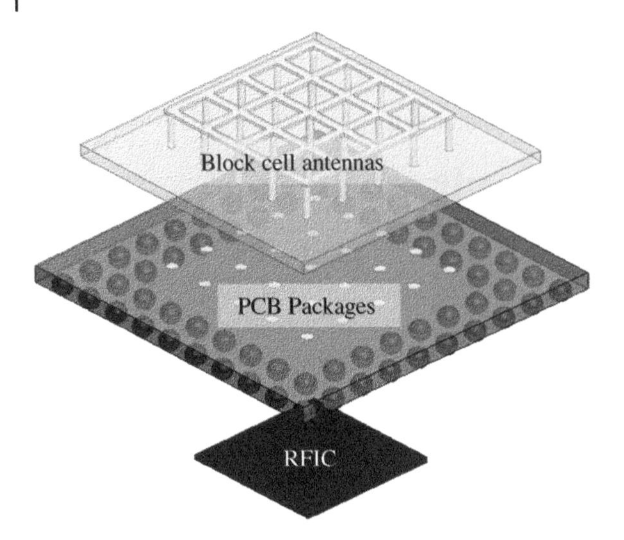

Figure 6.26 Conceptual figure of the block cell antenna (BCA) fully integrated with the mmWave 5G RFIC. *Source:* From [35]. ©2021 IEEE. Reproduced with permission.

6.2.1 Block Cell Antennas

The unit BCA consists of four connected monopole loop antennas and two GSG ports as illustrated in Figure 6.27. In this section, the proposed BCA is designed using an LTCC technology process. The antenna consists of 10 stacked layers with a thickness of 100 μm, relative permittivity of 5.9, and loss tangency of 0.002 at 28 GHz. A 10-μm-thick silver is inserted between each stacked layer. The optimized dimensions in Figure 6.27 are determined by considering the impedance matching at 28 GHz and the maximum radiation efficiency in a two-dimensional expanded BCA. The unit BCA contains four basic modes that utilize the phase difference between two ports with a phase resolution of 90°. Detailed analysis is required to understand the operation principle and background of the BCA.

The BCA originates from a single half-loop antenna with a half wavelength built on plane ground [36]. In Figure 6.28, step 1 shows that the current distribution is the same as that of a conventional loop antenna in accordance with the image theory. As $I_1(z)$ is the excited current to port 1, the arrow indicates the direction of the current at a single time domain, assuming a sinusoidal current is excited to the feed. In Figure 6.28, z is the distance from the feed, h is the height of the loop, and s is the length between the two vertical components. Throughout this manuscript, the loop length l denotes $s+2h$ as s and h denote the horizontal and vertical lengths of the loop, respectively. Step 2 in Figure 6.28 shows that the

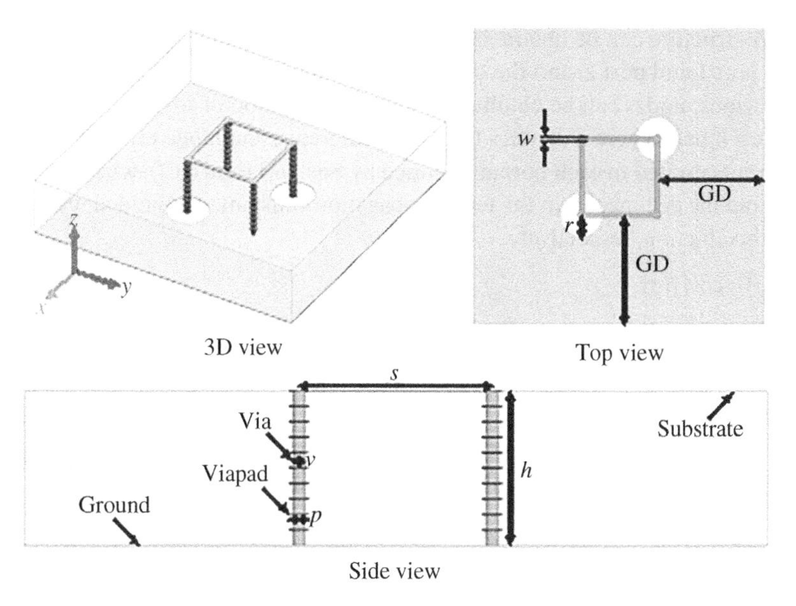

Figure 6.27 Schematic of the proposed unit BCA. Optimized dimensions (in mm): $w = 0.1$, GD = 2, $s = 1.4$, $h = 1.11$, $r = 0.4$, $v = 0.1$, $p = 0.15$. *Source:* From [35]. ©2021 IEEE. Reproduced with permission.

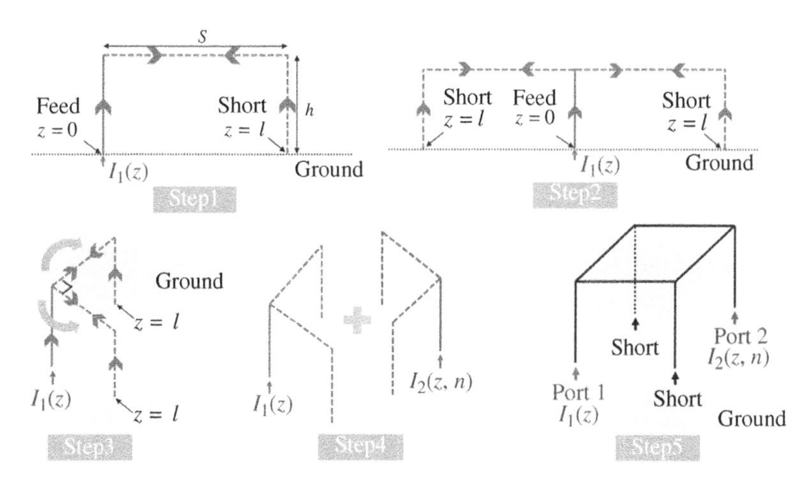

Figure 6.28 Derivation of the unit BCA from one loop antenna. *Source:* From [35]. ©2021 IEEE. Reproduced with permission.

two loops are connected in both directions from a single feed. From step 3 of Figure 6.28, the loops are folded to 90° along with the feed axis. The current distribution in the structure is the same as that in step 2. Finally, as shown in Figure 6.28, if a folded two-way loop structure is combined as illustrated in step 4, a unit BCA structure can be obtained as displayed in step 5. $I_1(z)$ and $I_2(z, n)$ are applied to port 1 and port 2, and the other vertical components are shorted to the ground. Various modes can be obtained by the superposition of $I_1(z)$ and $I_2(z, n)$ excited to each port where n denotes the phase difference and mode number.

The currents excited to each port are defined by Eqs. (6.1) and (6.2), where z, β, and λ_g denote the distance from the feed, propagation constant, and guided wavelength in the antenna, respectively.

$$I_1(z) = \cos(\beta z) \tag{6.1}$$

$$I_2(z, n) = \cos\left(\beta z + \frac{(n-1)\pi}{2}\right) \tag{6.2}$$

Figure 6.29a shows the location of each port and the names of the branches in alphabetical order. Figure 6.29b shows the current intensity in each branch, as

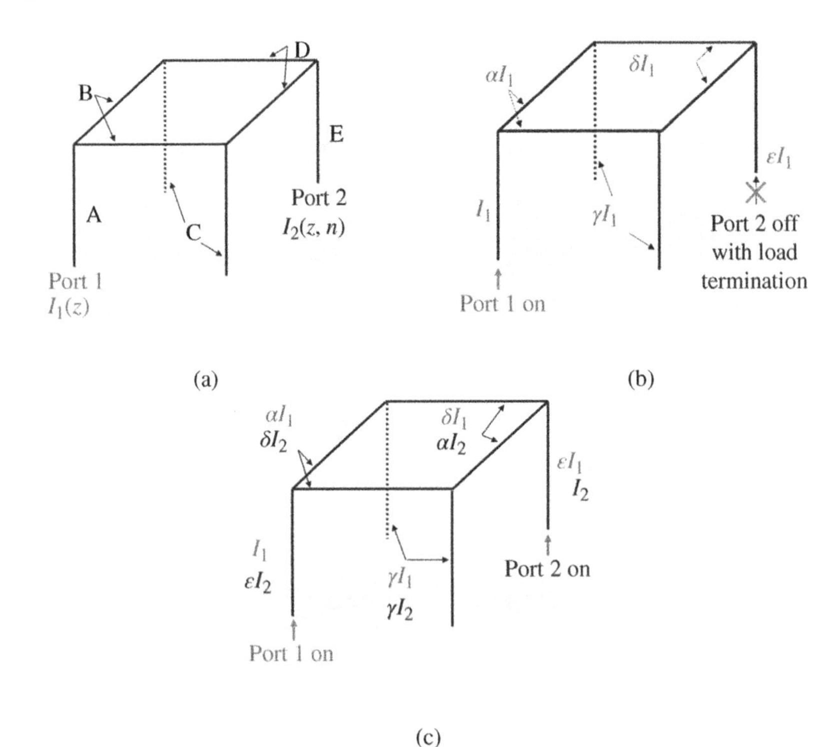

Figure 6.29 Derivation of the unit BCA from one loop antenna. Schematics of unit BCA. (a) Unit block cell with the notation of each branch and two ports. (b) Magnitude of the currents in each branch when the current $I_1(z)$ is excited to port 1. (c) The magnitude of the currents in each branch when the currents $I_1(z)$ and $I_2(z, n)$ are excited to port 1 and port 2, respectively. *Source:* From [35]. ©2021 IEEE. Reproduced with permission.

port 1 is ON and port 2 is OFF, where each Greek letter represents a coefficient indicating the magnitude of the current flowing through each branch. Figure 6.29c shows the current flowing in each branch when port 1 and port 2 are both ON at the same time. The current distribution of the unit BCA in $n = 3$ is illustrated in Figure 6.30a. As illustrated in Figure 6.30b, the effective current in $n = 3$ implies that the phase difference between the two ports is 180° and the radiation pattern in mode 3 will have a polarization direction in ($\phi = 135°$).

The simulation results of 3-D radiation patterns in the four modes of the unit BCA ($n = 1, 2, 3, 4$) are illustrated in Figure 6.31. An omnidirectional mode and a broadside mode are obtained when n is 1 and 3, respectively. Furthermore, the tilted beams are obtained when n is 2 and 4, respectively. As shown in Figure 6.32, the unit BCA exhibits the impedance characteristic of $|S_{11}| < -10$ dB at 24.3–29.7 GHz.

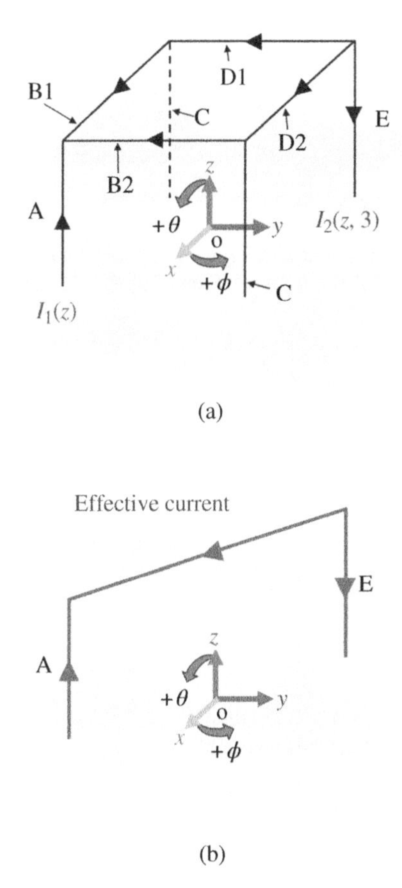

Figure 6.30 Current distribution of unit BCA in $n = 3$. (a) Current distribution on each branch. (b) The effective current in the far field. *Source:* From [35]. ©2021 IEEE. Reproduced with permission.

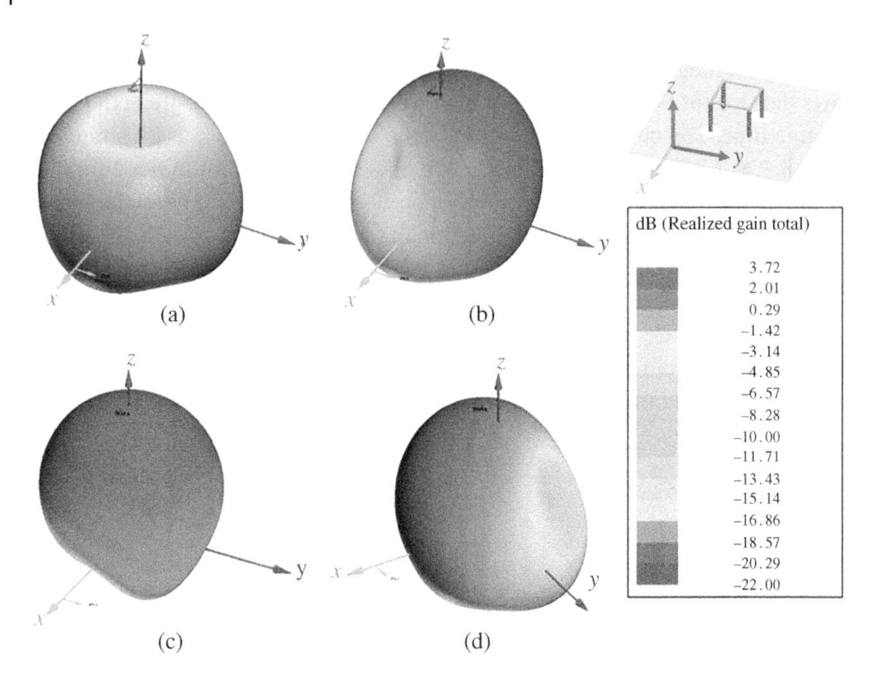

Figure 6.31 Simulated radiation patterns of the unit BCA. (a) $n = 1$ (0° diff.), (b) $n = 2$ (90° diff.), (c) $n = 3$ (180° diff.), and (d) $n = 4$ (270° diff.). *Source:* From [35]. ©2021 IEEE. Reproduced with permission.

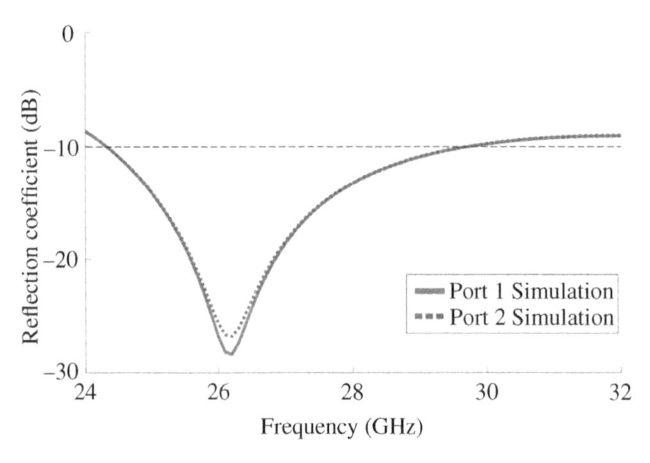

Figure 6.32 Simulated reflection coefficients of the unit BCA. *Source:* From [35]. ©2021 IEEE. Reproduced with permission.

6.2.2 Modular, Controllable Block Cell Arrays

According to the current distribution of the unit BCA in mode 3, the current distribution of one-dimensionally expanded 1×2 BCA with diagonal port configuration is shown in Figure 6.33. As illustrated in Figure 6.33a, the currents cancel each other in the branches in the expanded direction. Thus, the effective current appears as an array consisting of unit monopole loop antennas, as illustrated in Figure 6.33b. In addition, it is anticipated that if the block cell is expanded to the *y*-axis, it will have an *x*-axis polarization. The directivity and gain of the antenna are studied as a function of the number of the unit BCA. It is confirmed in Figure 6.34 that the proposed BCA structure can be modularly expanded for an antenna array configuration without any modification of the design. It is worth mentioning that the 1×13 BCA can achieve a peak realized gain of more than 10 dBi, as illustrated in Figure 6.35.

The proposed BCA is expanded to two-dimensionally expanded 4×4 BCA. The 4×4 BCA features 13 ports as illustrated in Figure 6.36. The weighting matrix defined by A represents the magnitude and delayed phase of the signals applied to each port. In Eq. (6.3), a_{mn} is the element of weighting matrix A where M and θ denote the magnitude and the delayed phase, respectively.

$$a_{\mathrm{mn}} = M\angle\theta\left(0 \le M \le 1, 0 \le \theta \le 2\pi\right) \tag{6.3}$$

A one-dimensionally expanded BCA has a single linear polarization. Therefore, if the BCA is expanded in the two-dimensional direction while maintaining the symmetry, it can be expected that multi-polarization can be obtained. For example, as the first (a_{11}, a_{12}, a_{13}) and fifth rows (a_{51}, a_{52}, a_{53}) of the weighting matrix A have a 180° phase difference, a high gain broadside mode having the *x*-axis polarization can be obtained. Likewise, by activating the first and fifth columns with 180° phase differences, *y*-axis polarization can be obtained. By activating

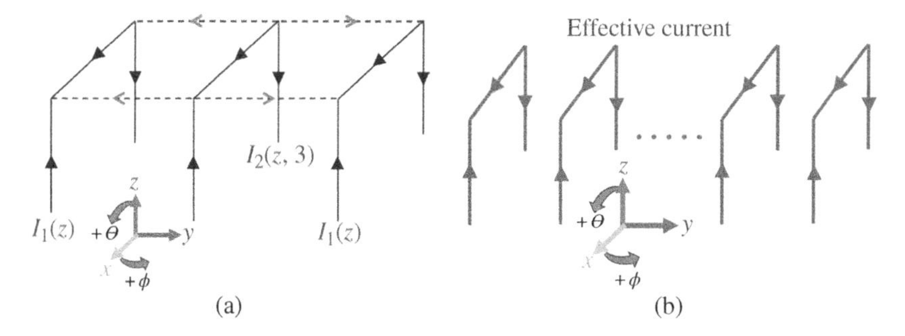

Figure 6.33 Current distribution of 1 × 2 BCA in mode 3 ($n = 3$). (a) Current distribution on each branch. (b) The effective current in the far field. *Source:* From [35]. ©2021 IEEE. Reproduced with permission.

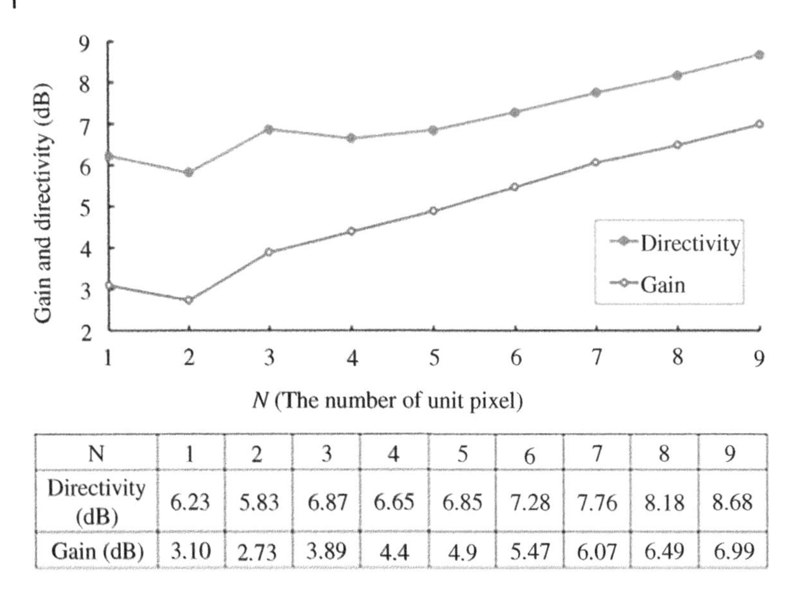

N	1	2	3	4	5	6	7	8	9
Directivity (dB)	6.23	5.83	6.87	6.65	6.85	7.28	7.76	8.18	8.68
Gain (dB)	3.10	2.73	3.89	4.4	4.9	5.47	6.07	6.49	6.99

Figure 6.34 Simulated gain and directivity of $1 \times N$ pixels with the phase difference between two adjacent ports is 180° ($n = 3$) as a function of the number of a unit pixel at 28 GHz. *Source:* From [34]. ©2021 IEEE. Reproduced with permission.

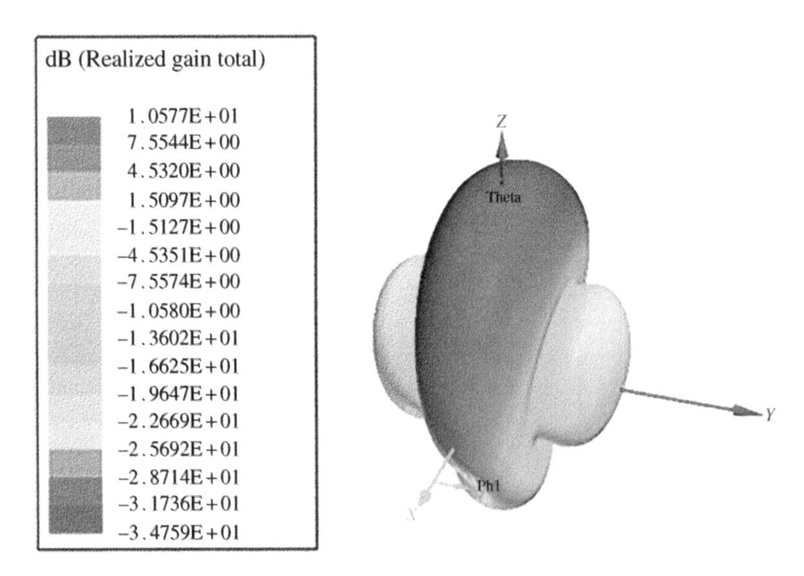

Figure 6.35 Simulated 3-D far-field radiation patterns of the 1×13 BCA. *Source:* From [35]. ©2021 IEEE. Reproduced with permission.

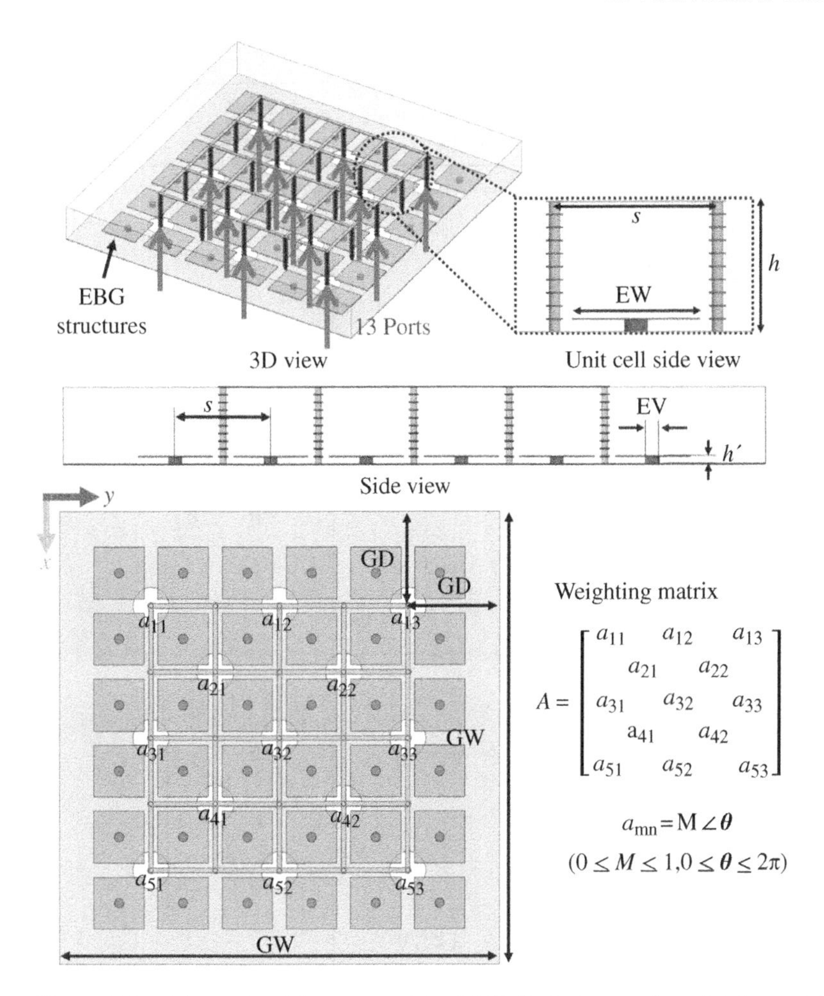

Figure 6.36 Schematic of the proposed 4×4 BCA. Each element in weighting matrix *A* is excited to each port. Optimized dimensions (in mm): *s* = 1.4, *h* = 1.11, EW = 1.1, EV = 0.2, *h'* = 0.11, GD = 2, GW = 9.6. *Source:* From [35]. ©2021 IEEE. Reproduced with permission.

diagonal ports, such as a_{11}, a_{21}, a_{31}, a_{42}, and a_{53}, a $\varphi = 45°$ polarization can be obtained. In addition, as with a unit BCA, a tilted-beam and dipole-like mode can be obtained using 90° and 0° phase differences, respectively. Using the weighting matrices in Table 6.2, the detailed 3-D simulated far-field radiation patterns of the 4×4 BCA are illustrated in Figure 6.37.

In Figure 6.38, the four ports located at the corners have two paths of current leaving each port, as in the unit BCA. However, since the other ports have three or four paths of current leaving each port, the impedance characteristic

Table 6.2 The mode and the weighting matrix of 4×4 BCA.

Mode	Polarization direction	Gain (dB)	Peak angle (θ, φ)	Weighting matrix
Low-gain broad-side	x-axis	−3.6	$\theta = 0°$	$\begin{bmatrix} 1 & & 1 & & 1 \\ & 0 & & 0 & \\ 0 & & 0 & & 0 \\ & 0 & & 0 & \\ 0 & & 0 & & 0 \end{bmatrix}$
	y-axis	−3.7	$(-5°, 0°)$	$\begin{bmatrix} 1 & & 0 & & 0 \\ & 0 & & 0 & \\ 1 & & 0 & & 0 \\ & 0 & & 0 & \\ 1 & & 0 & & 0 \end{bmatrix}$
High-gain broad-side	x-axis	−0.3	$(-2°, 0°)$	$\begin{bmatrix} 1 & & 1 & & 1 \\ & 0 & & 0 & \\ 0 & & 0 & & 0 \\ & 0 & & 0 & \\ -1 & & -1 & & -1 \end{bmatrix}$
	y-axis	0	$\theta = 0°$	$\begin{bmatrix} 1 & & 0 & & -1 \\ & 0 & & 0 & \\ 1 & & 0 & & -1 \\ & 0 & & 0 & \\ 1 & & 0 & & -1 \end{bmatrix}$
Tilted-beam (+direction)	x-axis	−5.5	$(25°, 0°)$	$\begin{bmatrix} 1 & & 1 & & 1 \\ & 0 & & 0 & \\ 1\angle\pi/2 & & 1\angle\pi/2 & & 1\angle\pi/2 \\ & 0 & & 0 & \\ 0 & & 0 & & 0 \end{bmatrix}$
	y-axis	−5.9	$(23°, 90°)$	$\begin{bmatrix} 1 & & 1\angle\pi/2 & & 0 \\ & 0 & & 0 & \\ 1 & & 1\angle\pi/2 & & 0 \\ & 0 & & 0 & \\ 1 & & 1\angle\pi/2 & & 0 \end{bmatrix}$
Tilted-beam (−direction)	x-axis	−6.0	$(-28°, 0°)$	$\begin{bmatrix} 0 & & 0 & & 0 \\ & 0 & & 0 & \\ 1\angle\pi/2 & & 1\angle\pi/2 & & 1\angle\pi/2 \\ & 0 & & 0 & \\ 1 & & 1 & & 1 \end{bmatrix}$

Table 6.2 (Continued)

Mode	Polarization direction	Gain (dB)	Peak angle (θ, φ)	Weighting matrix
	y-axis	−5.5	(−27°, 90°)	$\begin{bmatrix} 0 & & 1\angle\pi/2 & & 1 \\ & 0 & & 0 & \\ 0 & & 1\angle\pi/2 & & 1 \\ & 0 & & 0 & \\ 0 & & 1\angle\pi/2 & & 1 \end{bmatrix}$
Fan-beam	$\varphi = 45o$	−4.2	$\theta = 0°$	$\begin{bmatrix} 1 & & 0 & & 0 \\ & -1 & & 0 & \\ 0 & & 0 & & 0 \\ & 0 & & 1 & \\ 0 & & 0 & & -1 \end{bmatrix}$
	$\varphi = 135o$	−4.1	(3°, 135°)	$\begin{bmatrix} 0 & & 0 & & 1 \\ & 0 & & -1 & \\ 0 & & 0 & & 0 \\ & 1 & & 0 & \\ -1 & & 0 & & 0 \end{bmatrix}$
Bidirectional	x-axis beam direction	−2.9	(−74°, 0°)	$\begin{bmatrix} 1 & & 1 & & 1 \\ & 0 & & 0 & \\ 0 & & 0 & & 0 \\ & 0 & & 0 & \\ 1 & & 1 & & 1 \end{bmatrix}$
	y-axis beam direction	−3.5	(35°, 90°)	$\begin{bmatrix} 1 & & 0 & & 1 \\ & 0 & & 0 & \\ 1 & & 0 & & 1 \\ & 0 & & 0 & \\ 1 & & 0 & & 1 \end{bmatrix}$
	$\varphi = 45°$ beam direction	−4.9	(−27°, 45°)	$\begin{bmatrix} 1 & & 0 & & 0 \\ & 0 & & 0 & \\ 0 & & 0 & & 0 \\ & 0 & & 0 & \\ 0 & & 0 & & 1 \end{bmatrix}$
	$\varphi = 135°$ beam direction	−4.8	(−40°, 135°)	$\begin{bmatrix} 0 & & 0 & & 1 \\ & 0 & & 0 & \\ 0 & & 0 & & 0 \\ & 0 & & 0 & \\ 1 & & 0 & & 0 \end{bmatrix}$
Omnidirectional	z-axis	−6.1	(−48°, 0°)	$\begin{bmatrix} 1 & & 0 & & 1 \\ & 0 & & 0 & \\ 0 & & 0 & & 0 \\ & 0 & & 0 & \\ 1 & & 0 & & 1 \end{bmatrix}$

Source: From [35]. ©2021 IEEE. Reproduced with permission.

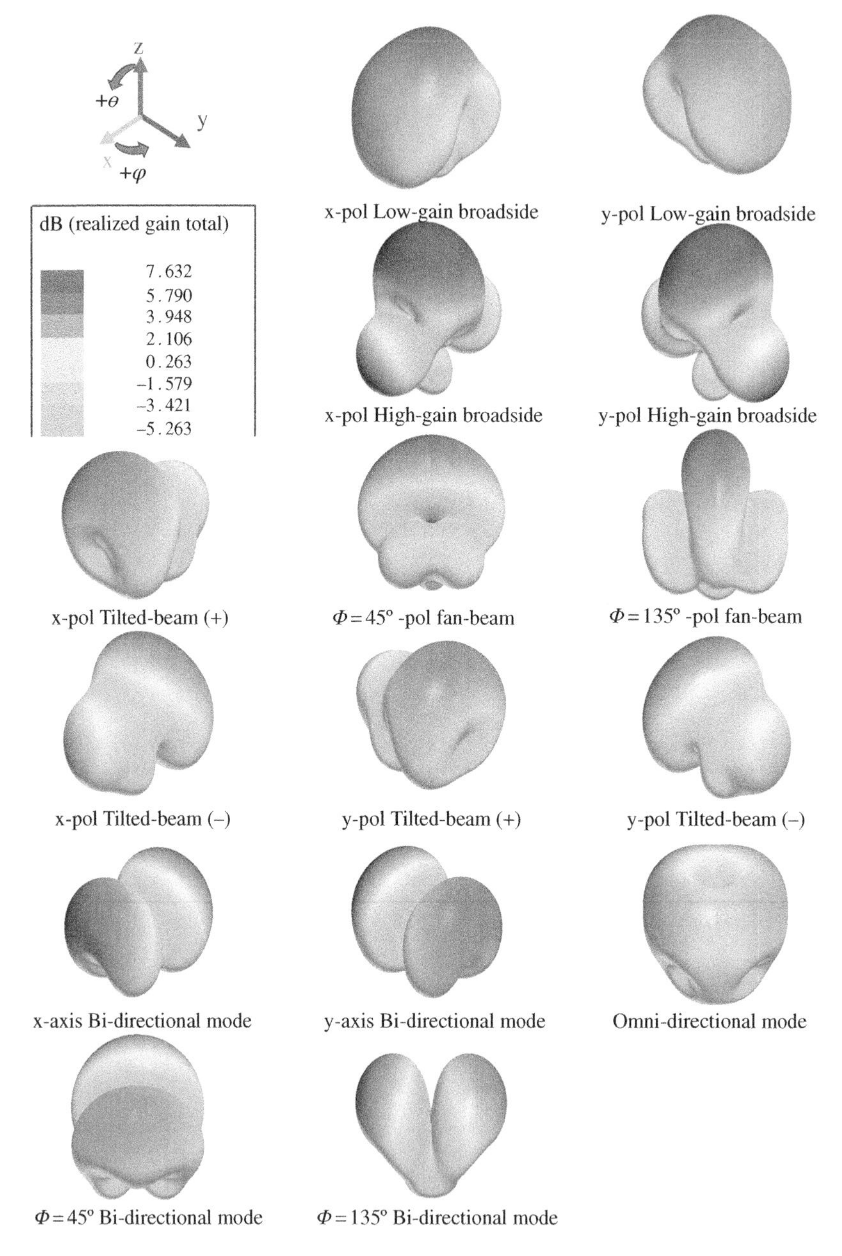

Figure 6.37 Simulated 3-D far-field radiation patterns of 4×4 BCA in all modes. *Source:* From [35]. ©2021 IEEE. Reproduced with permission.

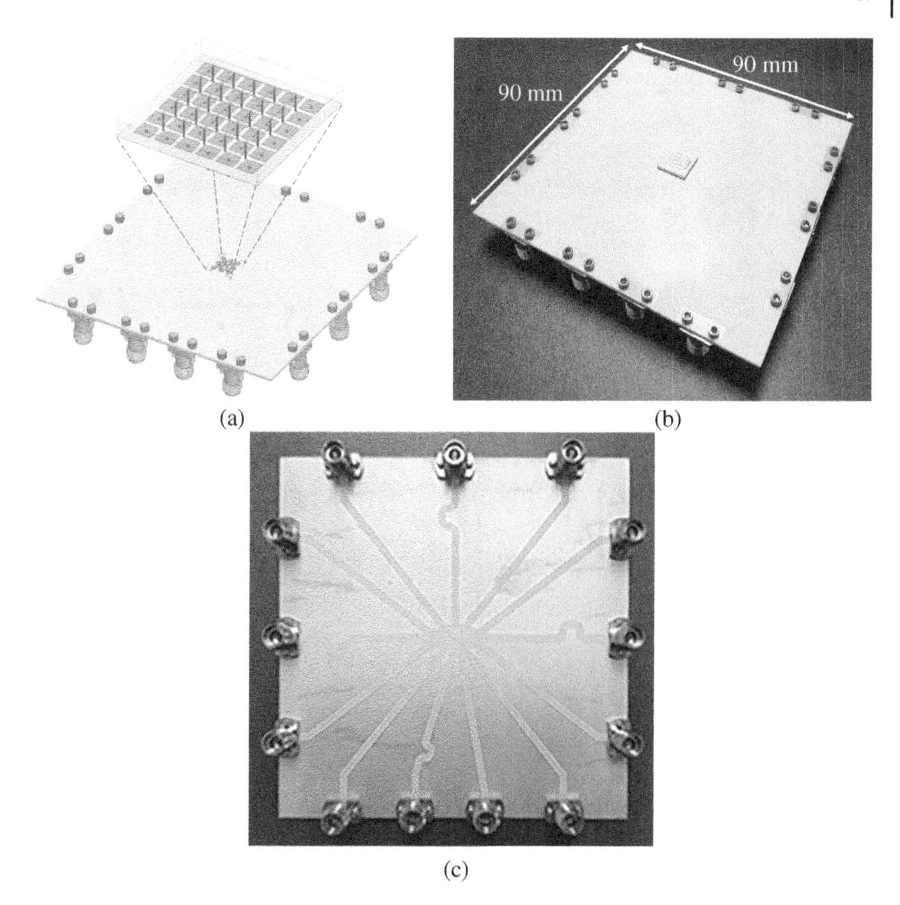

Figure 6.38 Schematic of the proposed 4×4 BCA and fan-out board, (a) conceptual schematic, (b) and (c) fabricated result (3D and bottom view). *Source:* From [35].

deteriorates at the target frequency. To mitigate this problem, a mushroom-type EM band gap (EBG) structure is used to improve the impedance characteristic of the proposed 4×4 BCA [37, 38]. Since the guided wavelength in LTCC is approximately 4.4 mm at 28 GHz, a square-shaped EBG structure with a width of 1.1 mm is added. As shown in Figures 6.38 and 6.39, the antenna is placed on top of the fan-out board using a surface mount technology (SMT) in order to link the proposed 4×4 BCA with a 5G phased-array module. As illustrated in Figure 6.39a, the signal and ground of the BCA are soldered to the signal and

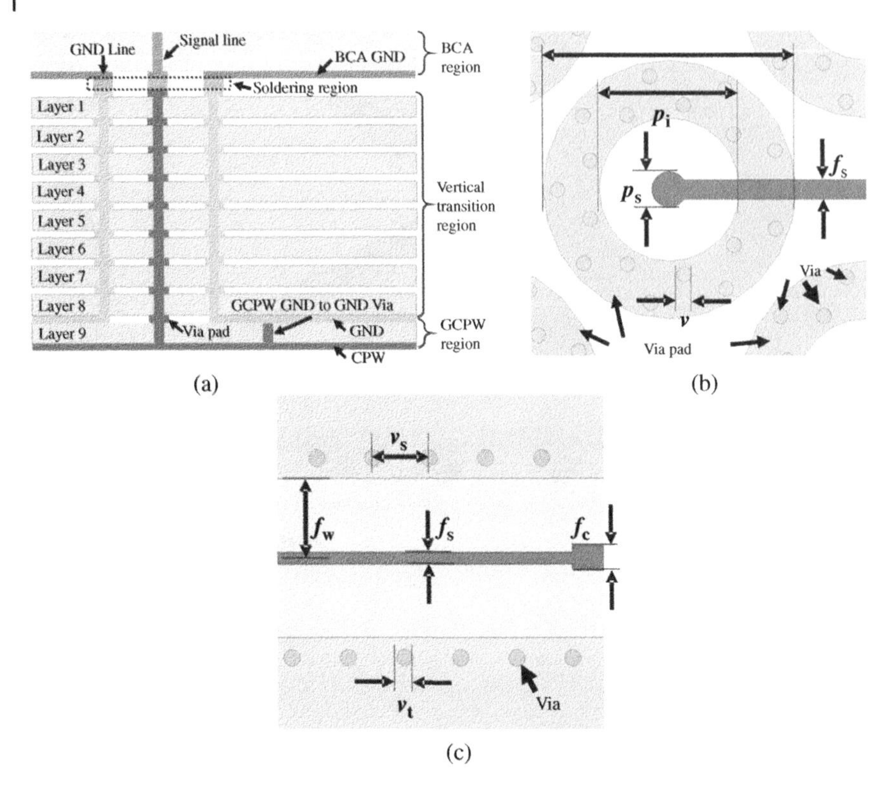

Figure 6.39 Schematic of the proposed fan-out board for 4×4 BCA. (a) Conceptual side view. (b) Bottom view of signal to the vertical transition. (c) GCPW structure for the transmission line; the convex part marked f_c is the part connected to the connector. The optimized dimensions (in mm): $p_i = 1.0, p_o = 1.8, p_s = 0.25, v = 0.05, f_s = 0.14, f_w = 0.97, f_c = 0.3, v_t = 0.2, v_s = 0.7$. *Source:* From [35]. ©2021 IEEE. Reproduced with permission.

ground of the board, respectively. As illustrated in Figures 6.39b,c, connectors are connected to the GCPW structures, and the GCPW structures are connected to the vertical transitions. It is confirmed from the EM simulation that each signal line of the fan-out board features equal phase delay, which alleviates the phase calibration issues.

The *S*-parameter of the fabricated BCA shown in Figure 6.38 is measured using the KEYSIGHT PNA-X, and the far-field radiation pattern is measured using a Ka-band RFIC module and mmWave far-field chamber at POSTECH. The reflection coefficients in the four different ports of the fabricated BCA are illustrated in Figure 6.40. Due to the structural symmetry of BCA, A_{11}, A_{13}, A_{51}, and A_{53} have the same impedance characteristics as A_{mn} denotes the port where the element a_{mn} of the weighing matrix is applied. Similarly, ports are classified as

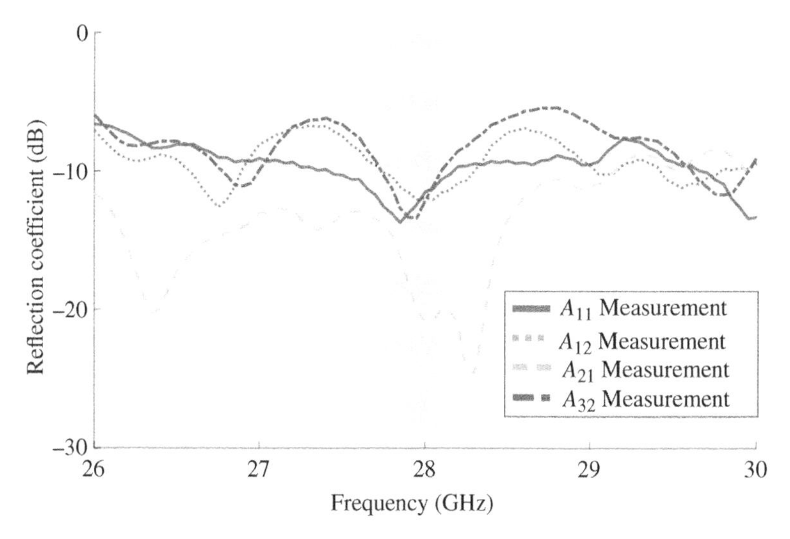

Figure 6.40 Measured reflection coefficients of the 4×4 BCA. A_{mn} denotes the port where the element a_{mn} of the weighing matrix is applied. *Source:* From [35]. ©2021 IEEE. Reproduced with permission.

$(A_{12}, A_{31}, A_{33}, A_{52})$, $(A_{21}, A_{22}, A_{41}, A_{42})$, and A_{32}. As shown in Figure 6.40, the four ports are below −10 dB around 28 GHz. The far-field radiation patterns are measured at 28 GHz using a mmWave RF module and the weighting matrices are depicted in Table 6.2. Table 6.2 shows measured normalized gains, peak gain directions, and weighting matrices in each mode. In Table 6.2, there are slight peak angle errors due to measurement environments, such as antenna under test (AUT) placement and sample fabrication error.

Several representative modes are depicted in Figures 6.41 and 6.42 to ascertain pattern and polarization reconfiguration, respectively. Simulation and measurement results of the 4×4 BCA with the fan-out board are also compared in Figures 6.41 and 6.42 to confirm consistency with intention of the design. In Figure 6.41a, the bidirectional mode exhibits a higher peak gain than the omnidirectional mode. In Figure 6.41b, the fan-beam mode and bidirectional mode feature peak gain and null point in the broadside point, respectively. In Figure 6.41c, the tilted-beam mode can be adjusted in both positive and negative directions. As depicted in Figure 6.42a,b, $E\theta$ of the x-axis polarized high gain mode in the xz plane corresponds to $E\theta$ of the y-axis polarized high gain mode in the yz plane. Likewise, the low gain modes depicted in Figure 6.42c can be achieved with orthogonal polarizations in the yz plane. Simulation and measurement differences occur in the local areas due to measurement environments, such as errors in RF cables and fabrication.

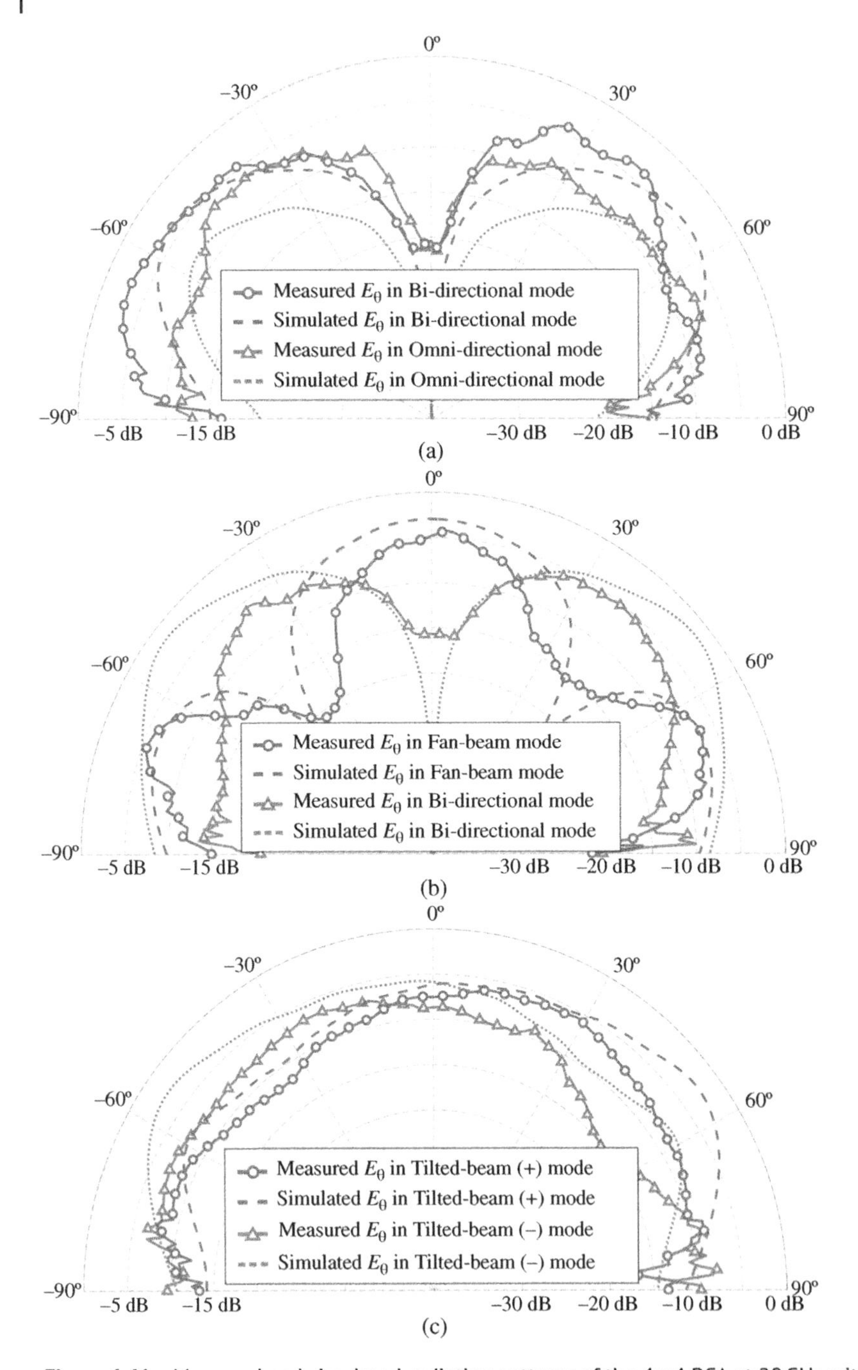

Figure 6.41 Measured and simulated radiation patterns of the 4×4 BCA at 28 GHz with pattern reconfiguration. (a) *x*-axis bidirectional mode and omnidirectional mode in the *xz* plane. (b) $\varphi = 135°$-polarized fan-beam mode and $\varphi = 135°$ bidirectional mode in $\varphi = 135°$ plane. (c) *x*-axis polarized tilted-beam modes in the *xz* plane. *Source:* From [35]. ©2021 IEEE. Reproduced with permission.

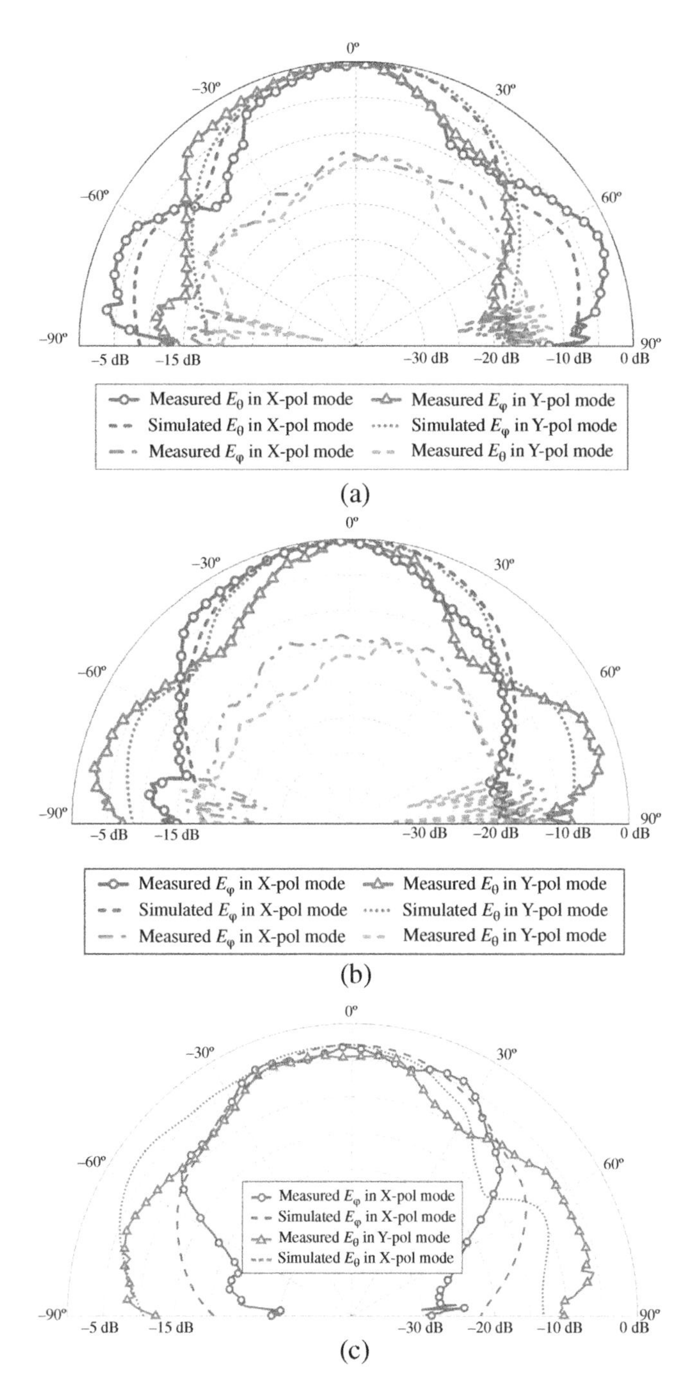

Figure 6.42 Measured and simulated radiation patterns of the 4×4 BCA at 28 GHz with polarization reconfiguration. (a) *x*- and *y*-axes polarized high-gain broadside mode in the *xz* plane. (b) *x*- and *y*-axes polarized high-gain broadside mode in the *yz* plane. (c) *x*- and *y*-axes polarized low-gain broadside mode in the *yz* plane. *Source:* From [35]. ©2021 IEEE. Reproduced with permission.

6.3 Conclusion

In Section 6.1, an extremely thin AiP with an integrated 5G vertically polarized (V-pol) endfire antenna array is proposed at 37–40 GHz for compact UE applications. The proposed AiP features a height profile <0.1 λ_0 which is the smallest height profile in the literature and therefore enables it to be integrated within compact devices with ease. Flip-chip-based packaging of mmWave beamforming AiPs is demonstrated using the V-pol endfire array and optimization methodology of vertical transition. The fabricated module achieves a measured EIRP of 18.2 dBm and a beam scanning range of $\pm 50°$ in azimuth at 28 GHz. The proposed thin AiP technology can provide an apparent guideline for designing miniaturized mmWave antenna systems within a cellular device.

In Section 6.2, a passive pattern- and polarization-reconfigurable BCA concept for mmWave 5G phased-array architectures is proposed. The proposed BCA only requires phase shifters and attenuators to control the current distribution on the antenna structure without using active components inside the antenna structure. The expanded 4×4 BCA can be utilized as a proof-of-concept model featuring theoretically available modes. Due to its two-dimensional symmetry, various types of polarization can be obtained depending on the port activation configuration. In the demonstration using the mmWave 5G beamforming reference board, the exemplified fabricated antenna exhibits 15 distinct far-field radiation patterns with reflection coefficients of −10 dB at 28 GHz. It should be noted that the existence of theoretical infinite modes requires an excessive resource allocation for mode selection in the baseband unit at present. Naturally, future works will focus on the development of optimization methodology based on machine learning algorithms.

References

1 W. Roh, J. Seol, J. Park, B. Lee, J. Lee, Y. Kim, J. Cho, K. Cheun and F. Aryanfar, "Millimeter-wave beamforming as an enabling technology for 5G cellular communications: theoretical feasibility and prototype results," *IEEE Commun. Magn.*, vol. 52, no. 2, pp. 106–113, Feb. 2014.

2 Samsung, "Analysis of mmWave performance: feasibility of mobile cellular communications at millimeter wave frequency," July 2017. http://www.samsung.com/global/business-images/insights/2017/Analysis-of-mmWave-Performance-0.pdf.

3 Qualcomm, Nokia, "Making 5G a reality: addressing the strong mobile broadband demand in 2019 and beyond," Sept. 2017. http://www.qualcomm.com/documents/making-5g-reality-addressing-strong-mobile-broadband-demand-2019-beyond.

4 V. Raghavan, A. Partyka, A. Sampath, S. Subramanian, O. H. Koymen, K. Ravid, J. Cezanne, K. Mukkavilli, and J. Li, "Millimeter-wave MIMO prototype: measurements and experimental results," *IEEE Commun. Magn.*, vol. 56, no. 1, Jan. 2018, pp. 202–209.

5 V. Raghavan, M. Chi, M. A. Tassoudji, O. H. Koymen, and J. Li, "Antenna placement and performance tradeoffs with hand blockage in millimeter wave systems," *IEEE Trans. Commun.*, vol. 67, no. 4, Apr. 2019, pp. 3082–3096.

6 J. Park, H. Seong, Y. N. Whang, and W. Hong, "Energy-efficient 5G phased arrays incorporating vertically polarized endfire planar folded slot antenna for mmWave mobile terminals," *IEEE Trans. Antennas Propag.*, vol. 68, no. 1, pp. 230–241, Jan. 2020.

7 W. Hong, "Solving the 5G mobile antenna puzzle: assessing future directions for the 5G mobile antenna paradigm shift," *IEEE Microw. Magn.*, vol. 18, no. 7, pp. 86–102, Nov.–Dec. 2017.

8 W. Hong, K. H. Baek, and S. Ko, "Millimeter-wave antennas for smartphones: overview and experimental demonstration," *IEEE Trans. Antennas Propag.*, vol. 65, no. 12, pp. 6250–6261, Dec. 2017.

9 T. Lin, T. Chiu, and D. Chang, "Design of dual-band millimeter-wave antenna-in-package using flip-chip assembly," *IEEE Trans. Compon. Packag. Manuf. Technol.*, vol. 4, no. 3, pp. 385–391, March 2014.

10 A. Vahdati, A. Lamminen, M. Varonen, J. Säily, M. Lahti, K. Kautio, M. Lahdes, D. Parveg, D. Karaca and K. A. I. Halonen, "90 GHz CMOS phased-array transmitter integrated on LTCC," *IEEE Trans. Antennas Propag.*, vol. 65, no. 12, pp. 6363–6371, Dec. 2017.

11 C. Mao, S. Gao, and Y. Wang, "Broadband high-gain beam-scanning antenna array for millimeter-wave applications," *IEEE Trans. Antennas Propag.*, vol. 65, no. 9, pp. 4864–4868, Sept. 2017.

12 B. Yu, K. Yang, C. Sim, and G. Yang, "A novel 28 GHz beam steering array for 5G mobile device with metallic casing application," *IEEE Trans. Antennas Propag.*, vol. 66, no. 1, pp. 462–466, Jan. 2018.

13 S. Zhang, X. Chen, I. Syrytsin, and G. F. Pedersen, "A planar switchable 3-D-coverage phased array antenna and its user effects for 28-GHz mobile terminal applications," *IEEE Trans. Antennas Propag.*, vol. 65, no. 12, pp. 6413–6421, Dec. 2017.

14 T. Zhang, L. Li, M. Xie, H. Xia, X. Ma, and T. J. Cui, "Low-cost aperture-coupled 60-GHz-phased array antenna package with compact matching network," *IEEE Trans. Antennas Propag.*, vol. 65, no. 12, pp. 6355–6362, Dec. 2017.

15 S. J. Park, D. H. Shin, and S. O. Park, "Low side-lobe substrate-integrated-waveguide antenna Array using broadband unequal feeding network for millimeter-wave handset device," *IEEE Trans. Antennas Propag.*, vol. 64, no. 3, pp. 923–932, March 2016.

16 Q. Zhu, K. B. Ng, C. H. Chan, and K. Luk, "Substrate-integrated-waveguide-fed Array antenna covering 57–71 GHz band for 5G applications," *IEEE Trans. Antennas Propag.*, vol. 65, no. 12, pp. 6298–6306, Dec. 2017.

17 D. Liu, X. Gu, C. W. Baks, and A. Valdes-Garcia, "Antenna-in-package design considerations for Ka-band 5G communication applications," *IEEE Trans. Antennas Propag.*, vol. 65, no. 12, pp. 6372–6379, Dec. 2017.

18 R. Alhalabi and G. Rebeiz, "High-efficiency angled-dipole antennas for millimeter-wave phased array applications" *IEEE Trans. Antennas Propag.*, vol. 56, no. 10, pp. 3136–3142, Oct. 2008.

19 Q. Chu, X. Li, and M. Ye, "High-gain printed log-periodic dipole array antenna with parasitic cell for 5G communication," *IEEE Trans. Antennas Propag.*, vol. 65, no. 12, pp. 6338–6344, Dec. 2017.

20 S.-S. Hsu, K.-C. Wei, C.-Y. Hsu, and H. Ru-Chuang, "A 60-GHz millimeter-wave CPW-fed Yagi antenna fabricated by using 0.18-μm CMOS technology," *IEEE Electron Device Lett.*, vol. 29, no. 6, pp. 625–627, Jun. 2008.

21 F. Taringou, D. Dousset, J. Bornemann, and K. Wu, "Broadband CPW feed for millimeter-wave SIW-based antipodal tapered slot antenna," *IEEE Trans. Antennas Propag.*, vol. 61, no. 4, pp. 1756–62, Apr. 2013.

22 W. Hong, K. H. Baek, Y. Lee, Y. Kim, and S. T. Ko, "Study and prototyping of practically large-scale mmWave antenna systems for 5G cellular devices," *IEEE Commun. Mag.*, vol. 52, no. 9, pp. 63–69, Sept. 2014.

23 Y. Li and K.-M. Luk, "A multibeam end-fire magnetoelectric dipole antenna array for millimeter-wave applications," *IEEE Trans. Antennas Propag.*, vol. 64, no. 7, pp. 2894–2904, Jul. 2016.

24 R. Suga, H. Nakano, Y. Hirachi, J. Hirokawa, and M. Ando, "Cost effective 60-GHz antenna package with end-fire radiation for wireless file-transfer system," *IEEE Trans. Microwave Theory Tech.*, vol. 58, no. 12, pp. 3989–3995, Dec. 2010.

25 W. El-Halwagy, R. Mirzavand, J. Melzer, M. Hossain, and P. Mousavi, "Investigation of wideband substrate-integrated vertically-polarized electric dipole antenna and arrays for mm-wave 5G mobile devices," *IEEE Access*, vol. 6, pp. 2145–2157, 2018.

26 Y. Hsu, T. Huang, H. Lin, and Y. Lin, "Dual-polarized Quasi Yagi–Uda antennas with endfire radiation for millimeter-wave MIMO terminals," *IEEE Trans. Antennas Propag.*, vol. 65, no. 12, pp. 6282–6289, Dec. 2017.

27 W. Heinrich, "The flip-chip approach for millimeter wave packaging," *IEEE Microwave Magn.*, vol. 6, no. 3, pp. 36–45, Sep. 2005.

28 R. Valois, D. Baillargeat, S. Verdeyme, M. Lahti, and T. Jaakola, "High performances of shielded LTCC vertical transitions from dc up to 50 GHz," *IEEE Trans. Microwave Theory Tech.*, vol. 53, no. 6, pp. 2026–2032, Jun. 2005.

29 T. Kangasvieri, M. Komulainen, H. Jantunen, and J. Vähäkangas, "High performance vertical interconnections for millimeter-wave multichip modules," *in 35th Eur. Microw. Conf.*, 2005, pp. 169–172.

30 A. Panther, C. Glaser, M. G. Stubbs, and J. S. Wight, "Vertical transitions in low temperature co-fired ceramics for LMDS applications," *IEEE MTT-S Dig.*, vol. 3, pp. 1907–1910, 2001.

31 D. G. Kam and J. Kim, "40-Gb/s package design using wire-bonded plastic ball grid array," *IEEE Trans. Adv. Packag.*, vol. 31, no. 2, pp. 258–266, May 2008.

32 S. Wu, X. Chang, C. Schuster, X. Gu, and J. Fan, "Eliminating Via-Plane Coupling Using Ground Vias for High-Speed Signal Transitions," in Proc. IEEE Conf. Electr Perform. Electron. Packag., San Jose, CA, USA, 2008, pp. 247–250.

33 M. A. Towfiq, I. Bahceci, S. Blanch, J. Romeu, L. Jofre, and B. A. Cetiner, "A reconfigurable antenna with beam steering and beamwidth variability for wireless communications," *IEEE Trans. Antennas Propag.*, vol. 66, no. 10, pp. 5052–5063, Oct. 2018.

34 J. Park, D. Choi, and W. Hong, "A software-defined mmWave radio architecture comprised of modular, controllable pixels to attain near-infinite pattern, polarization, and beam steering angles IMS," *in Proc. IEEE MTT-S Int. Microw. Symp. Dig.*, US, Jun. 2020.

35 J. Park, M. Choo, S. Jung, D. Choi, J. Choi, and W. Hong, "Software-programmable directivity, beamsteering, and polarization reconfigurable block cell antenna concept compatible with millimeter-wave 5G phased-array architectures," *IEEE Trans. Antennas Propag.*, vol. 69, no. 1, pp. 146–154, Jan. 2021.

36 C. A. Balanis, Antenna Theory: Analysis and Design. Hoboken, NJ, USA: Wiley, 2016, pp. 138–139.

37 D. Sievenpiper, L. Zhang, R. F. J. Broas, N. G. Alexopolous, and E. Yablonovitch, "High-impedance electromagnetic surfaces with a forbidden frequency band," *IEEE Trans. Microwave Theory Tech.*, vol. 47, no. 11, pp. 2059–2074, Nov. 1999.

38 P. Deo, A. Mehta, D. Mirshekar-Syahkal, P. J. Massey, and H. Nakano, "Thickness reduction and performance enhancement of steerable square loop antenna using hybrid high impedance surface," *IEEE Trans. Antennas Propag.*, vol. 58, no. 5, pp. 1477–1485, May 2010.

7

Multi-Physical Approach for Millimeter-Wave 5G Antenna-in-Package

7.1 Background and Current Challenges

The discrete packaging approach of the antenna and active RF components in the Sub-6 GHz antenna system provided the advantage of effectively dissipating the generated heat within the device. However, in the mmWave antenna modules, the increased number of RF channels and the high packaging density of RF/analog front-end circuitry not only accelerate the generation of trapped hot spots in the device but also make the integration of cooling systems challenging. Especially, in the case of state-of-the-art power amplifiers (PAs) for mmWave applications, the power added efficiency (PAE) is typically below 20% [1, 2], implying that the majority of the DC power will be converted into convection heat. This becomes significant for large-scale phased arrays containing a massive number of RF channels can generate a very large amount of heat from the high dissipated power of the entire system. It is recognized that the performance of the entire system degrades under high temperatures and can ultimately lead to malfunctions and mechanical damage, such as warpage or delamination within the package [3]. In addition, the temperature non-uniformity can cause the amplitude and phase variations of each antenna element excitation, which significantly effects the radiation pattern of the highly integrated phased-array system [4, 5]. Therefore, thermal management should be thoroughly addressed during the design process to guarantee the reliability of phased array systems that operate under various environmental conditions.

Considering that the heat generated from the conventional phased array configuration is concentrated in the center of the array [6], the excess heat should be transferred to a heat exchanger or to ambient air. Active cooling systems have

Microwave and Millimeter-Wave Antenna Design for 5G Smartphone Applications,
First Edition. Wonbin Hong and Chow-Yen-Desmond Sim.

Published 2023 by John Wiley & Sons, Inc.

received much attention due to their effective heat removal capability [4, 7]. In [4], the microchannel heat sink with a ladder-shaped inlet header is proposed and embedded in the highly integrated phased-array antenna module for cooling the internal heat and reducing temperature non-uniformity, as illustrated in Figure 7.1. To enhance the heat dissipation capacity of the microchannel heat sinks, a cylinder-shaped copper (Figure 7.1) is used to connect each amplifier and microchannel. Figure 7.2. shows a microchannel heat sink consisting of a ladder-shaped inlet head and a rectangular outlet head, eight small microchannels, an inlet, and an outlet. The heat generated by the amplifier is transferred to the coolant through a copper pillar, and the flow of coolant along the small micro-channel removes the heat from the phased array antenna [4]. Forced air cooling and forced liquid cooling are investigated for the Ka-band active phased array [7]. However, active cooling systems are not suitable for millimeter-wave mobile applications due to energy efficiency issues, cost issues, and increased complexity of the entire system.

Passive thermal management can be a more cost-effective and energy-efficient solution. The dissipated power is efficiently removed by a heat spreader connecting the transistor to the antenna [8, 9]. In [8], the multiphysical simulation demonstrates that most of the heat (~98%) is conducted down through the silicon and thermal interface material (TIM) toward the heat sink, as illustrated in Figure 7.3.

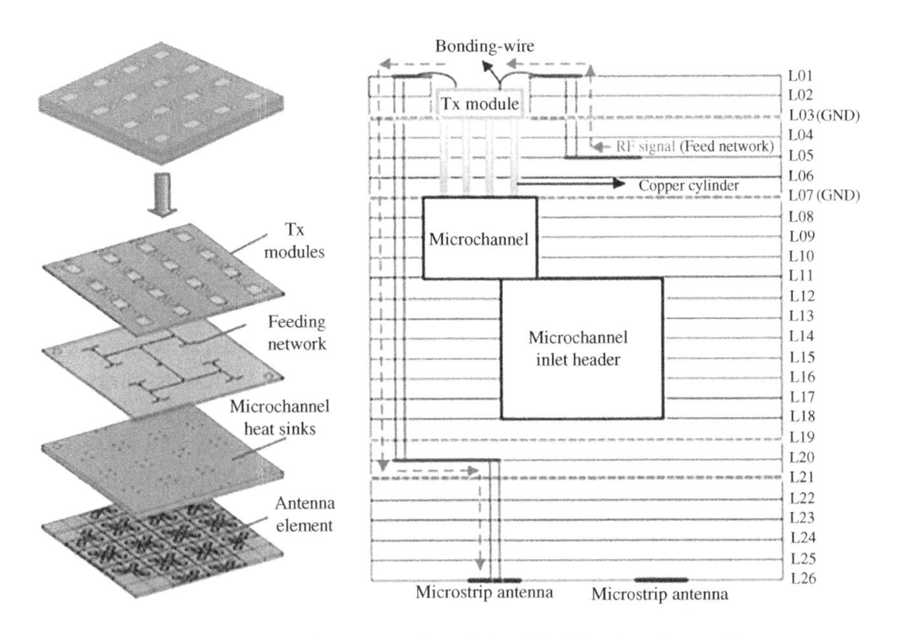

Figure 7.1 Exploded view and cross-section of the AiP with microchannel heat sinks. *Source:* From [4]. ©2019 IEEE. Reproduced with permission.

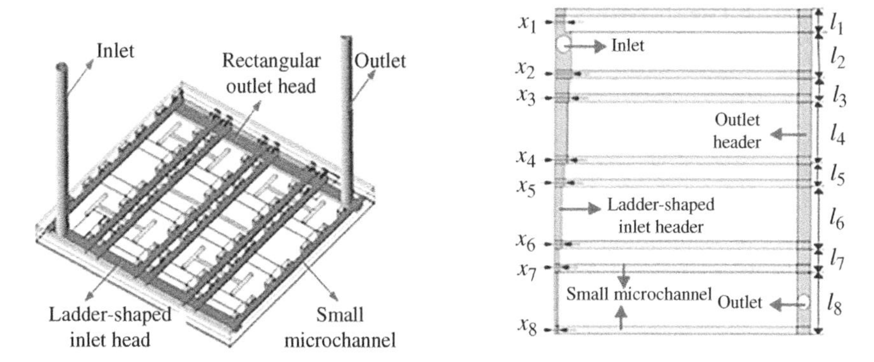

Figure 7.2 Detailed structure of the microchannel heat sinks with a ladder-shaped inlet head. *Source:* From [4]. ©2019 IEEE. Reproduced with permission.

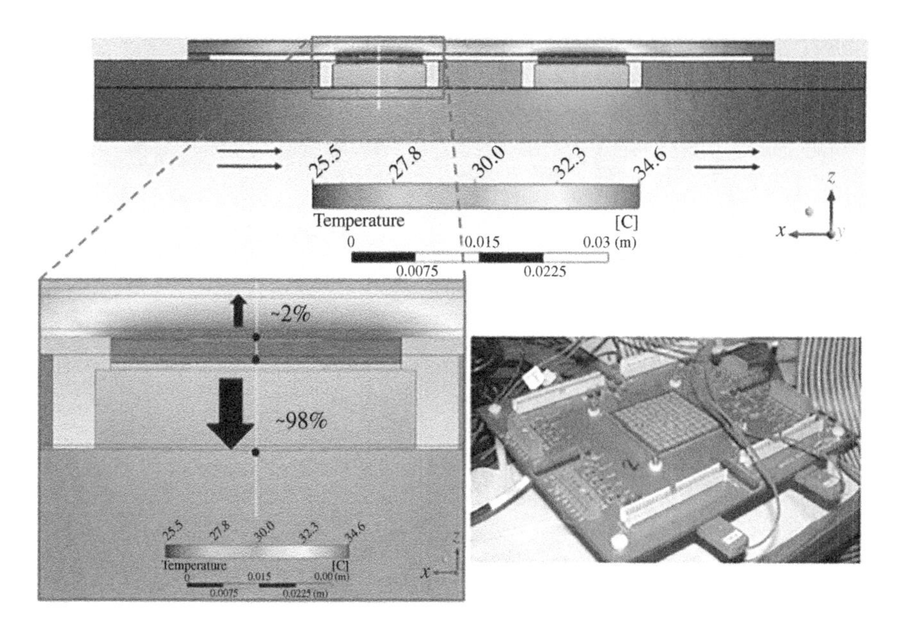

Figure 7.3 Configuration of multiphysical simulation and measurement for 5G mmWave applications. *Source:* From [8]. ©2019 IEEE. Reproduced with permission.

This result is also consistent with thermal experiments using the programmable thermal sensor chips on the thermomechanical test module. The 3-D fractal heatsink antenna is proposed to alleviate the thermal resistance in [10]. However, a relatively bulky heatsink configuration is inappropriate to be applied for compact mmWave phased arrays with flexible beam-scanning capability.

7.2 Heat Dissipation Strategies

7.2.1 Metal Stamped Antenna-in-Package Overview

This sub-section proposes a compact metal stamped antenna-in-package (AiP) structure that can be directly integrated with the transceiver module and expanded in the form of a modular configuration [11, 12]. The proposed AiP not only maximizes the heat removal capability at the passive level but also features good radiation performances. Figure 7.4a presents the conceptual diagram of the proposed AiP concept, which can be classified into several parts named the metal stamped

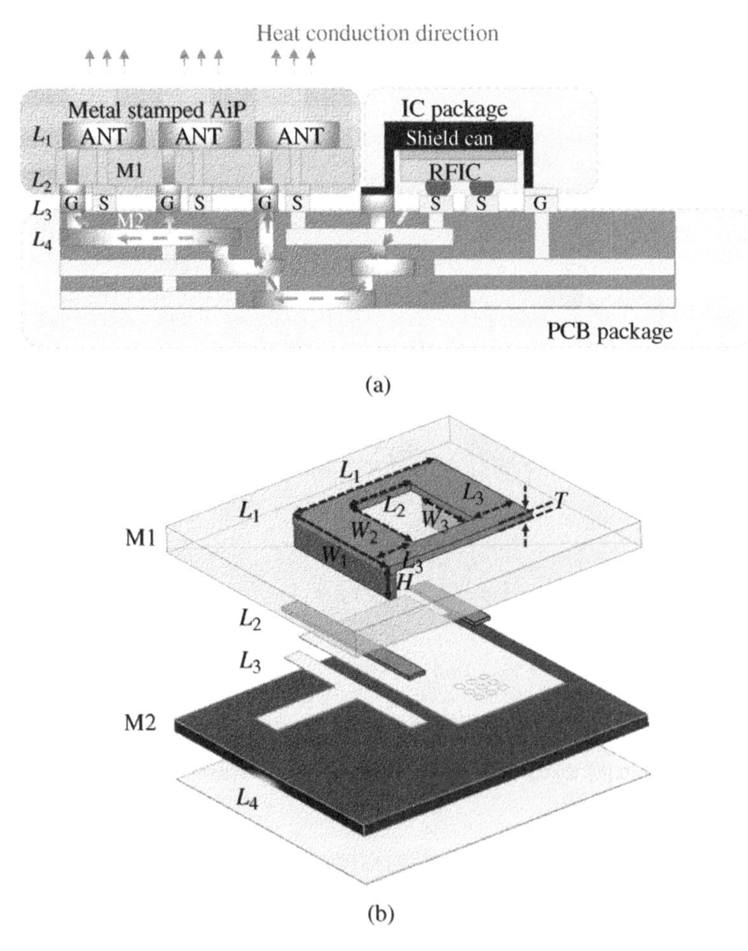

Figure 7.4 (a) The conceptual diagram of the proposed metal stamped AiP. (b) The detailed view of the antenna element. *Source:* From [12]. ©2020 IEEE. Reproduced with permission.

AiP, the IC package, and the PCB package. The metal stamped AiP can be directly connected to the RFIC mounted on the PCB using the vertical transition structure and soldering process. The metal stamped AiP dually functions as a phased-array antenna module and a heat spreader, which transfers heat from the IC to ambient air. The entire lamination is exemplified using Rogers RT 5880 dielectric substrate, which features a relative permittivity of $\varepsilon_r = 2.2$, loss tangent of $\tan \delta = 0.0009$, and single-layer thickness of $t = 0.254$ mm. The diameters of the vias and capture pads are configured to be 200 and 300 μm, respectively, throughout this chapter.

In the metal stamped AiP, the planar inverted-F antenna (PIFA) antenna is selected as a basic topology for the following reasons: (i) PIFA antennas can contain a large aperture and a wide feed structure, which maximizes the contact region between the antenna and the heat source. (ii) The shorting pin of the radiating structure can be electrically connected with the ground plane of the package and the shield can. This efficiently routes the internal convection heat to free space. (iii) PIFA antennas typically consist of orthogonal joints, which can be easily realized using metal stamping. The stainless steel ($\sigma = 1.4 \times 10^6$ [S/m]) is selected as a raw material for use in the stamping process primarily due to its corrosion resistance while maintaining formability and weldability. After being molded from the metal stamping, a plurality of antennas are deposited on the cooling medium (dielectric substrate) with relative permittivity of $\varepsilon_r = 3.5$, loss tangent of $\tan \delta = 0.001$ through thermal bonding. Ground signal pads, functioning as interconnection with the metal stamped antennas are situated at L_3. The PCB board and metal stamped AiP are individually fabricated and electrically connected using a soldering process.

7.2.2 Proof-of-Concept Model

The proof-of-concept (POC) model is designed and fabricated using standard PCB and press technology. As illustrated in Figure 7.4b, the radiating elements (L_1 to L_2) molded from the stamping process are deposited on the dielectric substrate (M_1) using thermal bonding. Ground and signal pads (L_3 to L_4) are printed on the dielectric substrate (M_2). The antenna element includes a radiating element with a volume of $L_1 \times W_1 \times H$ and a rectangular ground plane of PCB board with dimensions of 80 mm × 43 mm. Since the characteristics of PIFA antennas are affected by the dimensions of the ground and its location on the ground plane, the entire ground plane of the PCB board should be considered when designing the antenna elements. The feeding wall and the shorting wall are directly connected to the feedline pads and the ground pads of the PCB board, respectively, using the soldering process. It is well known that the resonance frequency of a conventional PIFA antenna can be calculated using the equation in [13]. However, this equation is an approximation and additional parameters, which contribute to the operating frequency of the PIFA antenna, need to be considered. The effect of the height (H) and width (W_3) of the shorting wall is studied in Figure 7.5a,b respectively. The 3-D electromagnetic (EM) simulator, ANSYS HFSS, is used to simulate the input impedance of the proposed antenna element. It is confirmed that the

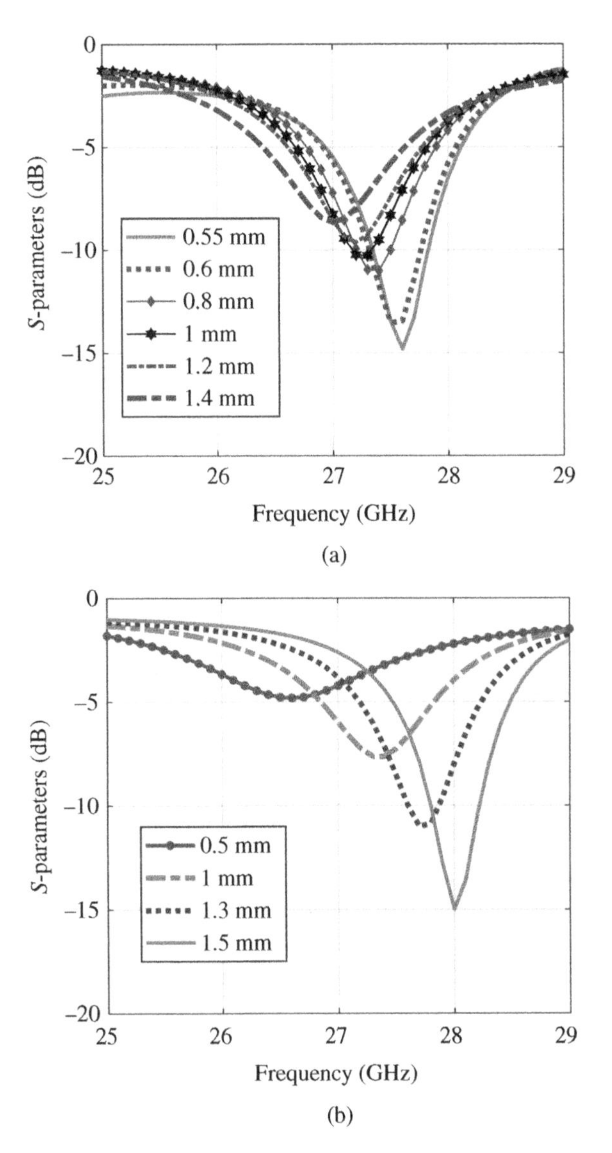

Figure 7.5 Simulated reflection coefficient of the radiating element: (a) As a function of the height of shorting wall (H). (b) As a function of the width of shorting wall (W_3). *Source:* From [11]. ©2019 IEEE. Reproduced with permission.

Table 7.1 Antenna parameters.

Parameter	Value (mm)	Parameter	Value (mm)
W_1	3	L_1	3.5
W_2	2	L_2	1.5
W_3	1.5	L_3	1
H	0.55	T	0.15

Source: From [12]. ©2020 IEEE. Reproduced with permission.

Table 7.2 Metal stamped AiP parameters.

Parameter	Value (mm)	Parameter	Value (mm)	Parameter	Value (mm)
O_A	2	D_T	49.7	D_C	2.6
S_{ANT}	5.6	D_L	7	D_S	4.5
L_1	3.5	D_A	5.15		
W_1	3	D_B	3.75		

Source: From [11]. ©2019 IEEE. Reproduced with permission.

width and height of the shorting wall affect the electrical length of the antenna. The optimized parameters are shown in Table 7.1.

The final POC model consists of the PCB package and the metal stamped AiP with eight antenna elements for emulating antenna array properties, as illustrated in Figure 7.6. The overall size of the POC model is 80 mm × 43 mm. The ground-signal-ground (GSG) pad for antenna evaluation is configured on the top side of the PCB package. Each antenna element of the metal stamped AiP is arranged at a distance (d_{ANT}) of 5.6 mm which is 0.53 λ_0 at 28.5 GHz. The metal stamped AiP is adhesively mounted on the PCB package using solder paste. The dummy shield can structure is used to align the two packages during the soldering process. The signal pad of each antenna is connected to the 1 : 8 power divider. The separation (S_{ANT}) between the elements is adjusted to 5.6 mm which is 0.53 λ_0 at 28.5 GHz, as shown in Figure 7.4a. The extracted S_{n1} for n = 2–9 of the planar 1×8 power divider is approximately −10.56 dB at 28.5 GHz which translates to approximately 0.27 dB/mm. The detailed dimensions are demonstrated in Table 7.2.

7.3 Multiphysical Analysis

7.3.1 Antenna Package Characterization

Figure 7.7 illustrates a photograph of the measurement setup for the radiation pattern of the fabricated metal stamped AiP POC. The metal stamped structures consisting of the antennas and the shield can are soldered to the PCB board. The

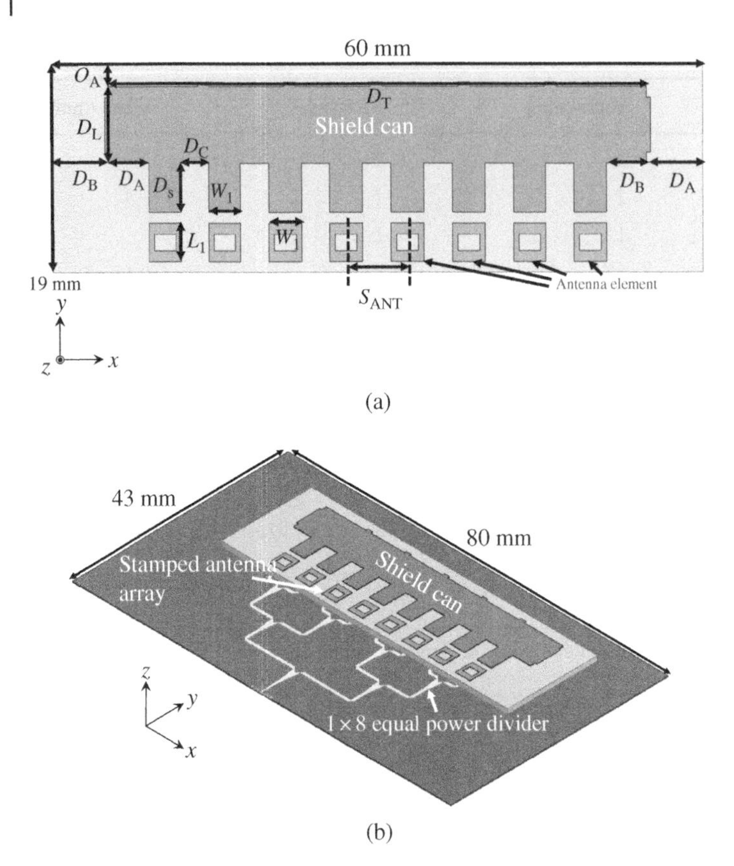

Figure 7.6 Illustration of the metal stamped AiP POC: (a) The detailed view of the antenna array. (b) 3-D view. *Source:* From [11]. ©2019 IEEE. Reproduced with permission.

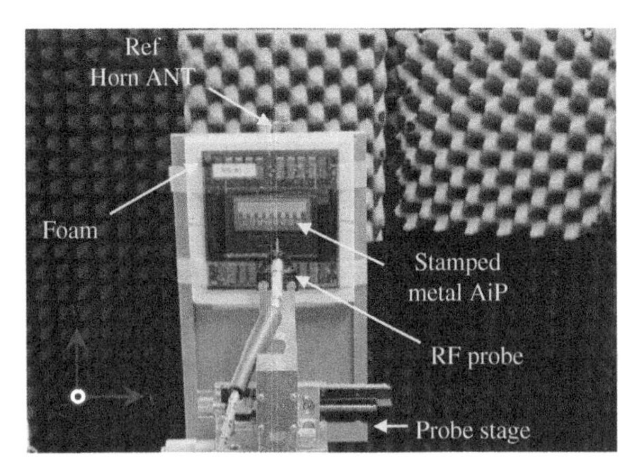

Figure 7.7 Photograph of the measurement setup for radiation patterns of the fabricated metal stamped AiP. *Source:* From [11]. ©2019 IEEE. Reproduced with permission.

S-parameters are measured using an N5247A KEYSIGHT PNA (10 MHz–67 GHz) at POSTECH. The GSG wafer probe with 350 μm pitch tips is employed to establish contact with the GSG pad. The ANSYS HFSS is used to extract the input impedance and the far-field properties of the metal stamped AiP POC. The input reflection coefficient |S11| of the metal stamped AiP POC is illustrated in Figure 7.8 and is ascertained to respectively feature 400 MHz (28.62–29.02 GHz) simulated and 700 MHz (28.27–28.97 GHz) measured impedance bandwidths. The discrepancy between the measurement and simulation results is due to fabrication errors in the stamping process, the rough surface part of the molded metal in the soldering region and deviation of the substrate electrical characteristics at the frequency of interest.

As illustrated in Figure 7.9, the far-field radiation pattern of the fabricated metal stamped AiP POC is measured using POSTECH probe-based anechoic chamber in both the *xz* plane and *yz* plane at 28.5 GHz. The measurement range of the elevation angle is limited to the range from −90° to 90° by clogging caused by surrounding metal objects and components, such as wafer probes, chucks, and microscopes. It can be seen in Figure 7.9a that the main lobe is tilted by approximately 30°, which is caused by the position and orientation of the PIFA antenna on a fixed-size ground plane [14]. It is confirmed in Figure 7.9b that the beam is synthesized by combining the 8-element linearly polarized antennas. The simulated and measured results feature a peak realized gain of 16.47 and 14.09 dBi, respectively. The gain deviation between the measured and simulated results is further investigated in the next sub-chapter 7.3.2.

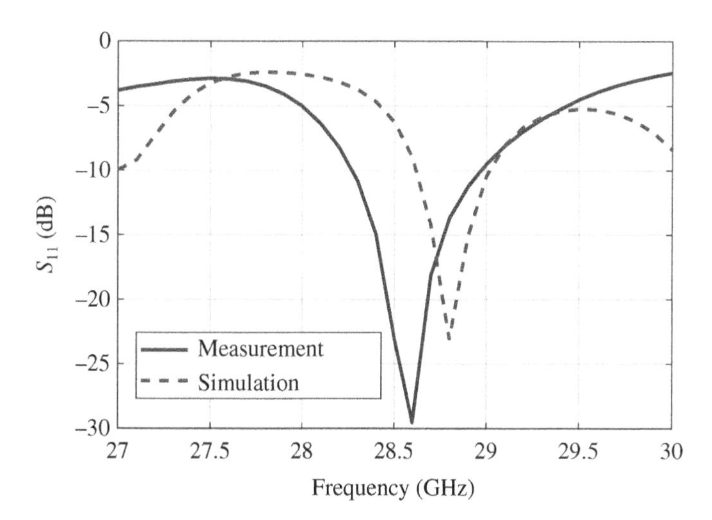

Figure 7.8 Measured and simulated input reflection coefficient of the metal stamped AiP POC. *Source:* From [12]. ©2020 IEEE. Reproduced with permission.

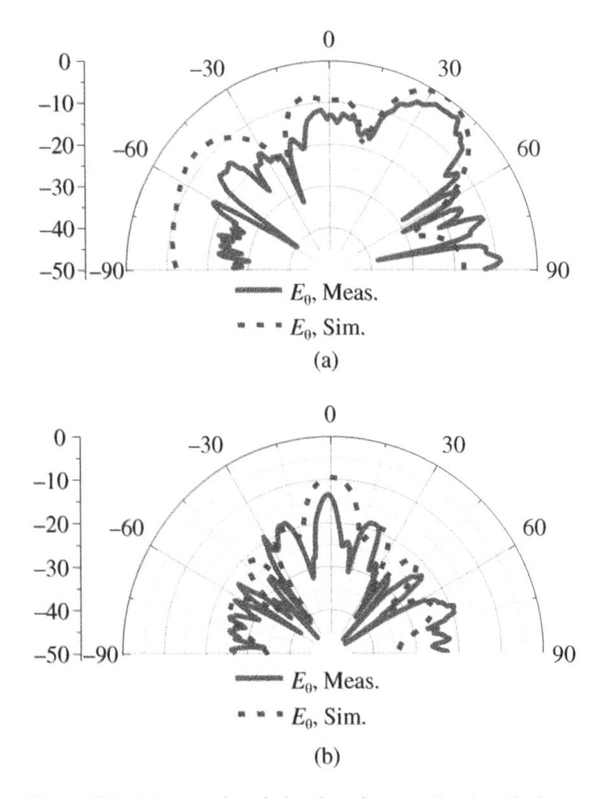

Figure 7.9 Measured and simulated normalized radiation patterns for the metal stamped AiP POC. (a) E_θ in the *yz* plane. (b) E_θ in the *xz* plane. *Source:* From [12]. ©2020 IEEE. Reproduced with permission.

7.3.2 Electrical Stability of the Fabrication Process

The antenna gain and the input impedance of the proposed AiP are significantly influenced by the misalignment between the PCB package and the metal stamped AiP during the soldering process, the machining error during the metal stamping process, and the rough surface of conductors. First, the effect of the misalignment is investigated. While $mis_{y\text{-axis}}$ is set to be 0 mm in an ideal condition, the misalignment (Figure 7.10a) exists in the fabrication process. It is confirmed in Figure 7.11a that the gain of the antenna deteriorates when $mis_{y\text{-axis}}$ gets larger. Second, the machining error during the stamping process is also one of the factors affecting the antenna properties. The parameter cut_{error} is depicted in Figure 7.10b. The effect of the machining error (cut_{error}) is studied in Figure 7.11b. The simulation results demonstrate that the gain of the antenna deteriorates when cut_{error} gets larger.

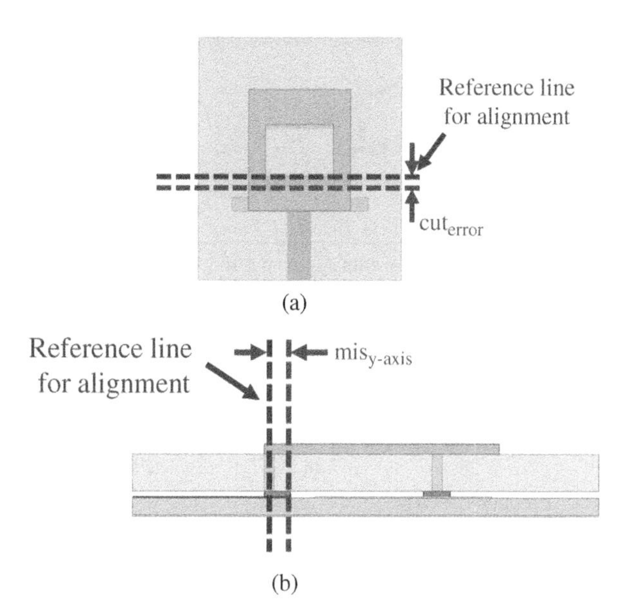

Figure 7.10 Geometry of the metal stamped AiP: (a) The top view after stamping. (b) The side view after soldering. *Source:* From [12]. ©2020 IEEE. Reproduced with permission.

Third, the effect of the rough surface on the antenna properties is simulated with the use of the average value of the rough surface (Ra). The range of Ra after the heat-treated stainless-steel finish is typically between 3.5 and 7.5 μm [15]. The Huray model [16] is used to simulate the effect of the rough surface on the antenna properties. Figure 7.12 demonstrates losses due to surface roughness by a conventional stainless metal surface finish. Therefore, the unavoidable loss caused by the fabrication error, conductive joints, and conductor surface roughness should be included in the design stage using the efficient numerical simulation methodology for accurate prediction of the antenna properties.

7.3.3 Thermal-Mechanical Analysis

To demonstrate the heat-dissipating capability of the proposed metal stamped AiP, a conventional 1×8 patch antenna array is designed at the center frequency of 28 GHz for the reference model, as illustrated in Figure 7.13a. For exact comparison, a conventional patch antenna array model featuring identical element spacing and stack-up to the proposed stamped AiP is designed. The RFICs are mounted on the bottom faces of each structure and electrically connected to the feed lines of each antenna element using vertically interconnected structures (vias). The heat emission generated from the RFIC is strongly correlated to the

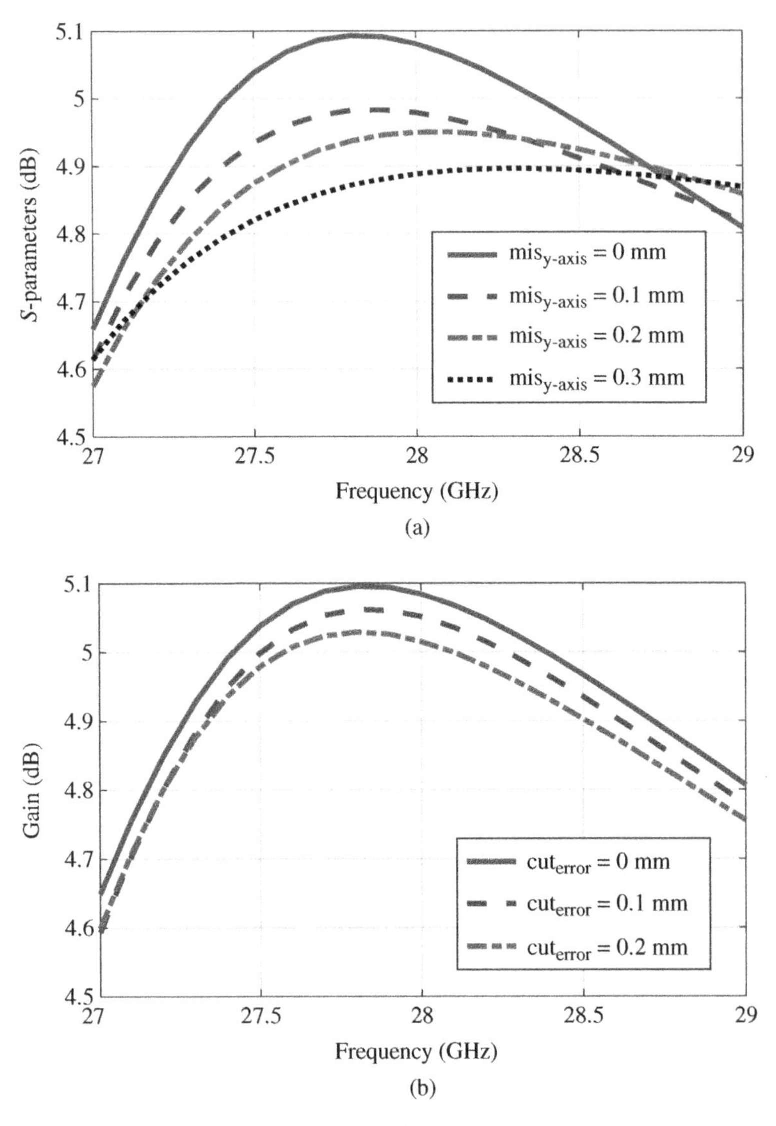

Figure 7.11 Simulated realized gain of the proposed metal stamped antenna element (a) as a function of $mis_{y\text{-axis}}$. (b) as a function of cut_{error}. *Source:* From [12]. ©2020 IEEE. Reproduced with permission.

equivalent isotropically radiated power (EIRP) and configured to be 20 dBm per each RF channel in this research [17]. The ambient temperature and the heat transfer coefficient of all materials in contact with air are set to 25 °C and 10 W/m^{2}K, respectively [18]. The temperature distributions (Figure 7.13) of the two structures

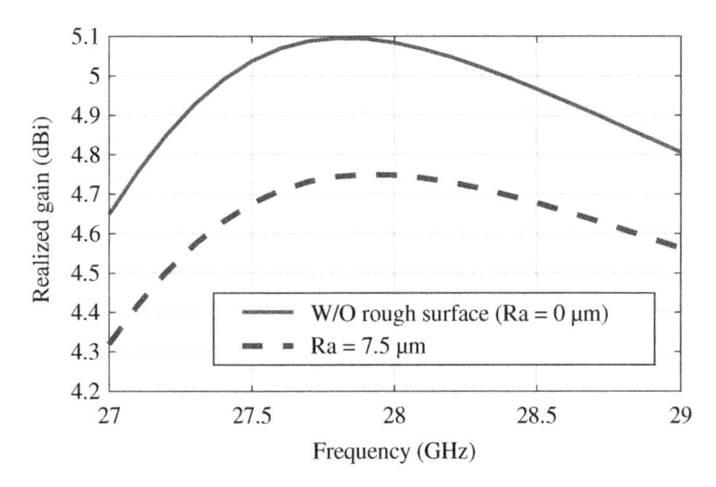

Figure 7.12 Simulated realized gain of the proposed metal stamped antenna element including the surface roughness model. *Source:* From [12]. ©2020 IEEE. Reproduced with permission.

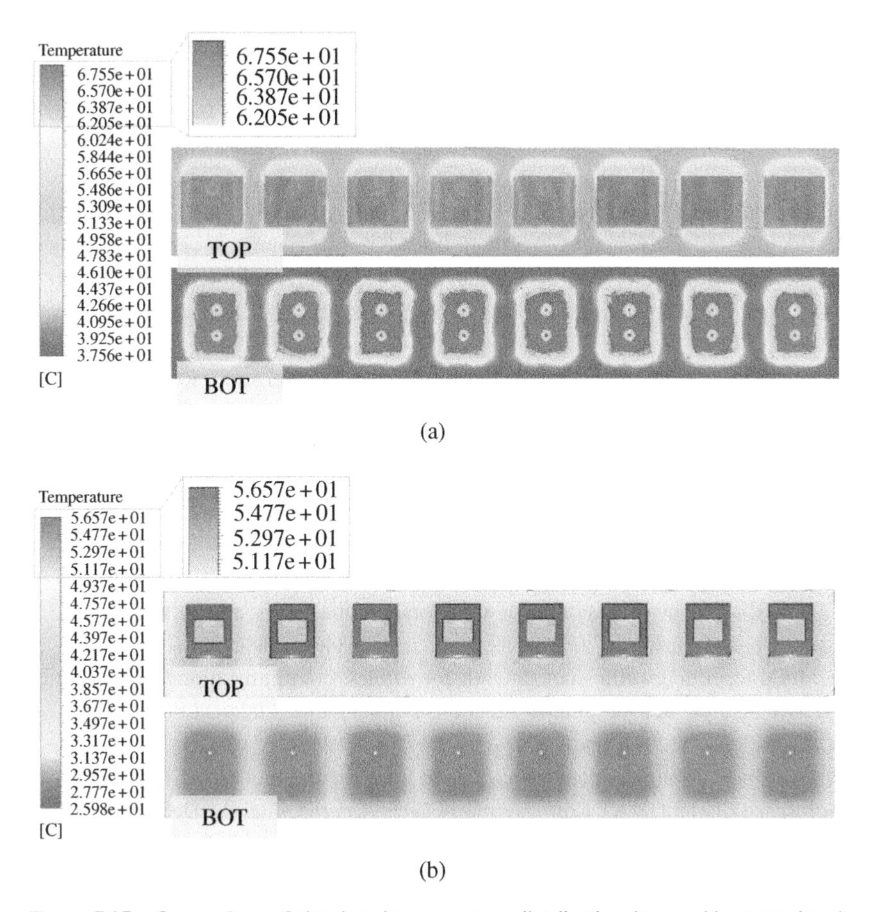

Figure 7.13 Comparison of simulated temperature distribution (top and bottom views). (a) The conventional patch antenna array. (b) The proposed stamped AiP.

are simulated and compared using the computational fluid dynamics (CFD) software, ANSYS Fluent. In the conventional patch array model, the maximum temperature of 67.6 °C is observed. In comparison, the proposed scheme features a maximum temperature of 56.6 °C, which is significantly lower than that of the conventional topology without using a standalone heatsink. In addition, the temperature distribution results ascertain that the heat emission generated from the RFIC is effectively discharged to the top surface of the package. This demonstrates the heat-dissipative functionality of the proposed stamped metal AiP.

References

1 B. Sadhu, Y. Tousi, J. Hallin, S. Sahl, S. Reynolds, O. Renström, K. Sjögren, O. Haapalahti, N. Mazor, B. Bokinge, G. Weibull, H. Bengtsson, A. Carlinger, E. Westesson, J.E. Thillberg, L. Rexberg, M. Yeck, X. Gu, D.Friedman and A. Valdes-Garcia, "A 28 GHz 32-element phased-array transceiver IC with concurrent dual polarized beams and 1.4 degree beam-steering resolution for 5G communication," *in IEEE Int. Solid-State Circuits Conf. (ISSCC) Dig. Tech. Papers*, San Francisco, CA, USA, Feb. 2017, pp. 128–129.

2 J. D. Dunworth, A. Homayoun, B-H. Ku, Y-C. Ou, K. Chakraborty, G. Liu, T. Segoria, J. Lerdworatawee, J. W. Park, H-C. Park, H. Hedayati, D. Lu, P. Monat, K. Douglas and V. Aparin, "A 28GHz Bulk-CMOS dual-polarization phased-array transceiver with 24 channels for 5G user and basestation equipment," *in IEEE Int. Solid-State Circuits Conf. (ISSCC) Dig. Tech. Papers*, San Francisco, CA, USA, Feb. 2018, pp. 70–72.

3 Z. Radivojevic, K. Andersson, J. A. Bielen, P. J. Van der Wel and J. Rantala, "Operating limits for RF power amplifier at high junction temperatures," *Microelectron. Reliabil.*, vol. 44, pp. 963–972, 2004.

4 J. Zhou, L. Yin, L. Kang, M. Wang and J. Huang, "Joint design and experimental tests of highly integrated phased-array antenna with microchannel heat sinks," *IEEE Antennas Wirel. Propag. Lett.*, vol. 18, no. 11, pp. 2370–2374, Nov. 2019.

5 Y. Aslan, J. Puskely, J. H. J. Janssen, M. Geurts, A. Roederer and A. Yarovoy, "Thermal-aware synthesis of 5G base station antenna arrays: an overview and a sparsity-based approach," *IEEE Access*, vol. 6, pp. 58868–58882, 2018.

6 E. McCune, "Energy efficiency maxima for wireless communications: 5G, IoT, and massive MIMO," *in Proc. of 2017 IEEE Custom Integrated Circuits Conference (CICC)*, TX, USA, 2017, pp. 1–8.

7 B. J. Döring, "Cooling system for a Ka band transmit antenna array," German Aerospace Center (DLR), Köln, Germany, *Tech. Rep. IB554-06/02*, Dec. 2005.

8 X. Gu, D. Liu, C. Baks, O. Tageman, B. Sadhu, J. Hallin, L. Rexberg, P. Parida, Y. Kwark and A. Valdes-Garcia, "Development, implementation, and characterization of a

64-element dual-polarized phased-array antenna module for 28-GHz high-speed data communications," *IEEE Trans. Microwave Theory Tech.*, vol. 67, no. 7, pp. 2975–2984, July 2019.

9. A. Alnukari, P. Guillemet, Y. Scudeller and S. Toutain, "Active heatsink antenna for radio-frequency transmitter," *IEEE Trans. Adv. Packag.*, vol. 33, pp. 139–146, Feb. 2010.
10. J. J. Casanova, J. A. Taylor and J. Lin, "Design of a 3-D fractal heatsink antenna," *IEEE Antennas Wirel. Propag. Lett.*, vol. 9, pp. 1061–1064, Nov. 2010.
11. J. Park, D. Choi and W. Hong, "Millimeter-wave phased-array antenna-in-package (AiP) using metal stamped process for enhanced heat dissipation," *IEEE Antennas Wirel. Propag. Lett.*, vol. 18, no. 11, pp. 2355–2359, Nov. 2019.
12. J. Park and W. Hong, "Metal stamped antenna-in-package for millimeter-wave large-scale phased-array applications using multiphysics analysis," *in Proc. 14th Eur. Conf. Antennas Propag. (EuCAP)*, London, DK, 2020.
13. K. Hirasawa and M. Haneishi, *Analysis, Design, and Measurement of Small Low Profile Antennas*. Norwood, MA: Artech House, 1992.
14. M. C. Huynh and W. Stutzman, "Ground plane effects on planar inverted-F antenna (PIFA) performance," *IEEE Proc. Microwave Antennas Propag.*, vol. 150, no. 4, pp. 209–213, Aug. 2003.
15. "Roughness measurements of stainless steel surfaces," 2014 [Online]. Available: http://www.worldstainless.org/Files/issf/non-image-files/PDF/Euro_Inox/RoughnessMeasurement_EN.pdf.
16. S. Hall and H. Heck, *Advance Signal Integrity for High-Speed Digital Designs*. Hoboken, NJ, USA: Wiley, 2009.
17. K. Kibaroglu, M. Sayginer and G. M. Rebeiz, "A low-cost scalable 32-element 28 GHz phased array transceiver for 5G communication links based on a 2×2 beamformer flip-chip unit cell," *IEEE J. Solid-State Circuits*, vol. 53, no. 5, pp. 1260–1274, May 2018.
18. "*Two-Resistor Compact Thermal Model Guideline*," JESD-1, JEDEC, Jul. 2008.

8

Frequency Tunable Millimeter-Wave 5G Antenna-in-Package

As shown in Figure 8.1a, an example of a mobile communication system is conceptually described. In a mobile communication system, a mobile device modulates an *N*-bit data stream, which is a digital signal, and converts the analog signal to the intermediate frequency (IF). Afterward, the IF signal is converted into a radio frequency (RF) carrier signal again, and data is transmitted in the form of electromagnetic waves through the air medium. The signal received through an air medium can be demodulated to an *N*-bit data stream in the reverse order of the transmission signal processing.

As described earlier, this signal processing is advanced at a remarkable pace with each generation. The second-generation (2G) communication, which started in the 1990s to migrate from voice to data, was the duration when digital communication started in earnest unlike first-generation (1G) when voice calls were everything. Since then, how quickly data can be downloaded has become a crucial topic. However, the existing method has limitations. Therefore, multiple access methods, such as code division multiple access (CDMA) are adopted to increase data speed and channel capacity. Furthermore, such a wideband-CDMA (W-CDMA) is started to use wider bandwidth. This is the beginning of what is called the third generation (3G) of communication.

A mobile communication device is no longer focused on voice communication but rather on data-oriented. The early period of the 3G started at approximately 384 kbps. However, as it grows to evolved high-speed packet access (HSPA+), the data speed increases by 40–50 Mbps. Nevertheless, whenever trying a video call or Internet through a mobile device, there are frequent disconnections and limitations in use due to slow data speed.

From the fourth generation (4G), what to do on a mobile device during a day is data service, not a voice call. The long-term evolution (LTE) service was started

Microwave and Millimeter-Wave Antenna Design for 5G Smartphone Applications,
First Edition. Wonbin Hong and Chow-Yen-Desmond Sim.

Published 2023 by John Wiley & Sons, Inc.

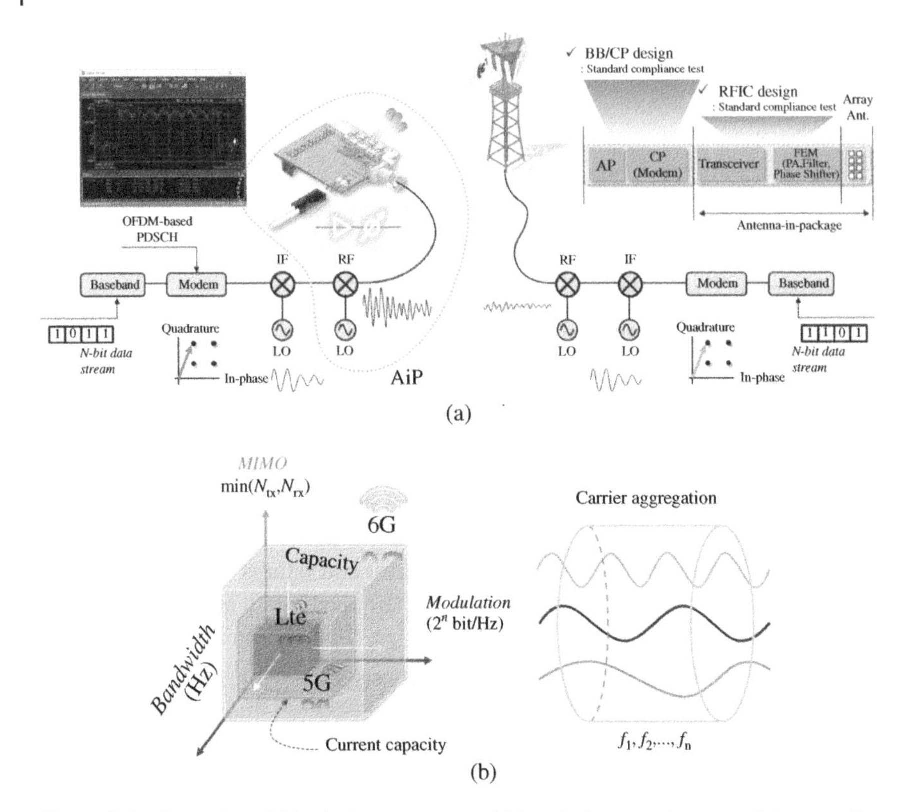

Figure 8.1 Examples of (a) wireless system and (b) techniques to increase data capacity.

with the primary objective of focusing on data in earnest and maximizing data speed. In other words, all the skills that can get the most out of your data are starting to come into demand. The initial goal for LTE was at least 100 Mbps on the downlink, which is the same as the internet speed used in the wired network at the time. Diverse techniques under a mobile device are exploited to achieve rapid speed. For instance, two or more antennas are utilized, and orthogonal frequency division multiplexing (OFDM) is adopted for the management of carriers efficiently under limited bandwidth.

Fifth-generation (5G) communication is developing to expand the data services as well, and its objectives are to receive more data faster than LTE communication. Increasing the data capacity and speed is intended to induce users to new experiences (e.g. VR/AR, etc.) completely different from previous services. An antenna can be considered as a gateway like a channel, which can transfer actual data through the air from the point of view of telecommunication. Since it is hard to increase the data transmission speed with a single antenna, as illustrated in

Figure 8.1b, technologies are introduced to increase the transmission speed by adding antennas and combining channels (e.g. multi-input multi-output, carrier aggregation), and transferring data in a wider bandwidth (e.g. mmWave spectrum).

In order to increase the number of antennas and bandwidth in mobile devices, sufficient space must be secured. However, it is not simple in reality. Therefore, a frequency reconfigurable method and an antenna-to-antenna switching technique are required in such mobile devices. Therefore, a frequency reconfigurable method and an antenna-to-antenna switching technique are required in such mobile devices. Briefly summarized, the purpose of frequency reconfigurable technology is similar between 4G and 5G, but requires a more cautious approach in mmWave spectrum. New approaches to a reconfigurable method for mmWave spectrum will be introduced in this chapter.

8.1 Background and Realistic Challenges for Mobile Applications

In wireless communication systems, various technologies are being introduced into mobile devices to increase data capacity. As mentioned earlier, a series of technologies, such as increasing the number of antennas and combining bands are being employed. However, the overall size of the mobile device is almost the same as that of the previous generation. So, the internal space for the antennas has been rather reduced. This trend is influenced by bandwidth and efficiency according to the theoretical limitations of McLean's (followed Chu's analysis) small antenna in the past decades as given by [1]

$$Q_r = \frac{1}{(ka)^3} + \frac{1}{ka}\left(k = \frac{2\pi}{\lambda}\right) \tag{8.1}$$

As shown in Figure 8.2a, the measured $\mathbf{B}\eta$ results regarding 110 related small antennas published in the IEEE transactions on antennas and propagation can also be ascertained. It is noted that the measured $\mathbf{B}\eta$ is based on the assumption of a self-matched antenna without additional matching circuits [2]. The results show that the bandwidth efficiency drops sharply in the region where $ka \leq 1$ or less.

Recently, the mmWave band for the 5G spectrum is being allocated along with the legacy 4G band as shown in Figure 8.2b. In other words, it is challenging to secure the required bandwidth in a limited internal space for many bands and channels and to maintain the radiation performance of the antenna. Furthermore, it seems that the expansion of the bands will be discussed continuously in 3GPP, such as mmWave $n259$ band added recently [3]. This is why frequency adjustment technology has emerged as an alternative to overcome technological limitations.

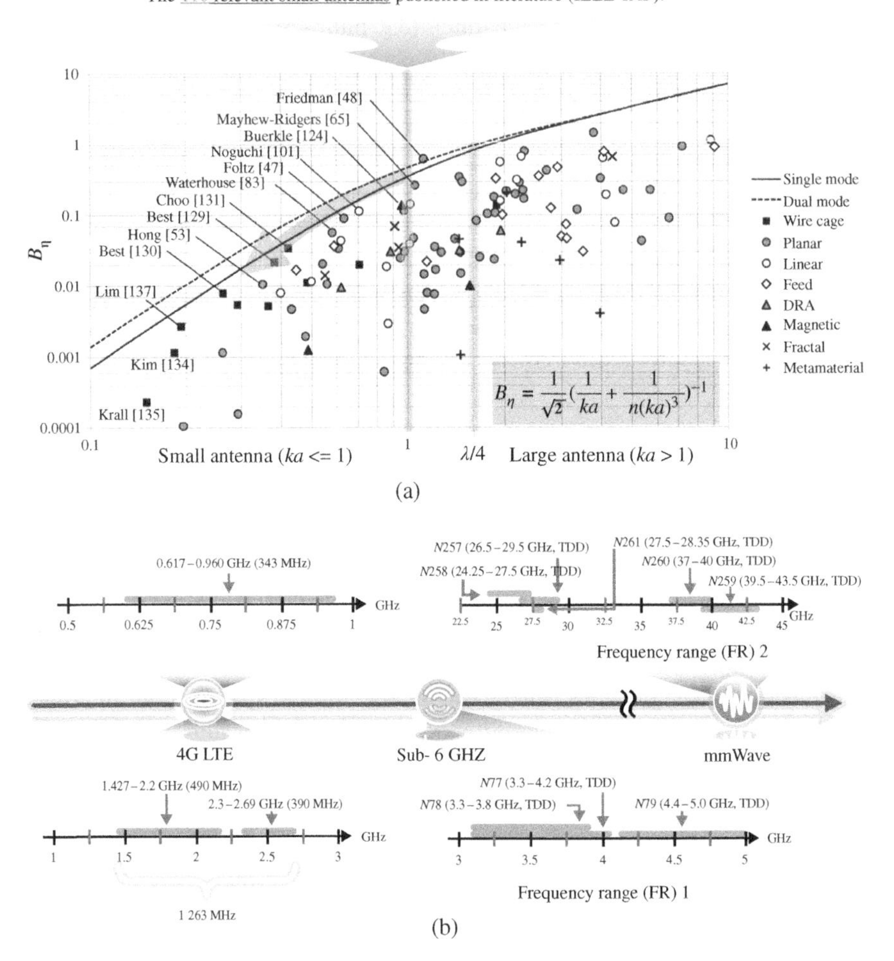

Figure 8.2 (a) Small antenna performance comparison. *Source:* From [2]. ©2020 IEEE. Reproduced with permission and (b) representative 4G and 5G spectrum.

In addition, the numerical variance, such as wavelength between the legacy 4G and millimeter wave spectrum results in unprecedented challenges for mobile antenna systems. As an example, the antenna-in-package consisting of an RF front-end module (FEM) and the phased array antenna is almost essential due to the short wavelength for the 5G antenna system. Therefore, the substrate properties and fabricating process resolution (e.g. line width, thickness, via size, etc.) become very important factors that result in substantial deviations since the RF performance can be affected by the substrate within a package. When an antenna is mounted on a substrate, it usually causes the wave velocity to be lowered depending on the material properties of the substrate. The velocity of electromagnetic waves is actually as follows [4]:

$$C = 1/\sqrt{\mu\varepsilon} = f\lambda \tag{8.2}$$

where C is the velocity of light, and μ and ε are permeability and permittivity. Transceivers that communicate wirelessly must be equipped with antennas. A mismatch between the antenna and the transmit/receive circuit causes performance degradation in a wireless device, such as a mobile phone. In general, a fixed matching network (FMN) method using an LC circuit has been exploited to solve the antenna mismatch, and it takes a lot of time to figure out an optimal matching value. Moreover, since the value of the LC circuit cannot be changed every time depending on the electric field conditions, there are performance problems, such as call drop and high talk current. As a consequence, a frequency tunable technology that can correct the characteristic impedance for the limitations of the fabricating process may be an alternative.

Furthermore, the velocity of propagation in the medium, and wave reflection depend on permittivity surrounding an antenna as well. Therefore, as shown in Figure 8.3, external factors, such as hand and protect cover case surrounding the

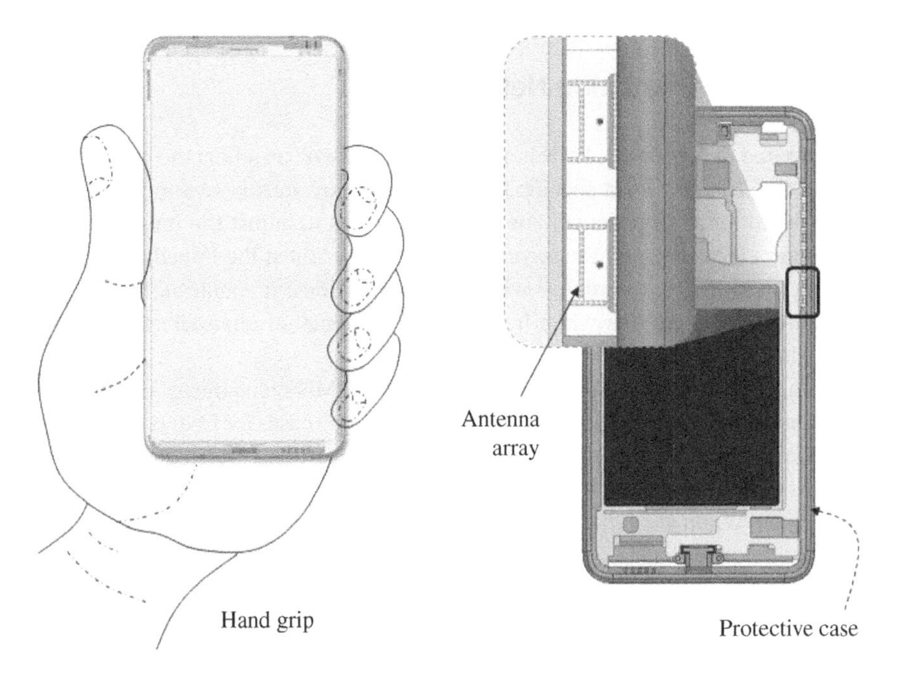

Figure 8.3 Example of tunable matching network (TMN) application for external factors.

mobile device can affect the resonance frequency or impedance of the antenna. In other words, since the antenna radiation performance is greatly affected as well by the external environmental interface of the mobile devices, the optimal antenna matching solution by considering various environments is required. That is why the frequency correction or antenna switching topology can be needed for mmWave module within the mobile device.

These frequency-adjustment and antenna-switching technologies are not first introduced in the 5G era but have been used in the previous generations as well. Although not much different from the matching principle used in conventional 4G, frequency tunable topology at millimeter-wave spectrum requires new approaches different from that of the previous generations. In a later sub-chapter 8.3, the limitations and considerations of frequency tunable technology at millimeter-wave will be discussed in more detail.

Considering mmWave tunable matching network (TMN) operation scenario for mobile devices, TMN topologies can be operated with the following functions as described in Figure 8.4 [5]: (i) Antenna impedance calibration owing to external causes (e.g. device fabrication errors and surrounding environments). (ii) Band selection and switching. The reasons for each scenario are numerous. However, the common purpose is to correct/adjust the antenna impedance status to maintain the optimum RF performance in diverse field conditions.

8.2 Tunable Matching Network

As illustrated in Figure 8.5, TMN is a kind of circuit that can adjust the impedance state of the antenna for a specific purpose. There are mostly two topologies to adjust the antenna impedance. The first approach is to adjust the impedance in the antenna feed line, and the second approach is to adjust the impedance at the antenna aperture. In accordance with the two solutions, it enables to adjust the impedance of an antenna when frequency discrepancy or channel switching is required.

Figure 8.6 shows the smith chart for a simplified TMN mechanism. The impedance transformation for TMN can be considered as two cases for frequency correction. The first is to put the impedance trajectory of the interesting frequency to the constant-conductance trajectory or the constant-resistance trajectory using the Smith chart. When $g = 1$, the interesting frequency can be transformed by either a shunt inductance (L_{shunt}) or capacitance (C_{shunt}). Otherwise, when $r = 1$, the frequency can be adjusted by either a series inductance (L_{series}) or capacitance (C_{series}). For the analysis and detailed mechanism of impedance matching using the Smith chart, please refer to Chapter 5 (Impedance Matching end Tuning) of Microwave Engineering written by David M. Pozar [6].

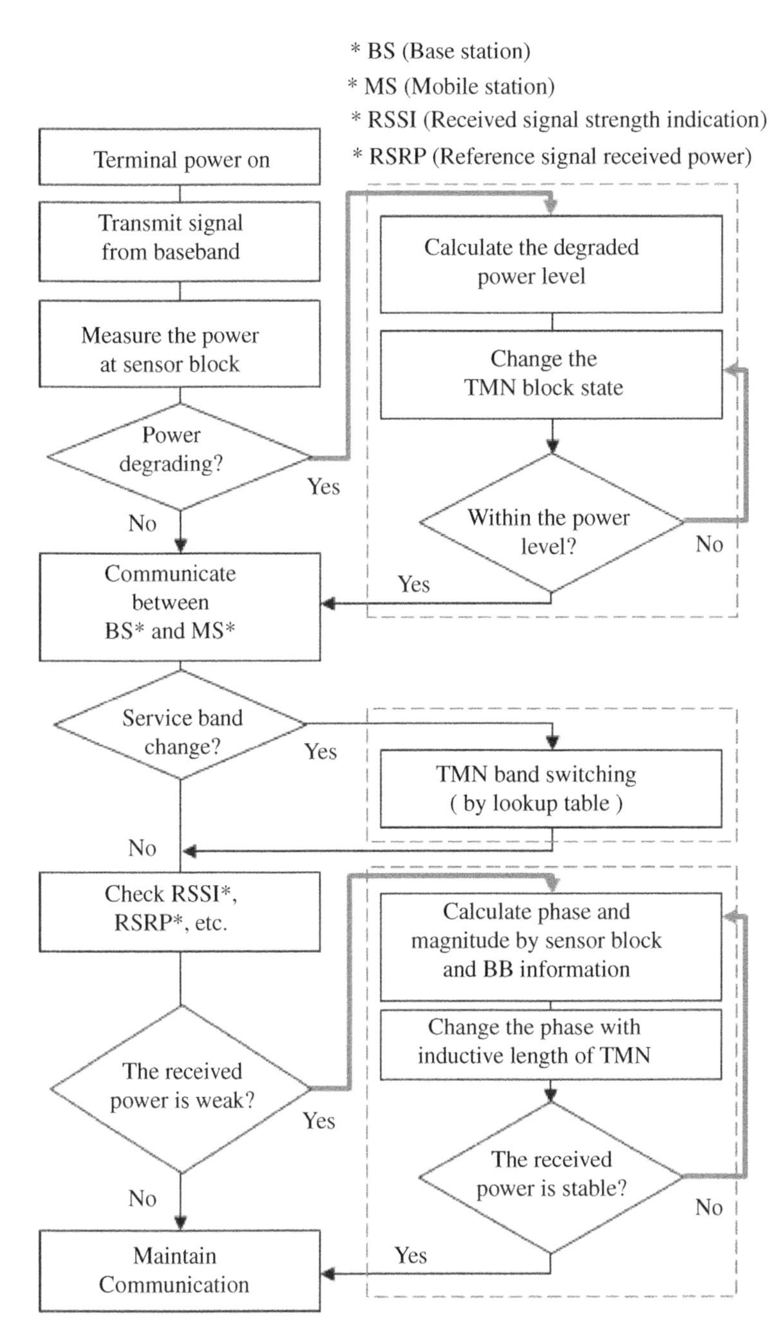

Figure 8.4 Flowchart of the proposed mmWave TMN operation scenario. *Source:* From [5]. ©2020 IEEE. Reproduced with permission.

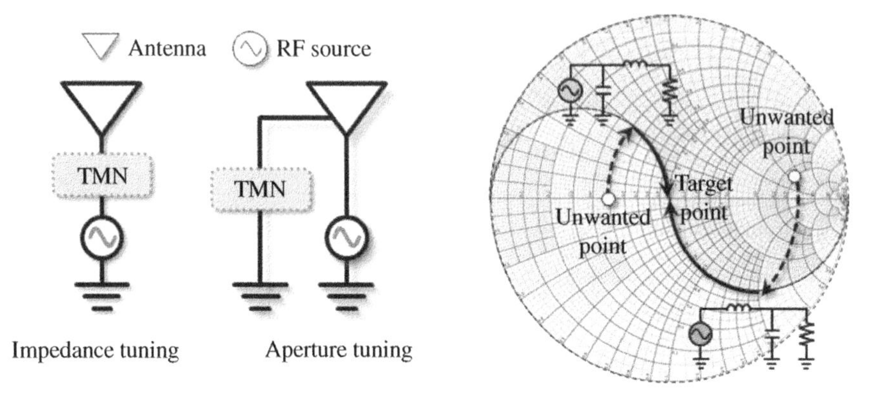

Figure 8.5 Representative cases of the tunable matching network.

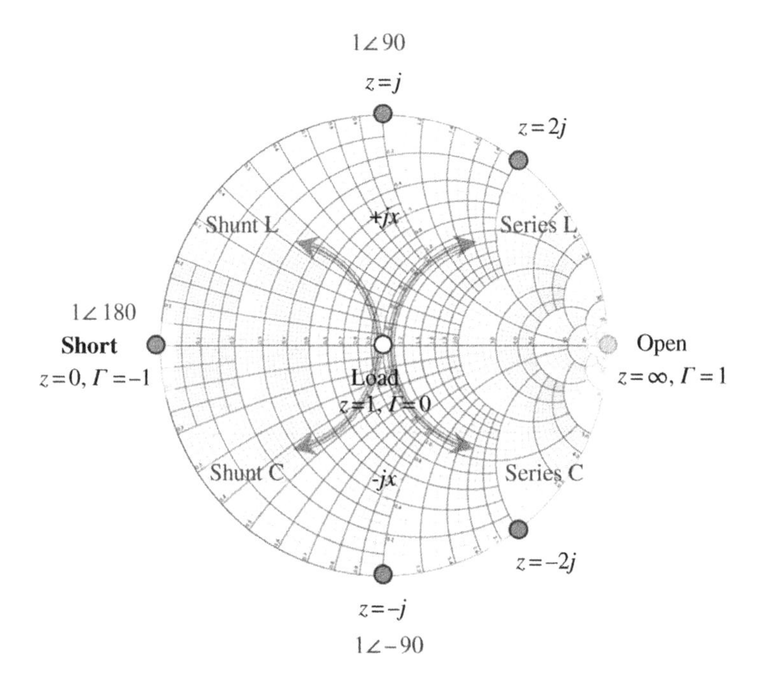

Figure 8.6 Smith chart for reconfigurable matching mechanism.

8.3 Topology and Design Considerations

The resonant frequency can be determined by the following equation in terms of the circuit.

$$\text{Resonance frequency} = \frac{1}{2\pi\sqrt{L \cdot C}} \tag{8.3}$$

That is, if the value of either inductance (L) or capacitance (C) is changed, the resonance frequency can also be altered. TMN components for real-time variation can be positive-intrinsic-negative (PiN) diodes, variable capacitor, micro electro mechanical system (MEMS), switches (e.g. single pole double throw [SPDT]), varactors, etc. Above all, frequency tunable antenna topology that can be easily employed can be configured by utilizing the properties of the capacitance.

However, the configuration of TMN is different from the legacy LTE. Since the impedance (Z) of the antenna which is given by [7]

$$Z = R + jX = R + j2\pi fL + j\frac{1}{2\pi fC} \tag{8.4}$$

is completely dependent on the frequency (f), which means that the inductance (L) and capacitance (C) of each frequency are different. That is, the required impedance level is different from that of 4G LTE. For an intuitive understanding, the range of the lumped element used in 4G LTE and the impedance required at the millimeter wave spectrum are summarized in Table 8.1. In addition, commercial components with the required impedance at millimeter waves go beyond manufacturing limitations.

Figure 8.7 shows an example of an aperture tuning approach. The antenna is based on a planar folded slot antenna (PFSA) [8, 9] with low-profile end-fire radiation characteristics. Capacitance variation at the center of the PFSA's slot exhibiting high impedance is employed to reconfigure the frequency. The PFSA can be equivalent to a parallel resonator circuit due to a slot-based antenna. Once again since the slot structure is a parallel resonator, so adding capacitance to the slot structure allows the frequency to be adjusted, and the impedance is determined by the earlier equation, which is the opposite of the series resonator. Therefore, adding additional capacitance to the slot of the PFSA increases the antenna's capacitance and the additional capacitance (Additional Cap + C) so that the resonance frequency can be adjusted. As depicted in Figure 8.8, although this approach is useful for frequency adjustment featuring wideband coverage, however, owing to the adjusted current distribution of the aperture, the electromagnetic wave direction is changed.

For short wavelengths, such as mmWave spectrum, the characteristics of the far-field radiation pattern, such as beamforming are very crucial to surpass the loss in the air. Therefore, an alternative that takes into account far-field radiation patterns is the impedance tuning method. While the aforementioned aperture tuning method can affect the current distribution of the radiator, impedance tuning can change the impedance in the transmission by keeping the current

Table 8.1 Equivalent reactance by frequencies.

	Lumped elements		Required equivalent reactance	Equivalent reactance	Required lumped elements	Equivalent reactance	Electronic component manufacturing company	Currently available value (minimum)
	Value	1 GHz	2 GHz	28 GHz	Value	28 GHz		
Capacitor	0.5 pF	318.31 Ω	159.15 Ω	11.37 Ω	20 fF	318.31 Ω	Murata	0.1 pF
	3 pF	53.05 Ω	26.53 Ω	1.89 Ω	110 fF	53.05 Ω	Samsung EM	0.2 pF
	10 pF	15.91 Ω	7.96 Ω	0.57 Ω	360 fF	15.92 Ω	Taiyo Yuden	0.2 pF
	100 pf	1.59 Ω	0.8 Ω	0.06 Ω	3.57 pF	1.59 Ω	TDK	0.2 pF

Source: From [5]. ©2020 IEEE. Reproduced with permission.

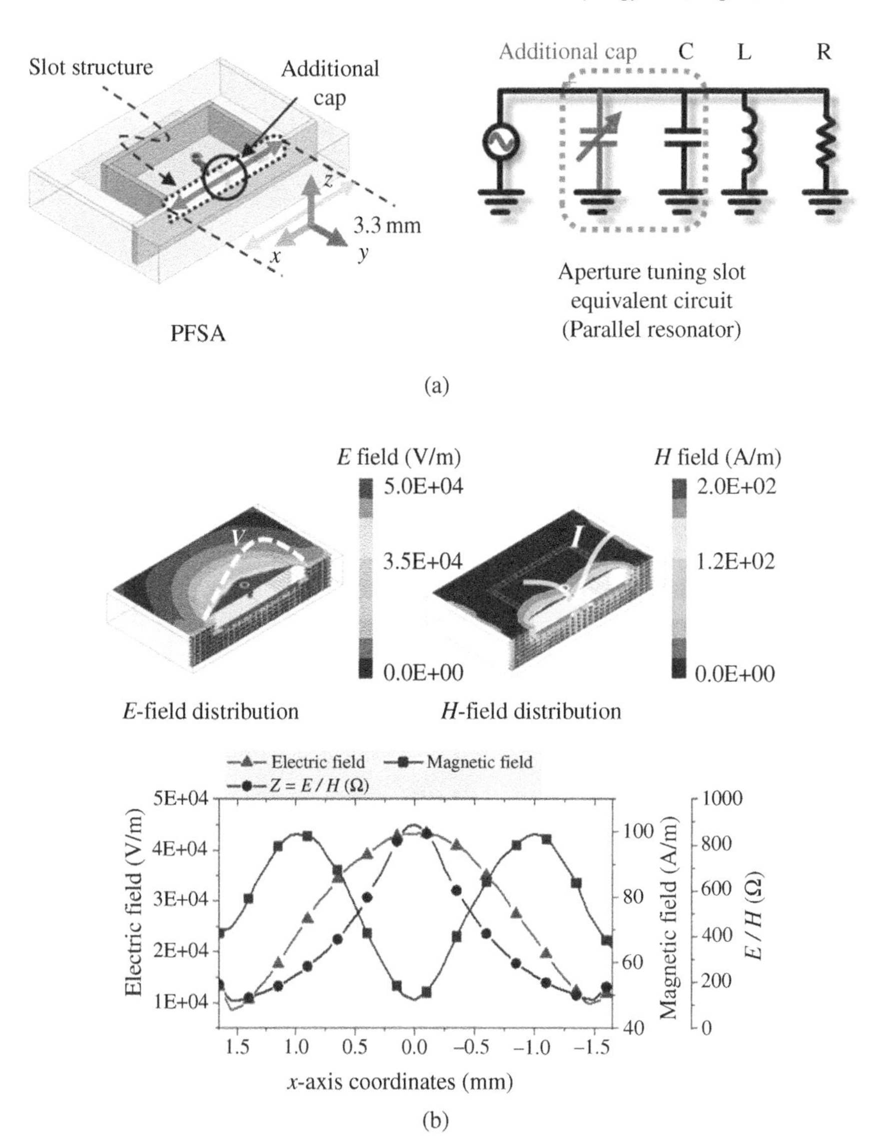

Figure 8.7 (a) PFSA with aperture tuning. (b) Simulated field distributions. *Source:* From [5]. ©2020 IEEE. Reproduced with permission.

distribution of the radiator. To adjust the resonance frequency with the impedance tuning method as illustrated in Figure 8.9, this approach is just effective when the impedance of the antenna is situated on the constant resistance circle. Thus, the transmission line may be needed to modify the phase while maintaining

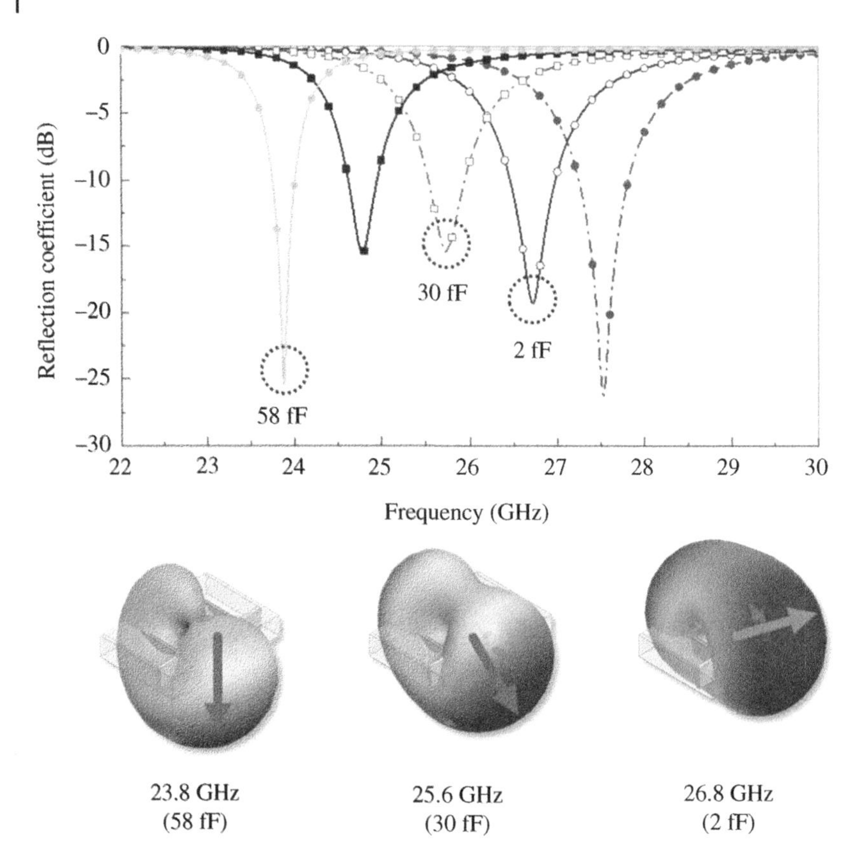

Figure 8.8 Simulated reflection coefficient and 3-D far-field radiation pattern of the PFSA with aperture tuning. *Source:* From [5]. ©2020 IEEE. Reproduced with permission.

the reflection coefficient. (Note that the phase can be changed to 180° at a 1/4 wavelength period.)

Once the antenna impedance has been configured within the required impedance region, the impedance can be adjusted via the series and shunt networks. For example, the matched frequency to the system impedance can be adjusted when the value of the series capacitance is lowered at $r = 1$. This further increases the resonant frequency as shown in Figure 8.10. Frequency-reconfigurable topology is possible via the shunt networks as well. However, this chapter only elaborates on TMN with the serial case.

As shown in Figure 8.11 [6], although there can be many ways to implement impedance, verification of the impedance tuning method for frequency adjustment is conducted based on physically distributed form. The equivalent capacitance employed in the transmission line of the PFSA is implemented as 63 and

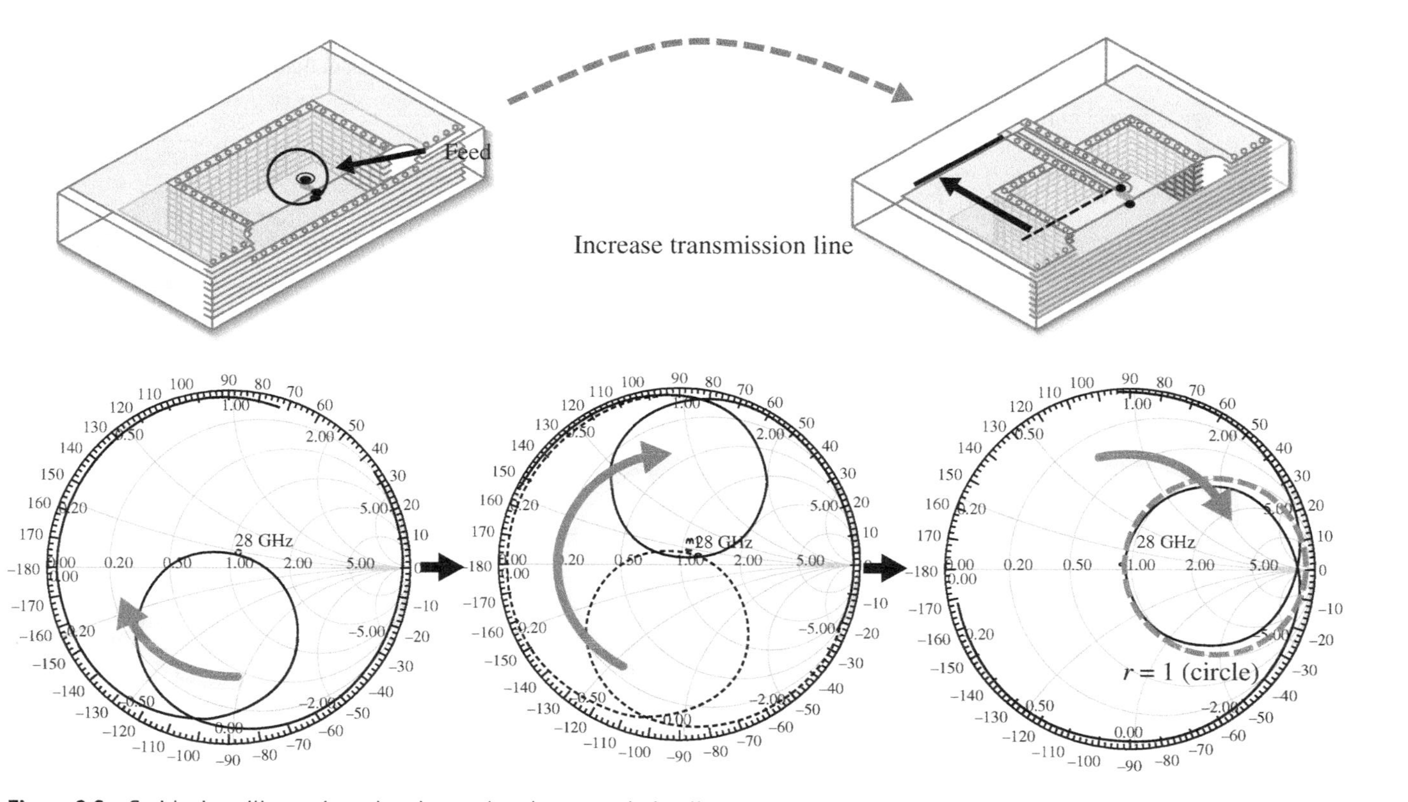

Figure 8.9 Smith chart illustration when increasing the transmission line.

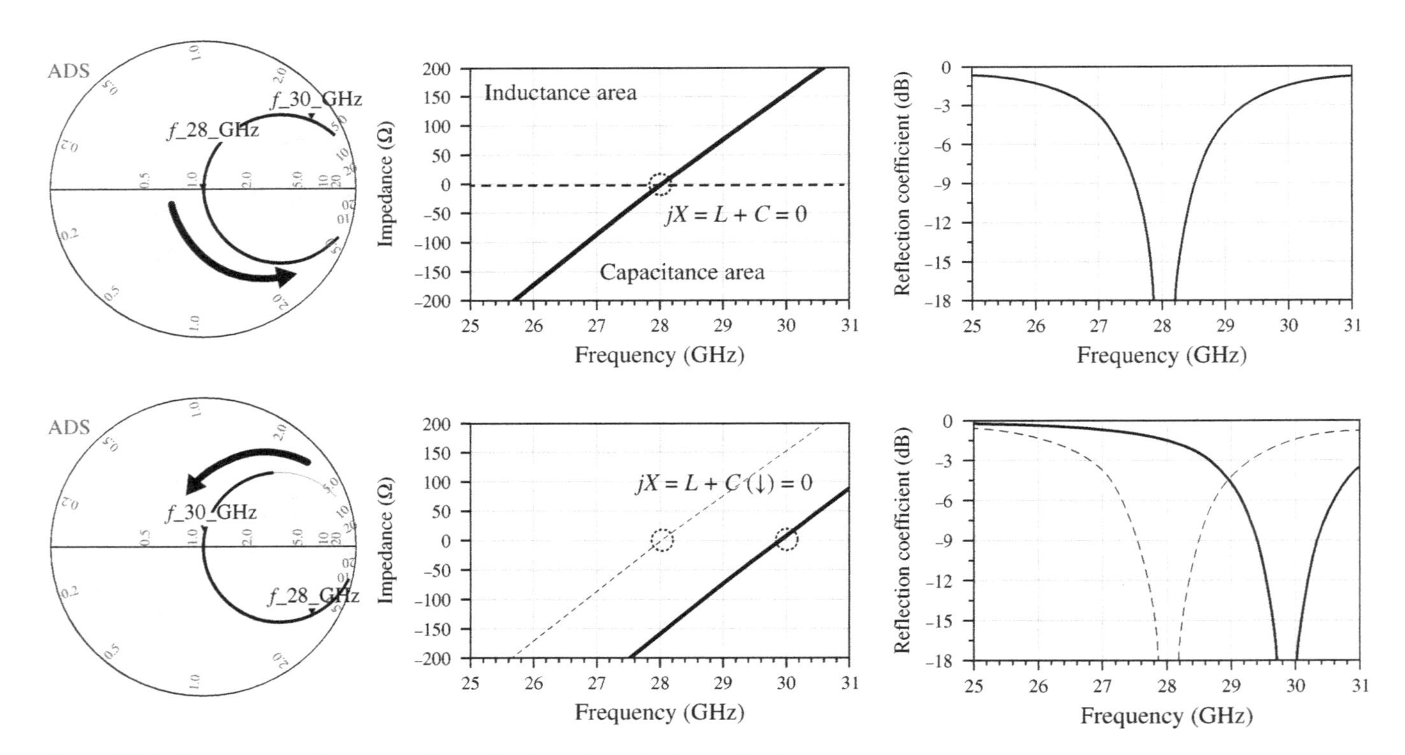

Figure 8.10 Frequency tuning mechanism using impedance tuning method.

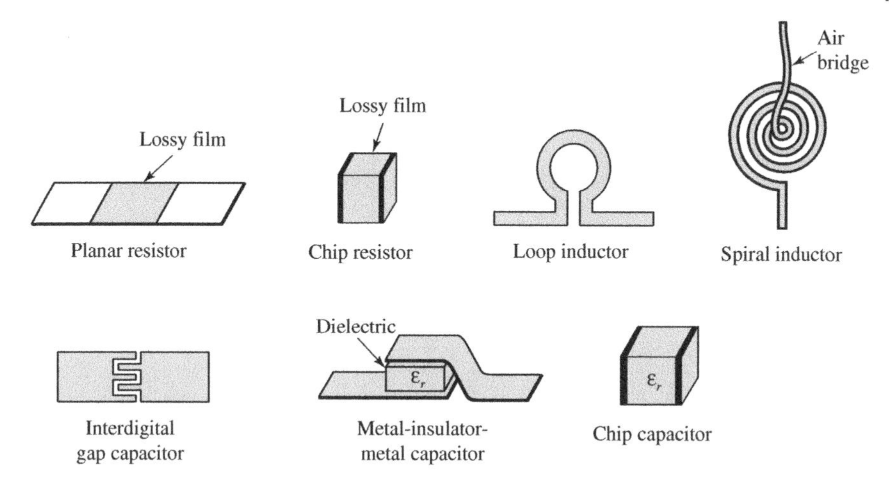

Figure 8.11 Examples of resistance and reactance which can be fabricated [6].

103 fF with reference to Table 8.1. As illustrated in Figure 8.12, the peak resonance frequency is able to be tuned between 27.8 and 28.8 GHz. Furthermore, this topology has an advantage to minimize the distortion of the radiation pattern if the beamforming characteristic is considered. Within the effective frequency range, the radiation patterns exhibit a negligible change in far-field radiation patterns as a function of operating frequency.

The reason why the frequency variable range is different between Figures 8.8 and 8.12 is due to the position of the capacitor. Figure 8.8 can be equivalently modeled as the sum of parallel capacitors while Figure 8.12 is equivalently modeled as the sum of series capacitors. This is because the value of the impedance is different due to the difference in equivalent capacitance according to the position of the capacitor as shown in the example.

RF choke circuit for DC bias is essential if active elements are considered for frequency tuning since RF choke circuit is for eliminating the interference between the RF and DC signals. However, If the RF choke is utilized through lumped elements, the self-resonant frequency (SRF) can be induced at the frequency of interest as shown in Figure 8.13. Therefore, an RF choke employing a microstrip line can be considered.

Since the frequency tunable antenna-in-package (AiP) may contain various unpredictable process errors, some differences, which can lead to unexpected results. At high frequencies, such as mmWave spectrum, the soldering effect on component mounting cannot be neglected as illustrated in Figure 8.14. In addition, if the process is not properly controlled, it may not be the desired result. Therefore, the reflection coefficient and the far-field radiation pattern should be designed and verified in real conditions under AiP form including such as several solder balls.

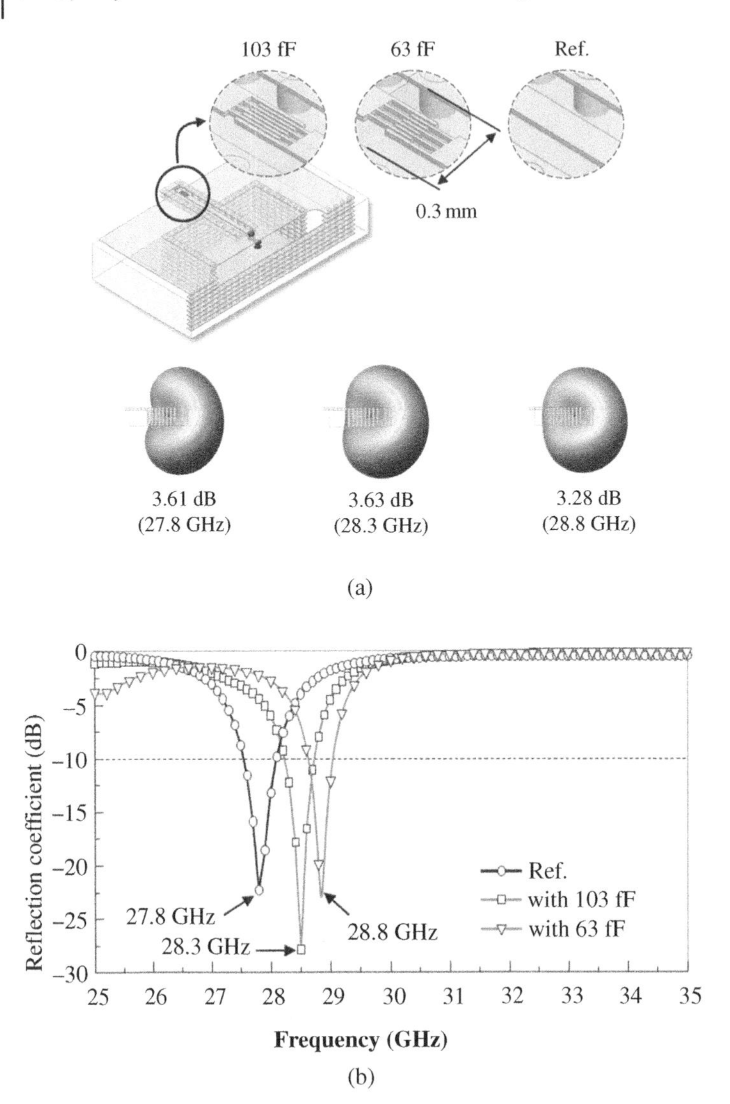

Figure 8.12 (a) PFSA examples and (b) simulated reflection coefficients. *Source:* From [5]. ©2020 IEEE. Reproduced with permission.

The unwanted parasitic impacts, such as the soldering of IC or electric components can result in discrepancies between fabrication and design. Figure 8.15 shows the effect of soldering for better understanding. The results noted that the capacitance can be varied depending on the amount of soldering (shown in

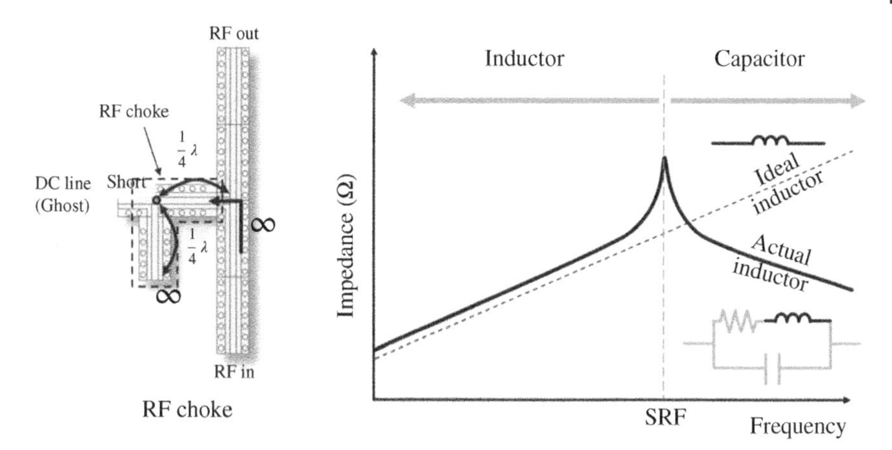

Figure 8.13 An example of RF choke and the self-resonant frequency (SRF). *Source:* From [5]. ©2020 IEEE. Reproduced with permission.

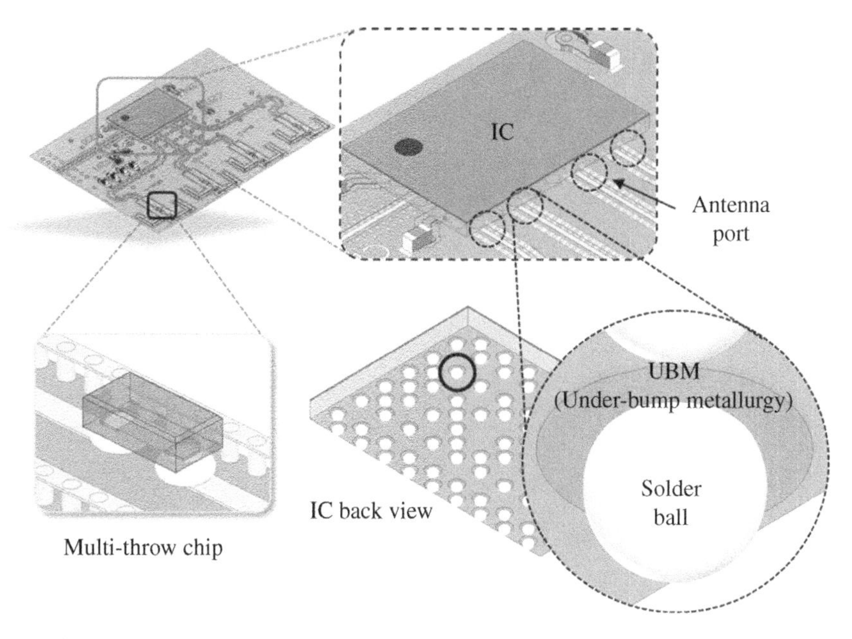

Figure 8.14 The effect of solder balls when implementing components and integrated circuits (ICs).

Figure 8.15 (1) and (3)). In addition, the real impedance can be altered even with similar capacitance (shown in Figure 8.15 (1) and (2)). That is, the actual impedance depends on the parasitic effects, resulting in the variation of the final impedance. Thus, the amount of soldering in the package must be well managed.

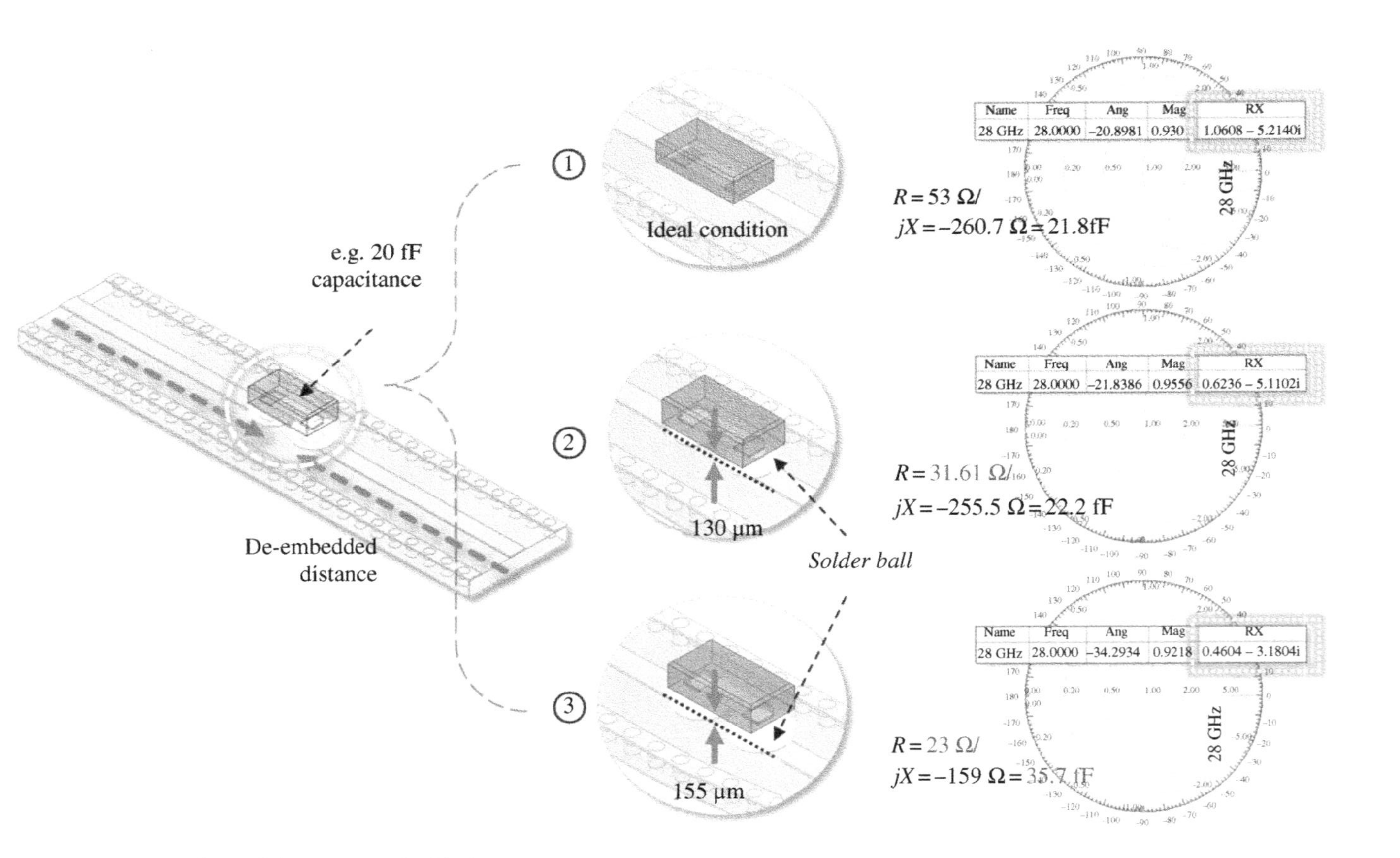

Figure 8.15 The effect of the amount of soldering.

In addition, since the RF signal is actually transmitted through the ball grid arrays (BGA) when the IC is mounted, a suitable transition structure between the BGA and microstrip line may be required. Several types of transition structures are introduced in the literature [10–12] as shown in Figure 8.16. In [10], the effects of a keep-out circle on the ground plane are investigated. Transmission performance can be improved by reducing parasitic capacitance. On the other hand, if the signal line as shown in Figure 8.16a and 8.16b passes through the bottom (vertical via transition) plane, the performance of the transmission line can be improved by adjusting the interval and number of ground vias. In [11], To improve the transmission signal, a stagger via (STV) structure and embedded air cavities are implemented. In [12], the effect of the number of layers and via clearance under a multilayer via transition is investigated by analytical equations.

8.4 Examples and Demonstrations

Figure 8.17 shows a new class of frequency tunable 5G architecture for mmWave AiP using surface mount technology (SMT). This type of concept and approach is introduced in [13]. This 5G AiP is implemented to adjust impedance at each channel by a half-wavelength transformer. The first salient point of this 5G AiP architecture is the use of TMN topology using an active component with arbitrary impedance ($1/2\ \lambda$ transformer) at each antenna feed line. This arbitrary impedance concept is introduced to adapt a commercial active component since the width of the commercial active components is larger than the $50\,\Omega$ impedance transmission line at mmWave spectrum. Therefore, any system can be used without any limitation of the size of active components.

The second salient point is its 100% compatibility with the current mmWave phased array 5G architectures through SMT. The mmWave 5G radio systems require complete integration of the antenna and the RF integrated circuit (RFIC). This architecture is implemented at each antenna element using a multi-throw switching chip with low capacitance (20 fF) suitable for impedance adjustment at mmWave spectrum. That is, the individual frequency can be adjusted for each antenna under the phased array antenna condition using a simplified and efficient SMT topology. Therefore, this architecture can simultaneously emit at 27 and 28 GHz to the selected antenna as illustrated in Figure 8.17. For this reason, it is also applicable to the INTRA-BAND carrier aggregation (CA) being discussed in the 3GPP [14] and beam squint due to the beam angle changing as a function of frequency.

In order to demonstrate the frequency tunable mmWave 5G AiP, it has a 9-layer configuration due to the numerous power lines, signal lines, clocks, and so on as

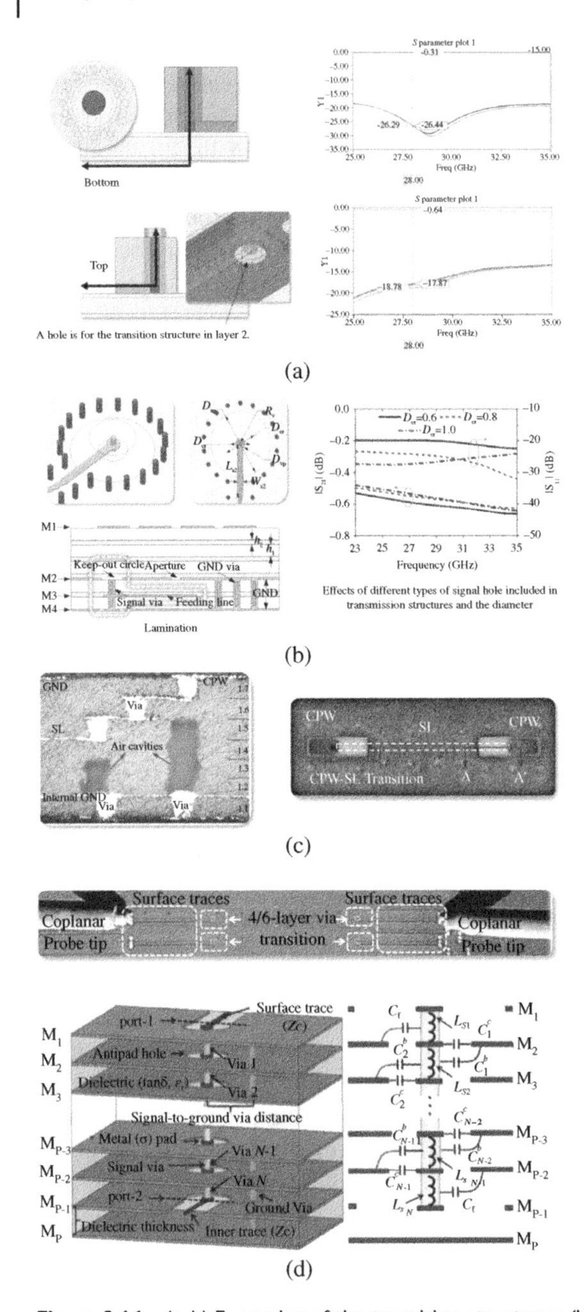

Figure 8.16 (a, b) Examples of the transition structures. (b) *Source:* From [10]. ©2020 IEEE. Reproduced with permission; (c) an aperture etched on a ground plane. *Source:* From [11]. ©2005 IEEE. Reproduced with permission; (d) multilayer via Transitions [12]. *Source:* From [11]. ©2011 IEEE. Reproduced with permission.

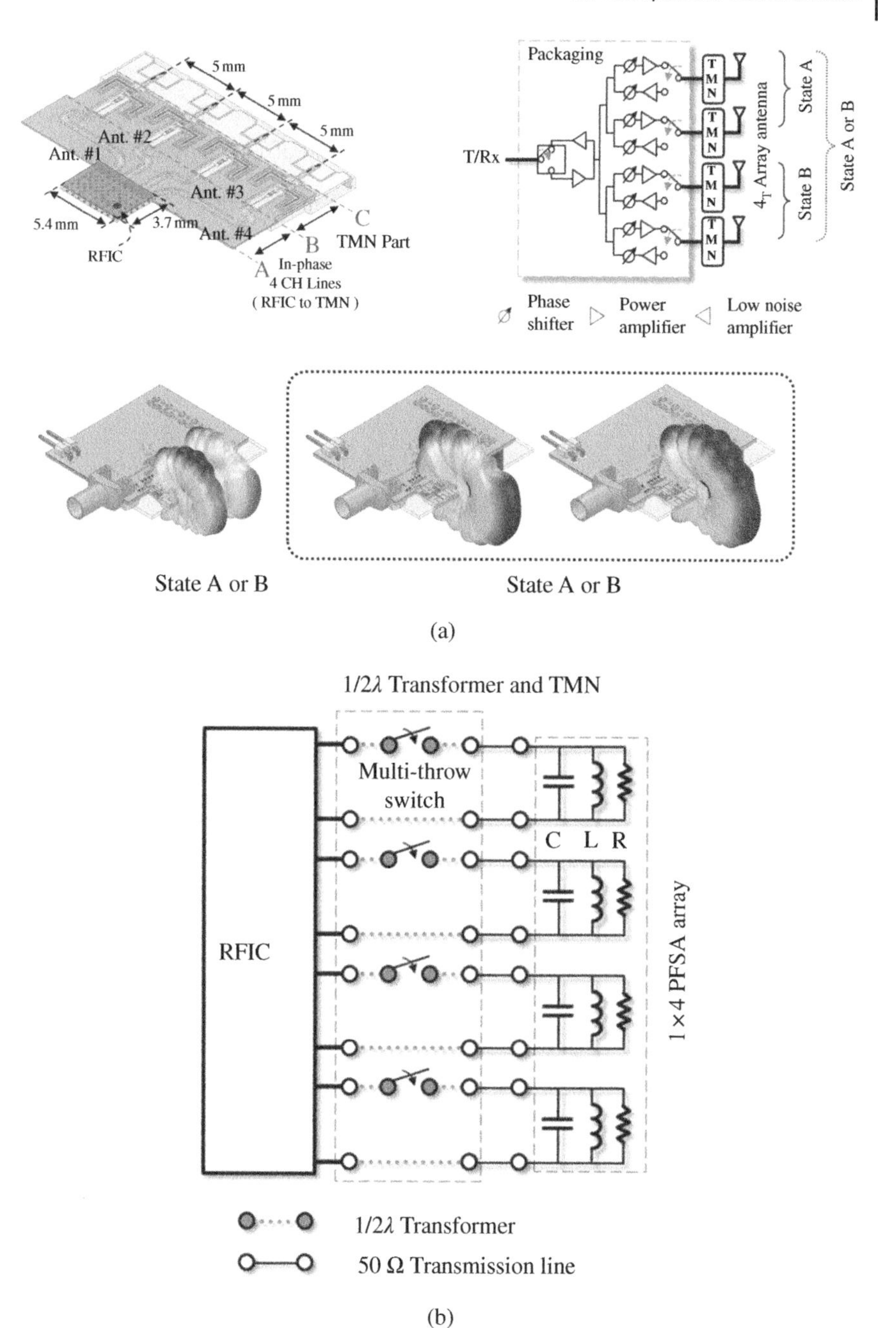

Figure 8.17 (a) The example of a frequency tunable AiP architecture and (b) equivalent circuit of its RF module. *Source:* From [13]. ©2020 IEEE. Reproduced with permission.

depicted in Figure 8.18. When the signal lines are stacked within a thin area, a stepped layer configuration is generated, and thus the overall height may vary. It can be a problem for RF impedance status. Therefore, the RF signal line needs to be routed very carefully. For demonstration, the circuit of TMN and RF signal line are designed in the top layer.

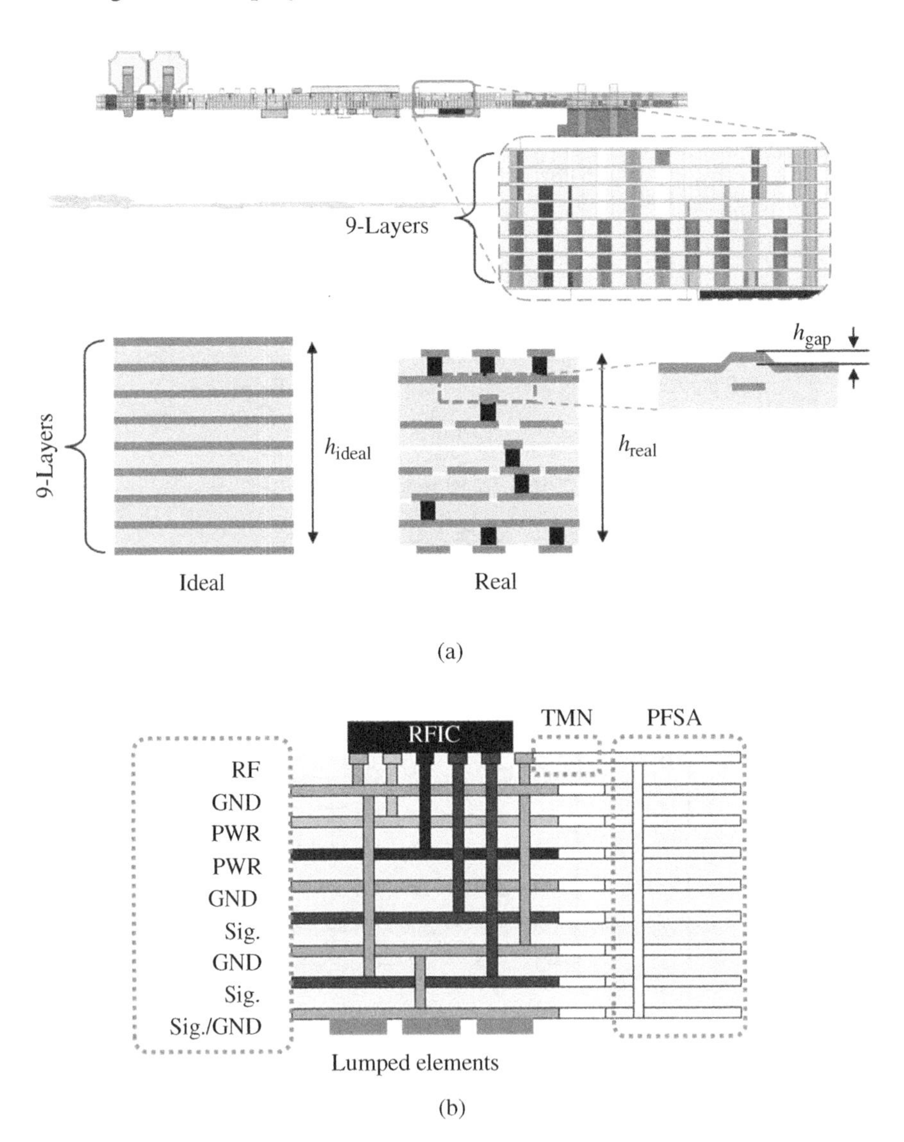

Figure 8.18 (a) Nine layers of designed AiP configuration for fabrication, and (b) detailed stack-up information of the substrate.

Figure 8.19 shows the working mechanism of an arbitrary impedance with a half-wavelength. The transmission line featuring an arbitrary impedance can be employed in any system. The ideal transmission line impedance can be returned to the original impedance status at every half-wavelength interval. The input impedance of a transmission line having a length (l) can be expressed by Eq. (8.4) [15],

$$Z_{in} = Z_o \left[\frac{Z_L + jZ_o \tan \beta \left(= \frac{2\pi}{\lambda} \right) \ell}{Z_o + jZ_L \tan \beta \left(= \frac{2\pi}{\lambda} \right) \ell} \right] (\text{lossless}) \tag{8.5}$$

that is, as can be seen from this equation, assuming that the length is 0, the input impedance (Z_{in}) and the load impedance (Z_L) are the same. Similarly, when the

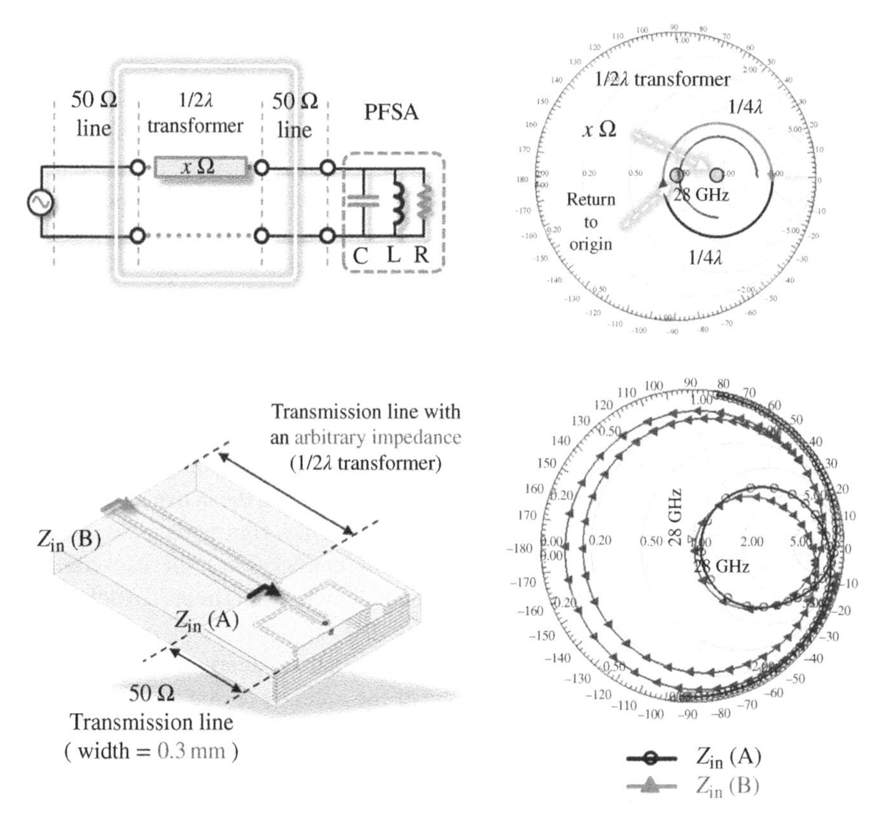

Figure 8.19 Working mechanism of the 1/2λ times transformer. *Source:* From [13]. ©2020 IEEE. Reproduced with permission.

guided wavelength (λ) becomes half-wavelength, the input impedance (Z_{in}) and the load impedance (Z_L) become the same. That is, the output impedance returns to the original position at half-wavelength intervals based on the above equation. Hence, this approach can be effectively implemented when a size of an active circuit, such as TMN is larger than the width of the transmission line.

The result of the half-wavelength transformer with the antenna section can be proved through simulation. Although the half-wavelength transformer is different from the system impedance, the effect exhibits similar impedance at the frequency of interest. In other words, it means that each section (e.g. antenna and tunable circuit) can be designed independently. Thus, it is expected that this method is applicable to any other application as well.

As explained earlier, the operating frequency is totally dependent on inductance (L) and capacitance (C). To realize a frequency tunable mmWave AiP, the impedance of the antenna must be adjusted through inductance or capacitance. Furthermore, antenna impedance must be situated on a constant resistance circle ($r = 1$) if the impedance matching topology is considered. That is why an arbitrary half-wavelength transformer is needed with active components.

However, if a single antenna structure, such as PFSA is integrated with a system ground plane, the three-dimensional far-field radiation shape or direction can be affected since the current on the extended ground can influence the radiation pattern as illustrated in Figure 8.20a. Therefore, when composing a module with a package, it is necessary to carefully investigate the effect on the radiation pattern of the antenna. Considering antenna-in-package, an additional reflector is introduced at the backside of the antenna allowing the end-fire radiation. The RF choke, switch chips, and a DC block capacitor are located to implement the TMN behind the reflector. DC signal lines are configured using conductive lines between layer 3 and layer 5 as illustrated in Figure 8.20b.

Once again, it is noted that the far-field radiation pattern is dependent on the current distribution including the current flowing on the ground plane around the signal transmission line. The capacitance ($1/j2\pi\omega C$) of the active switch chip corresponds to the imaginary part (jX) of the impedance. Hence, the phase of the current distribution on the ground plane may change since the capacitance changes according to the frequency adjustment, which results in partial deformation of the far-field radiation as shown in Figure 8.21.

Additionally, it is worthwhile noting that the radiation pattern may be distorted when the GND area is wider than the size of the antenna or due to surrounding metal. Some examples are shown in Figure 8.22 for better understanding. In [17], as shown on the left of Figure 8.22a, beamforming investigation with a ground plane similar to a smartphone board is conducted. It is ascertained that a wide GND size can give distortion of the radiation pattern. In addition, in the case of an antenna mounted on the side-view mirror of the vehicle on the right in

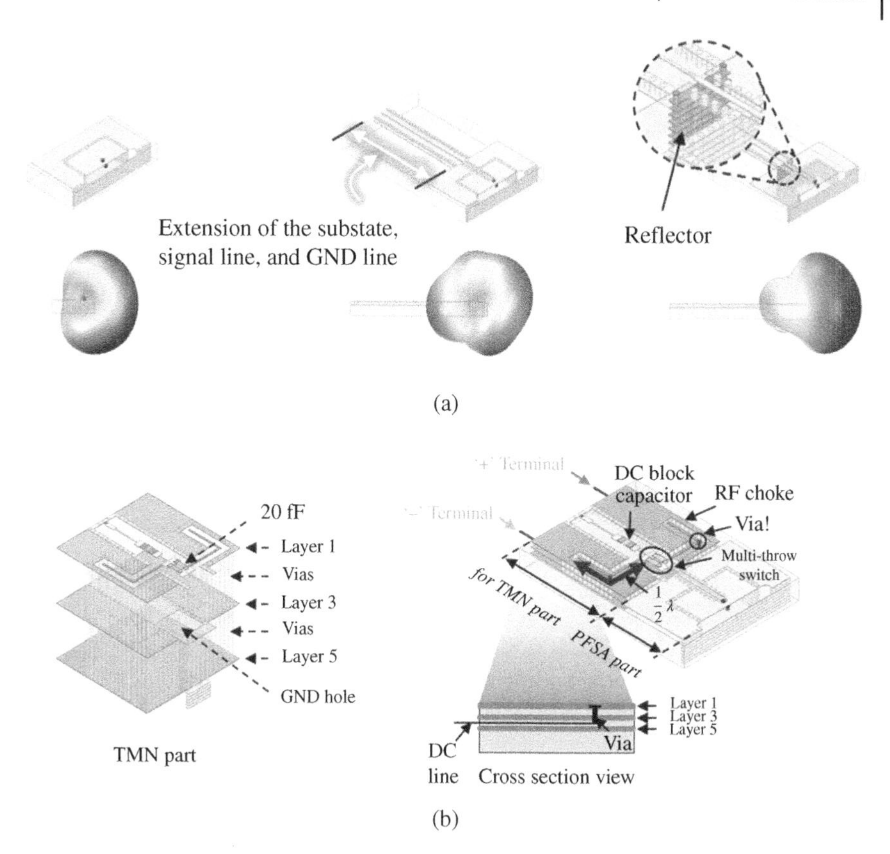

Figure 8.20 (a) The dielectric and ground plane effect, and (b) the overall configuration of the single antenna concept. *Source:* From [13]. ©2020 IEEE. Reproduced with permission.

Figure 8.22a, even when comparing the results of analyzing only a stand-alone side-view mirror with the analysis including a car body, the size of the GND can cause serious distortion in the radiation pattern.

In this demonstration as illustrated in Figure 8.22b, a slightly distorted pattern can be seen since it has a wider ground than the antenna, and the current distribution of the ground and metal cover is involved in the radiation pattern. Therefore, the surrounding environment should be included when designing mmWave module.

As depicted in Figure 8.23a, a low-temperature co-fired ceramic (LTCC) substrate with excellent loss characteristics (Df) and process capability is exploited for a tunable mmWave 5G AiP fabrication. The LTCC substrate designed for the demonstration contains nine layers including VDD (voltage drain drain), several signal lines, and the system ground plane. The actual circuit area including the

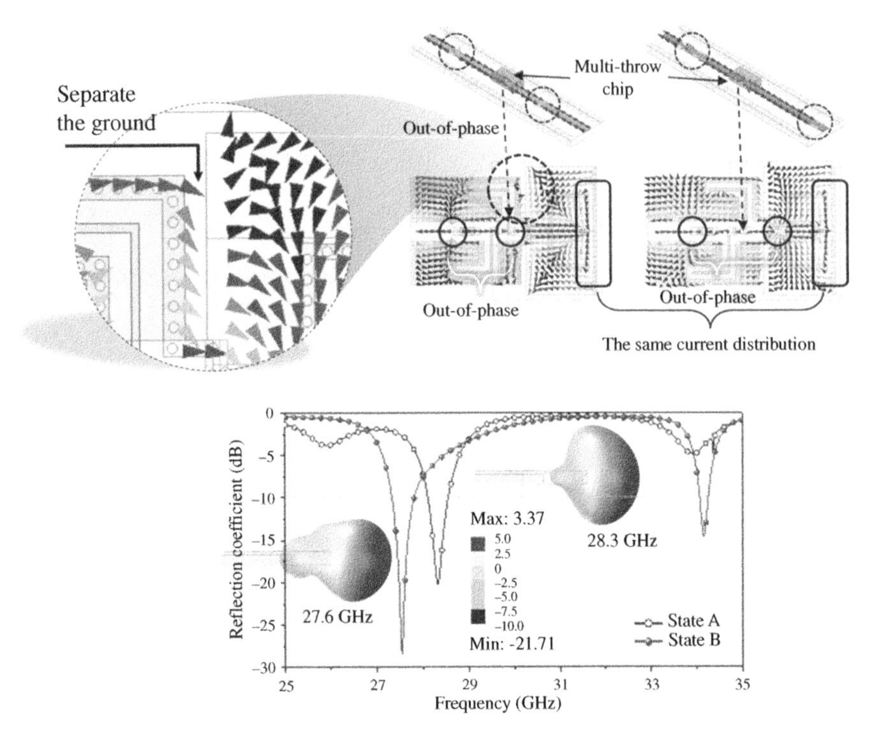

Figure 8.21 Simulated current distribution at each state. *Source:* From [16]. ©2020 IEEE. Reproduced with permission.

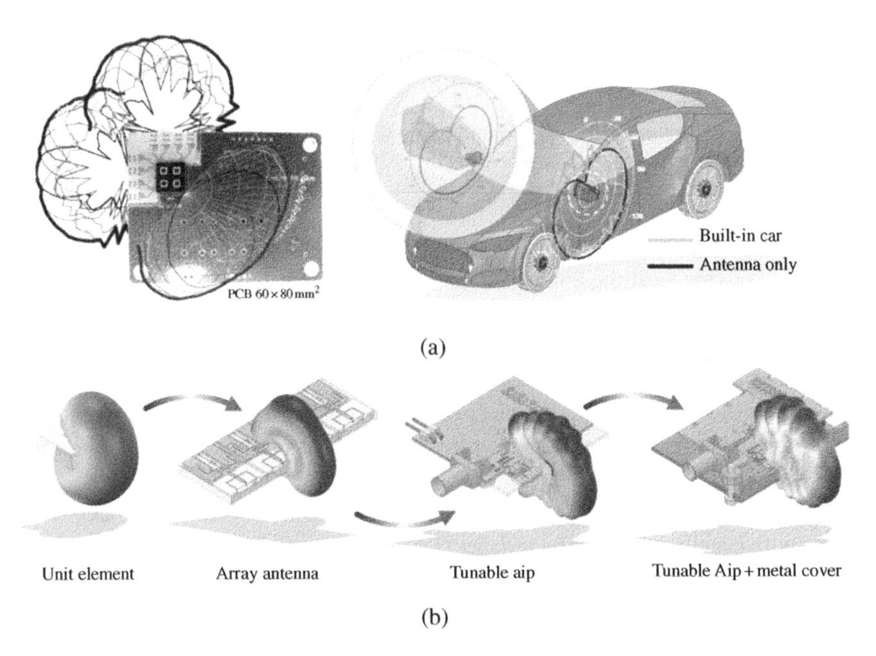

Figure 8.22 Example of the large metal ground effect. (a) Similar to smartphone board and radiation pattern with the car body. (b) Extended ground and metal effect. *Source:* From [13]. ©2020 IEEE. Reproduced with permission.

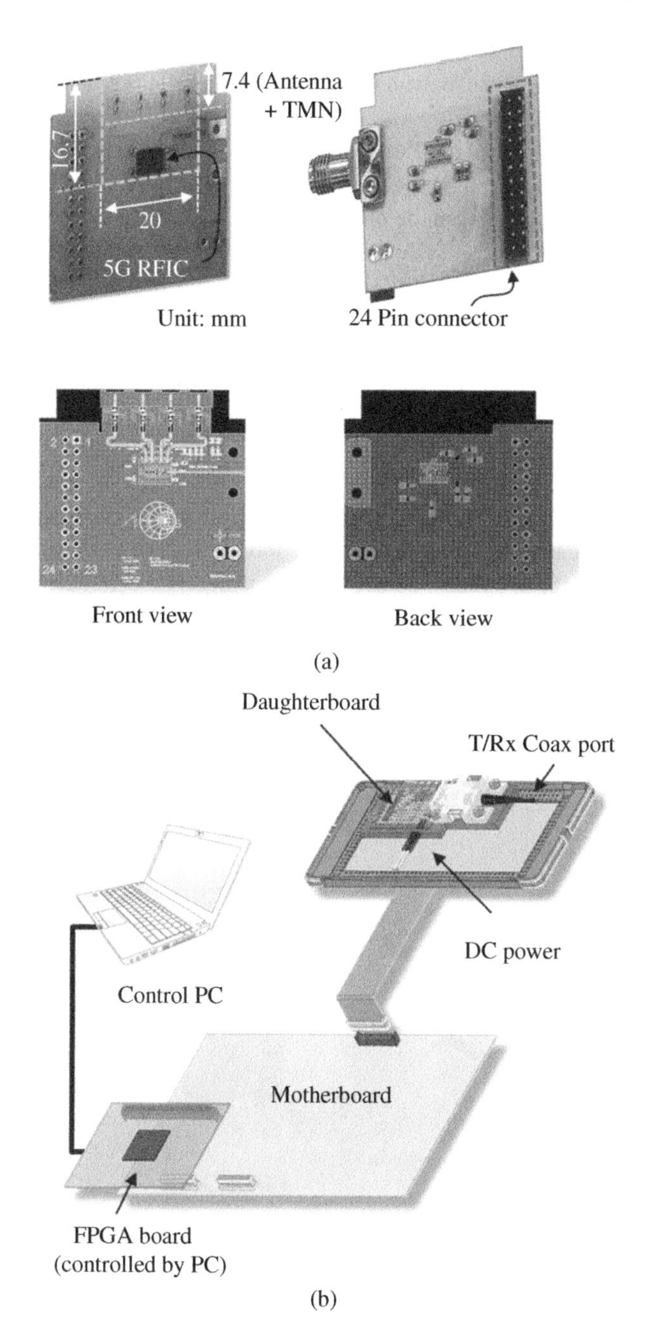

Figure 8.23 (a) The photo of the fabricated tunable 5G AiP and (b) setup for controlling the 5G RF module. *Source:* From [13]. ©2020 IEEE. Reproduced with permission.

TMN for frequency adjustment is 7.4 by 20 mm, and the area of the AiP package including RFIC is 16.7 by 20 mm. The experimental scenario is configured taking into account the mobile phone currently in use as shown in Figure 8.23b. Therefore, 4G LTE antennas are placed on the top and bottom, and the mmWave AiP module for demonstration is placed on the side. In addition, the back cover of the fabricated mobile phone mockup is made of metal material. Furthermore, a laptop, field programmable gate array (FPGA), and motherboard are required to control the AiP Module (daughterboard).

The far-field radiation pattern of the manufactured tunable 5G AiP is measured by a distributed axis chamber as depicted in Figure 8.24. The fabricated 5G AiP module and K-connector (2.92 mm) are fixed to the mobile phone mockup through a 3D printing jig. Then, the active switch chip is controlled by a DC power line connected to AAA-sized battery power.

From the results, 5G AiP module featuring frequency switching exhibits peak radiation at each center frequency (27, 28 GHz) depending on the switch state. It is also ascertained that the measured peak gain of the AiP module exhibits 13.7 dBi at state A (27 GHz) and 12.6 dBi at state B (28 GHz), respectively. The conducted power of the RFIC is 5 dBm when the 3D far-field radiation pattern measurement is conducted. Figure 8.25 shows the normalized gain of beam steering for each

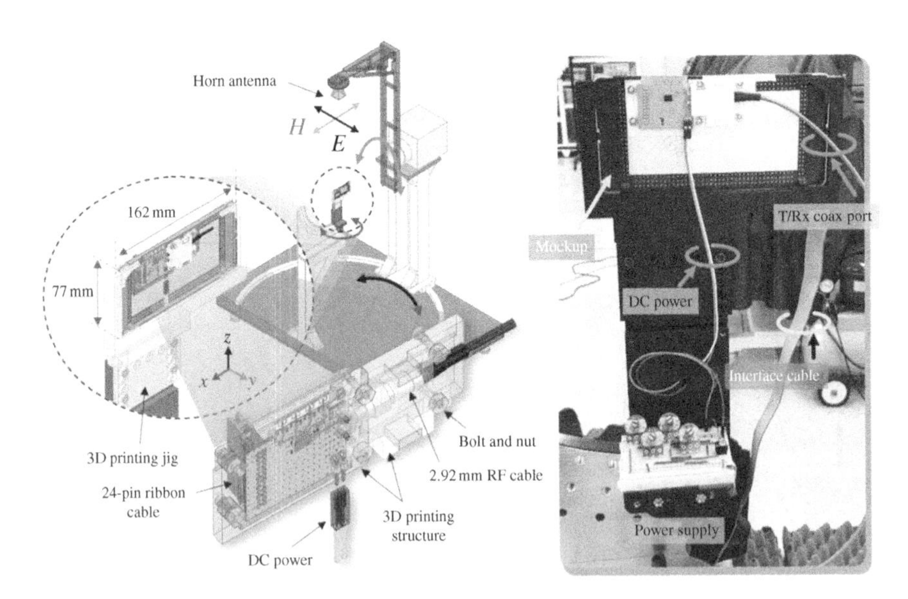

Figure 8.24 Measurement setup for the far-field radiation pattern. *Source:* From [13]. ©2020 IEEE. Reproduced with permission.

state. As mentioned above, it is worth noting that the beam steering pattern may be distorted owing to several reasons discussed in previous (e.g. module size including ground area, solder ball, the metal cover of device, etc.). Therefore, it is also worth noting that a slightly distorted radiation pattern can be observed if manufacturing process controls and the influence of surrounding AiP conditions are not taken into account. The beamforming steering range is from −35 to +50° at each switching state.

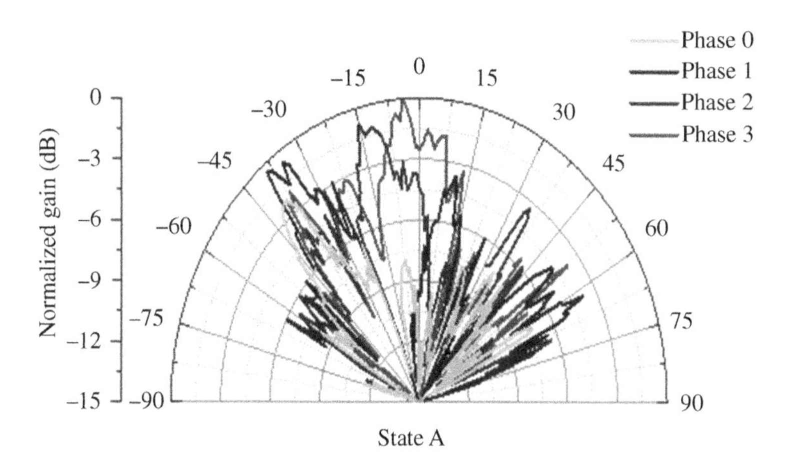

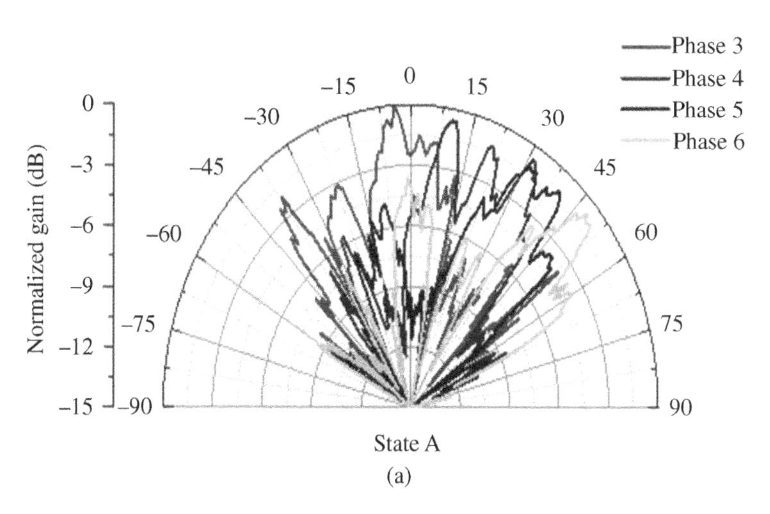

Figure 8.25 Measured beamsteering on mobile platform at (a) state A and (b) state B. *Source:* From [13]. ©2020 IEEE. Reproduced with permission.

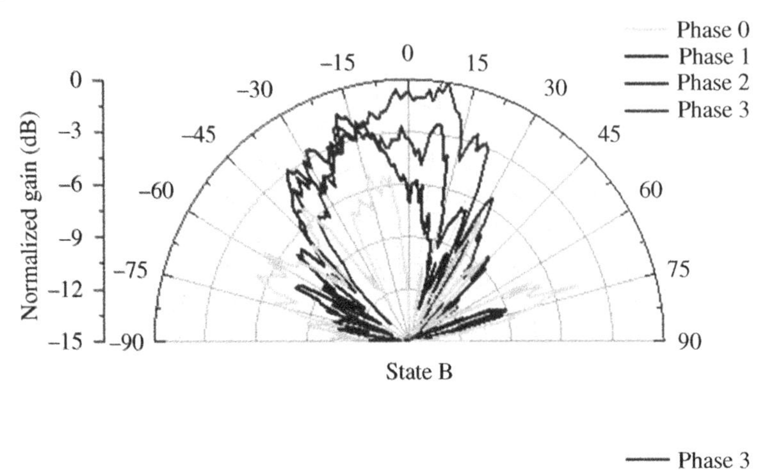

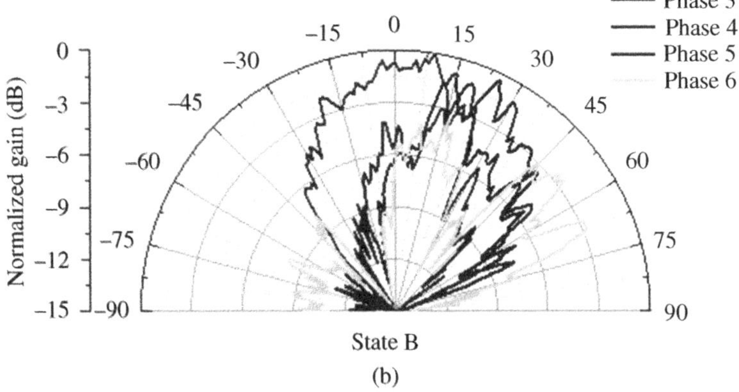

Figure 8.25 (Continued)

8.5 Upcoming Challenges

So far, the reason why the frequency reconfigurable topology at mmWave spectrum is needed has been described through working mechanism and examples of demonstrations. Recently, various tunable matching studies for mmWave spectrum have been conducted by several research groups. For brevity, some research cases will be briefly described here.

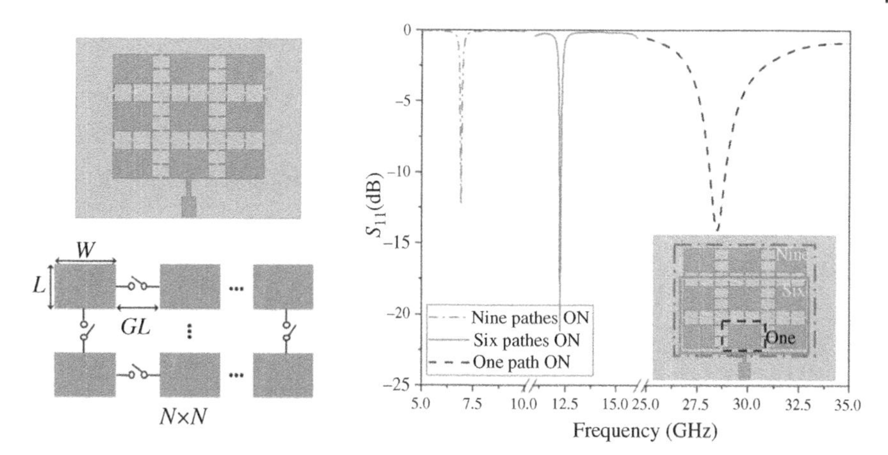

Figure 8.26 Scalable patch antenna and reflection coefficient results. *Source:* From [18]. ©2020 IEEE. Reproduced with permission.

1) Scalable patch antenna [18]:
   From the mmWave band (frequency range 2) to the CmWave (below 6 GHz), a reconfigurable antenna is presented. As illustrated in Figure 8.26, The fundamental idea is that the physical size of the patch determines the resonant frequency, so the resonant frequency can be adjusted through an electrical structure that changes the dimension of the actual patch. That is, when all nine individual elements are connected, the antenna resonates at a low frequency, and when only one of the nine elements is connected to the feed line, the antenna resonates at a high frequency (Figure 8.26).
2) Reconfigurable intelligent surface [19]:
   A study of reconfigurable intelligent surfaces (RIS) at mmWave spectrum is introduced by means of a capacitively coupled patch parasitic resonator as shown in Figure. 8.27. A binary reflection array is required to change the reflection phase, which can be achieved through the electrical length of the resonator. The proposed RIS is featured by a directivity close to 30 dBi and reaching angles of up to 60° (Figure 8.27).
3) Frequency-adjustable LC filter [20]:
   As illustrated in Figure 8.28, the nematic LC-based frequency adjustment method for filtering at mmWave spectrum can realize a frequency tunable range of 2 GHz, and the bandwidth of each resonant frequency is 3.3 GHz.

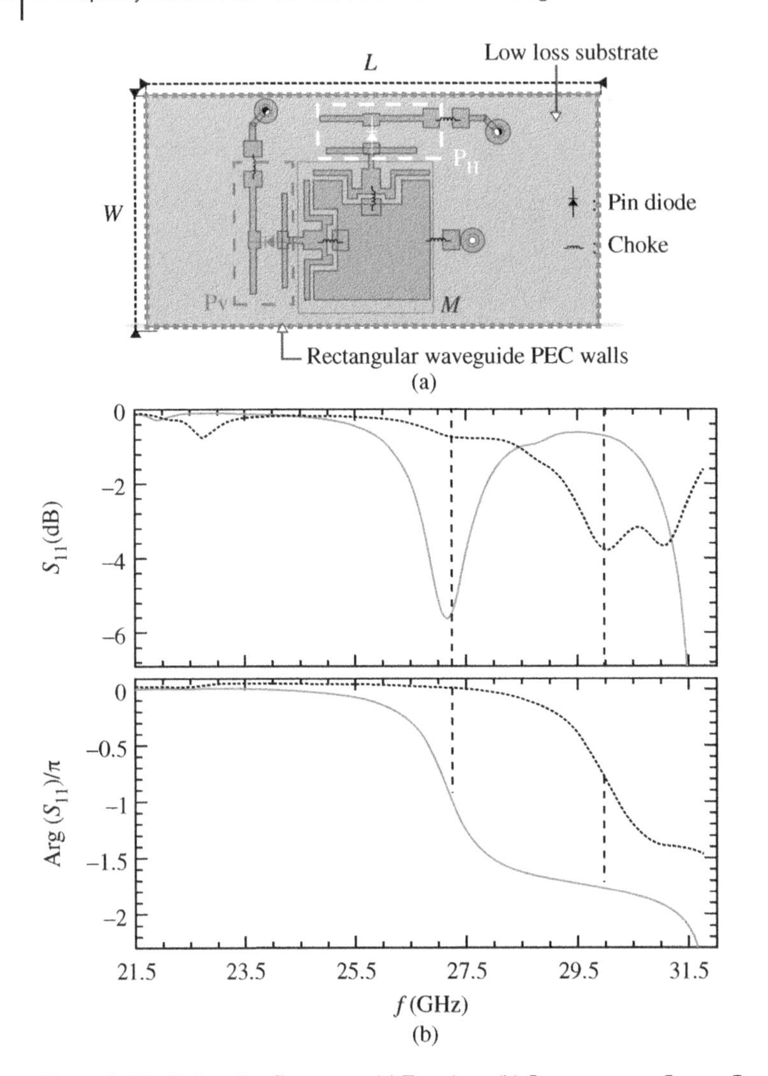

Figure 8.27 Unit cell reflect array. (a) Top view; (b) S-parameters. *Source:* From [19]. ©2021 IEEE. Reproduced with permission.

When a voltage is biased to the LC, the effective dielectric constant of the LC is changed, and which center frequency can be adjusted by this approach. The insertion loss of the proposed filter is approximately 4.5 dB (Figure 8.28).

4) Tunable reflective-type phase shifter [21]:
The reflective-type phase shifter (RTPS) utilizing silicon-on-insulator (SOI) RF MEMS is presented as shown in Figure 8.29. The tunable phase shifter featured

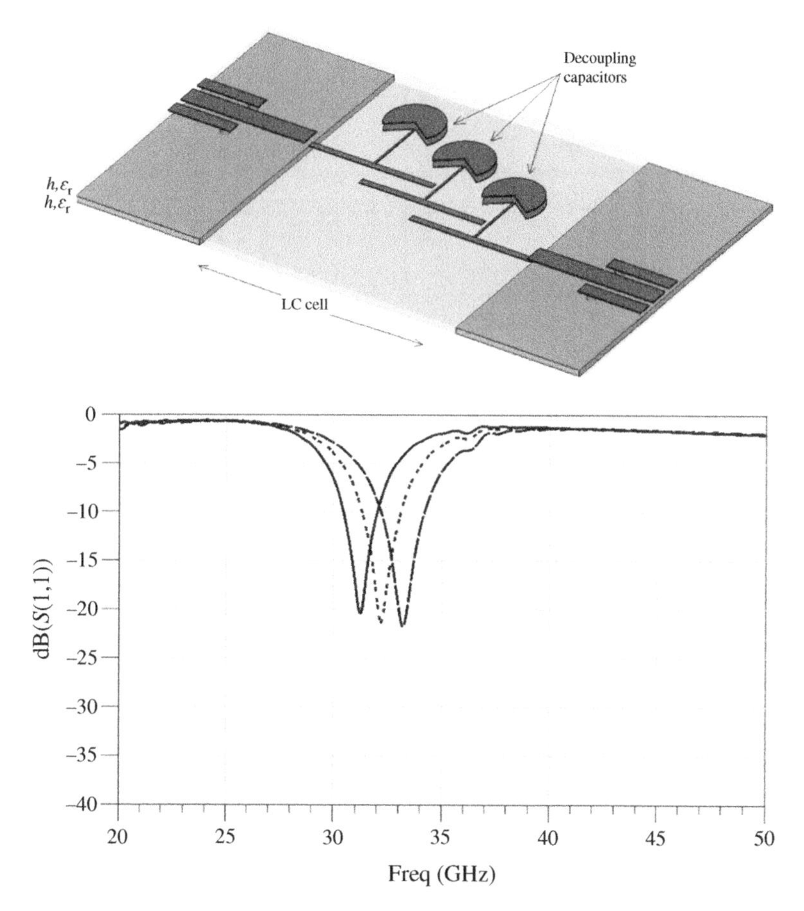

Figure 8.28 The LC-based tunable bandpass filter. *Source:* From [20]. ©2012 IEEE. Reproduced with permission.

a continuous 120° from 26 to 30 GHz. In MEMS-based phase shifters, the phase is controlled by the voltage connected to the pad of the actuator, and when the capacitance of each reflective load is adjusted, a phase shift between the input and output ports is tuned (Figure 8.29).

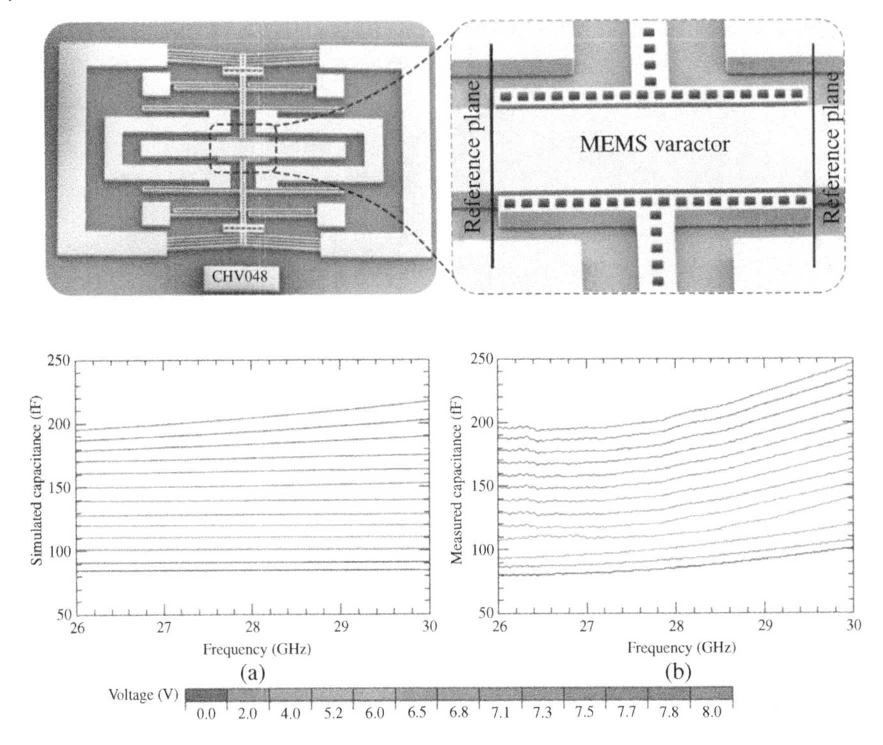

Figure 8.29 Measured capacitance over frequency range with the MEMS. *Source:* From [21]. ©2021 IEEE. Reproduced with permission.

References

1 J. S. McLean, "A re-examination of the fundamental limits on the radiation Q of electrically small antennas," *IEEE Trans. Antennas Propag.*, vol. 44, no. 5, pp. 672–676, May. 1996.

2 D. F. Sievenpiper, D. C. Dawson, M. M. Jacob, T. Kanar, S. Kim, J. Long and R. G. Quarfoth, "Experimental validation of performance limits and design guidelines for small antennas," *IEEE Trans. Antennas Propag.*, vol. 60, no. 1, pp. 8–19, Jan. 2012.

3 3GPP, "NR; User Equipment (UE) radio transmission and reception; Part 2: Range 2 standalone (Release 16)," *Technical Specification (TS) 38.101-2*, version 17.0.0, Dec. 2020.

4 W. L. Stutzman and G. A. Thiele, *Antenna Theory and Design*, 3rd Edition, John Wiley & Sons, Inc., 2012.

5 J. Choi, J. Park, Y. Youn, W. Hwang, H. Seong, Y. Whang, and W. Hong, "Frequency-adjustable planar folded slot antenna using fully integrated multithrow function for 5G mobile devices at millimeter-wave Spectrum," *IEEE Trans. Microwave Theory Tech.*, vol. 68, no. 5, pp. 1872–1881, May. 2020.
6 D. M. Pozar, *Microwave Engineering*, 4th Edition, John Wiley & Sons, Inc., 2011.
7 C. A. Balanis, *Antenna Theory*, 3rd Edition, John Wiley & Sons, Inc., 2005.
8 J. Park, D. Choi, and W. Hong, "28 GHz 5G dual–polarized end fire antenna with electrically-small profile," *in Proc. 12th Eur. Conf. Antennas Propag. (EuCAP)*, London, UK, 2018.
9 J. Park, H. Seong, Y. N. Whang, and W. Hong, "Energy efficient 5G phased–arrays incorporating vertically–polarized endfire planar folded slot antenna for mmWave mobile terminals," *IEEE Trans. Antennas Propag.*, vol. 68, no. 1, pp. 230–241, Jan. 2020.
10 M. Xue, W. Wan, Q. Wang and L. Cao, "Low-profile wideband millimeter-wave antenna-in-package suitable for embedded organic substrate package," *IEEE Trans. Antennas Propag.*, 2021. vol. 69, no. 8, pp. 4401–4411, Aug. 2021.
11 Y. C. Lee and C. S. Park, "A novel CPW-to-stripline vertical via transition using a stagger via structure and embedded air cavities for V-band LTCC SiP applications," *2005 Asia-Pacific Microwave Conference Proceedings*, Suzhou, China, 2005.
12 G. Hernandez-Sosa, R. Torres-Torres, and A. Sanchez, "Impedance matching of traces and multilayer via transitions for on-package links," *IEEE Microwave Wirel. Componen. Lett.*, vol. 21, no. 11, pp. 595–597, Nov. 2011.
13 J. Choi, D. Choi, J. Lee, W. Hwang, and W. Hong, "Adaptive 5G architecture for a mmWave antenna front-end package consisting of tunable matching network and surface-mount technology," *IEEE Trans. Componen. Packag. Manuf. Technol.*, vol. 10, no. 12, pp. 2037–2046, Dec. 2020.
14 3GPP, "NR; User Equipment (UE) radio transmission and reception; Part 2: Range 2 standalone (Release 16)," *3rd Generation Partnership Project (3GPP), Technical Specification (TS) 38.101-2*, version 16.3.1, Mar. 2020.
15 M. O. Sadiku, *Elements of Electromagnetics*, 5th Edition, Oxford University Press, USA, 2009.
16 J. Choi, J. Park, W. Hwang, and W. Hong, "Millimeter-wave 5G antenna-in-package for mobile devices featuring intelligent frequency correction using distributed surface mount technologies," *2021 15th European Conference on Antennas and Propagation (EuCAP)*, DE, 2021.
17 H. Kim, B. Sun Park, S. S. Song, T. S. Moon, S. H. Kim, J. M. Kim, J. Y. Chang and Y. C. Ho, "A 28-GHz CMOS direct conversion transceiver with packaged 2×4 antenna array for 5G cellular system," *IEEE J. Solid-State Circuits*, vol. 53, no. 5, pp. 1245–1259, May 2018.

18 J. Ge and G. Wang, "CmWave to MmWave reconfigurable antenna for 5G applications," *2020 IEEE International Symposium on Antennas and Propagation and North American Radio Science Meeting*, Montreal, QC, Canada, 2020.

19 J. -B. Gros, V. Popov, M. A. Odit, V. Lenets, and G. Lerosey, "A reconfigurable intelligent surface at mmWave based on a binary phase tunable metasurface," *IEEE Open J. Commun. Soc.*, vol. 2, pp. 1055–1064, 2021.

20 M. Yazdanpanahi and D. Mirshekar-Syahkal, "Millimeter-wave liquid-crystal-based tunable bandpass filter," *2012 IEEE Radio and Wireless Symposium*, Santa Clara, CA, USA, 2012.

21 T. Singh, N. K. Khaira, and R. R. Mansour, "Thermally actuated SOI RF MEMS-based fully integrated passive reflective-type analog phase shifter for mmWave applications," *IEEE Trans. Microwave Theory Tech.*, 69, no. 1, pp. 119–131, Jan. 2021.

9

Cost-Effective and Compact Millimeter-Wave 5G Antenna Solutions

9.1 Background

Millimeter-wave (mmWave) antennas are a core technology in 5G new radio (NR) mobile communication to address the increasingly severe bandwidth shortage issues being driven by the exponential growth in global wireless data traffic [1, 2]. Because each national government are assigned to various mmWave spectrums for interference avoidance with the occupied applications, the ultra-wide bandwidth 5G antenna in global roaming cellular devices should be required, as illustrated in Figure 9.1a [3, 4]. Moreover, in order to provide reliable mmWave communications in mobile user equipment (UE), some researches on phased array antennas featuring large scanning coverage with end-fire fan-beam have been reported [4–8]. Thus, these UE antennas for mmWave 5G global service should exhibit wide-angle scanning capability across broadband spectrum.

The mmWave antennas should be integrated into a traditional phased array architecture together with radio frequency integrated circuit (RFIC) chipsets and other components for coverage enhancement [3]. At the same time, multidisciplinary challenges associated with cost-effective fabrication technique, compatibility among design and fabrication technique, and integration with other electronic components can be arisen to realize mobile devices for 5G communication [3]. In particular, various commercial network components should also be integrated into mobile devices and the display panel has been occupied for the most part. Thus, mmWave antenna should be mounted within limited real estate and extreme propagation environments while maintaining sufficient antenna performance at a low fabrication cost [3, 7].

For realizing a compact mmWave antenna with cost-effective and wide-beam coverage in broadband, other critical design factors are to minimize mutual

Microwave and Millimeter-Wave Antenna Design for 5G Smartphone Applications,
First Edition. Wonbin Hong and Chow-Yen-Desmond Sim.

Published 2023 by John Wiley & Sons, Inc.

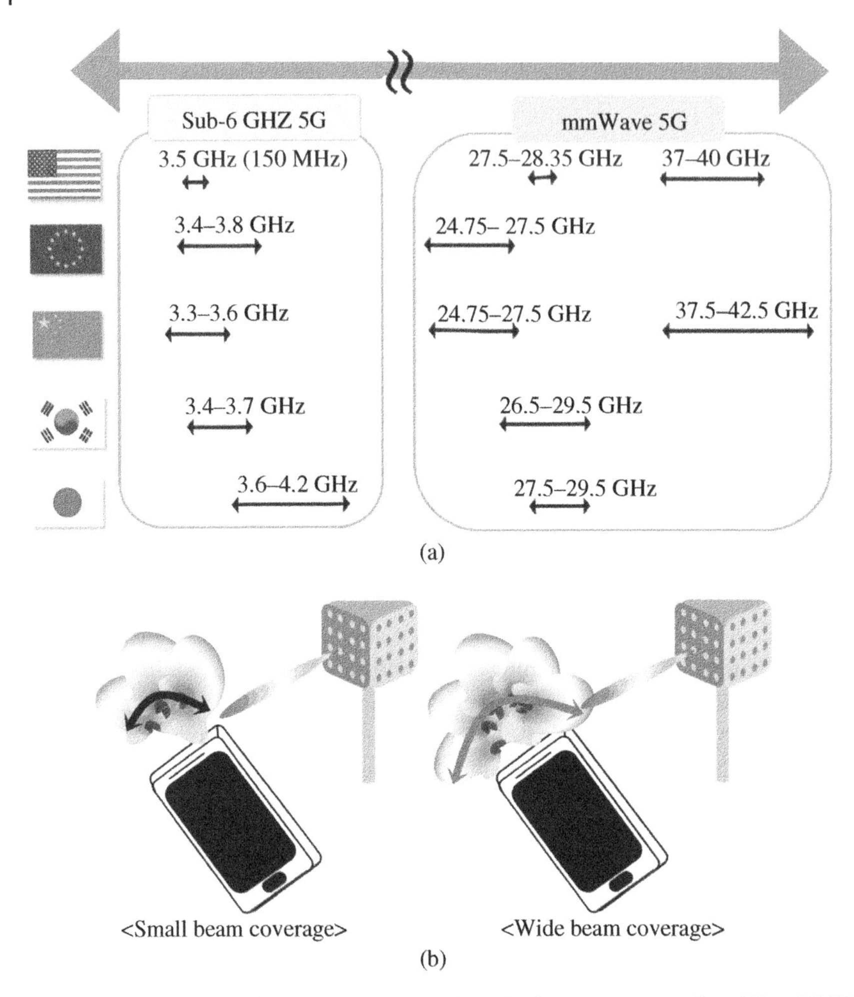

Figure 9.1 (a) Global 5G NR spectrum by regions. (b) Beam coverage of mmWave 5G UE antenna. *Source:* From [4]. ©2021 IEEE. Reproduced with permission.

coupling between each antenna element and insertion loss of feeding networks [1–4]. When multiple elements in mmWave antennas are placed within $\lambda_0/2$ spacing, mutual coupling can degrade the radiation performance of phased arrays. To investigate the reason for mutual coupling and reduce the mutual coupling in compact multiple antenna, several research efforts in multiple-input multiple-output (MIMO) systems and phased array architectures have been devoted, not only from antenna engineers but also from communication societies [9–14].

In this chapter, FR-4 PCB-based cost-effective antenna solutions at microwave and mmWave spectrum are proposed and discussed using planar high-impedance surfaces (HISs) in detail. First, in Section 9.1.1, the faced difficult issues of conventional printing and packaging technologies in mmWave antenna will be discussed in detail. In Section 9.1.2, the fundamental source of mutual coupling and the previous researches on compact antenna technologies featuring the reduction of mutual coupling with low-production cost will be described. Additionally, the previous researches on the design and verification methodologies of low-loss feeding networks will be described in Section 9.1.3. As the first implemented example of a cost-effective antenna element in this chapter, a compact inverted-L antenna (ILA) topology integrating with one-dimensional electromagnetic (EM) bandgap (1-D EBG) ground structures has been presented to enhance antenna performance on FR-4 printed circuit board (PCB) at microwave regime [15]. The 1-D EBG ground structures in ILA topology within a compact size ($0.21\lambda_0 \times 0.32\lambda_0$) can serve as HIS with slow-wave behavior controlling both space-wave and surface-wave. Afterward, by utilizing the ILA topology with 1-D EBG structures, a compact millimeter-wave (mmWave) phased array additionally incorporating via wall structures for HIS properties has been presented [4, 16]. By designing and measuring large-scaled phased arrays fabricated on single-layer FR-4 PCB without any additional packaging process, this phased array antenna in linear array configurations has been realized and investigated about wide-angle scanning capability in broadband, respectively.

9.1.1 Challenging Issues of Conventional Printing and Packaging Techniques in mmWave Antenna

As depicted in Figure 9.2, the conventional printing and packaging technologies in mmWave antenna, which are usually utilized in mobile devices, exist troublesome problems between production cost and antenna performance [4, 17, 18]. Figure 9.2a illustrates that these fabrication technologies mainly determine the production cost and applicable operating frequency as this is linked to the fabrication resolution, such as minimum metal line width-spacing and thickness [4]. Moreover, in Figure 9.2b, the material properties of the utilized substrate and metal line for the mmWave antenna can impact significantly the antenna performance, such as antenna gain and loss of feeding networks.

In low-frequency mmWave spectrum below 100 GHz, antenna-in-package (AiP) techniques have been mainly utilized [18]. The representative AiPs are low-layer PCB, high-density interconnect (HDI) PCB, low-temperature co-fired ceramic (LTCC), and fan-out wafer-level package (FOWLP) process [7, 18–20]. However, due to the fabrication resolution limit of 3-D feeding networks consisting of via hole and metal line in AiP, it can cause impedance mismatch and large insertion loss. At sub-terahertz (sub-THz) frequency, the antenna-on-chip (AoC) based on

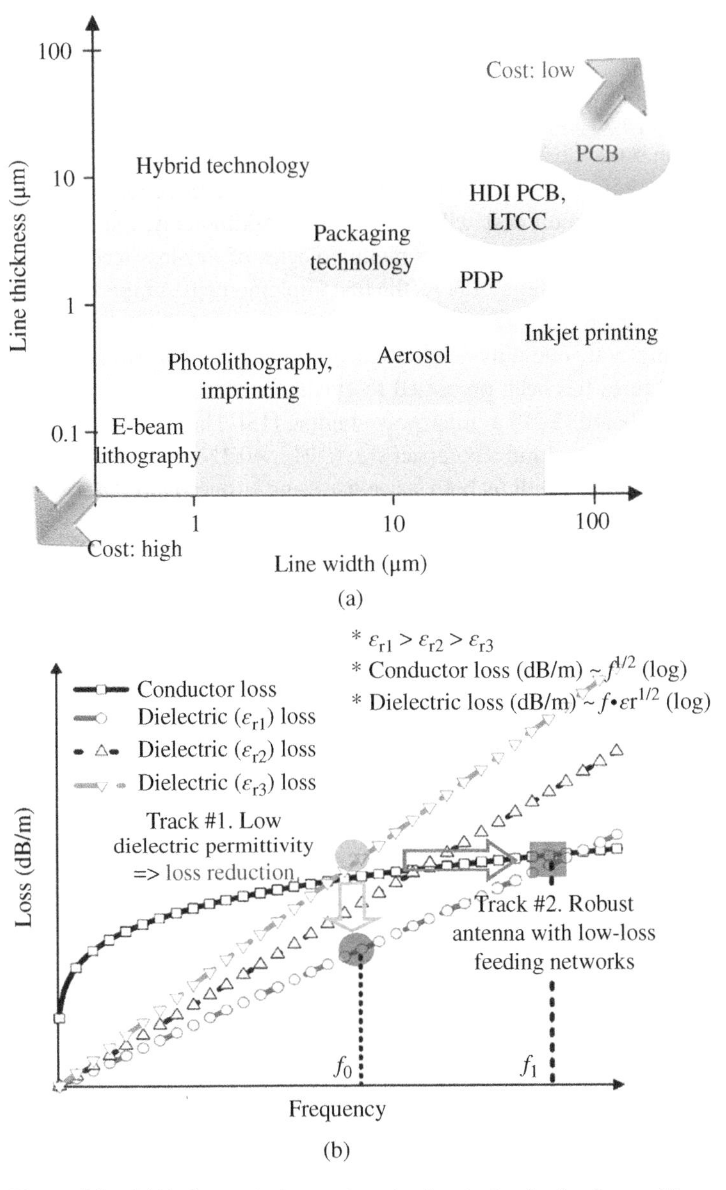

Figure 9.2 (a) Various printing and packaging technologies for mmWave antenna applications. *Source:* From [4]. ©2021 IEEE. Reproduced with permission. (b) Relation between operating frequency and loss in mmWave antenna by various material properties.

the standard photolithography process has been actively spotlighted due to easy integration with other Si CMOS components and no need for additional interconnect structures.

Nevertheless, AoC inherently exhibits low radiation performance due to extremely thin back-end-of-the-line (BEOL) dielectric layers (≤10 μm thickness) [21–23]. Recently, in order to achieve high performance with cheap production cost and insufficient fabrication resolution, single-layer PCB or multi-layer FR-4 substrate-based mmWave UE antennas have been presented [7, 24–27]. In particular, as illustrated in Figure 9.2b, two tracks of mmWave antenna solution have been described. First, the antenna fabricated on single-layer PCB exhibits high radiation performance using low dielectric constant and low dielectric lossy media, such as Teflon substrate [24, 25]. Second, a 60-GHz antenna module embedded in a multi-layer FR-4 substrate provides low-loss feeding networks with 3-D interconnect structures [7, 26, 27].

However, in order to further reduce the production cost with performance enhancement despite insufficient fabrication resolution (100 μm minimum metal width) and large loss materials (tanδ: 0.032 at 28 GHz), the research on single-layer FR-4 PCB process-based mmWave phased arrays should be required.

9.1.2 Previous Researches on Mutual Coupling in Planar Multiple Antenna

When each antenna element in planar multiple antenna (e.g. MIMO antenna and phased array antenna) has been tightly arranged, some of the EM energy of one antenna element is transferred to the other elements. The amount of the mutually coupled energy depends on the main radiation characteristics of each antenna in near-field or far-field, spacing between the antenna elements, and adjacent environment. An EM field exists around the antennas due to the nature of the current distribution on linear antennas, leading to radiation energy into free space, and some energy possibly coupling to adjacent elements as illustrated in Figure 9.3. This coupling energy is regarded as mutual coupling [9–14]. Although the antenna impedance can be tuned or optimized by additionally reconfigurable devices, the mutual coupling as the dominant reason that deteriorates the performance of multiple antenna systems still does not remove.

Due to the effect of mutual coupling causing the induced current, closely spaced antennas would induce the distorted current distribution and the unintended input antenna impedance in turn unlike single antenna element. In the case of a single antenna without mutual coupling, the input impedance of an antenna is equal to its self-impedance. However, with closely spaced antenna elements, the input impedance depends on the mutual impedance and the currents on

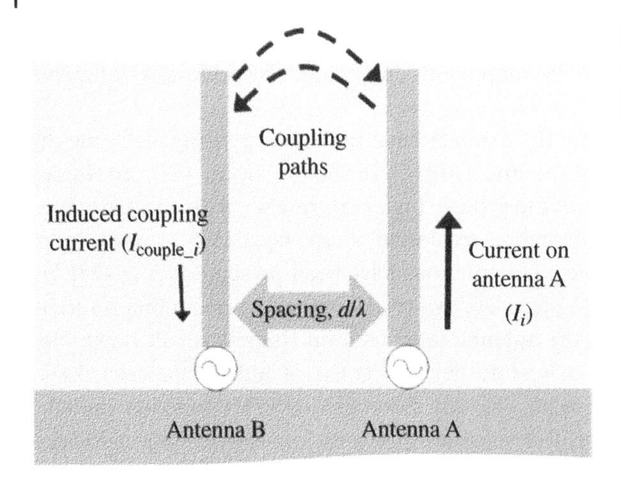

Figure 9.3 Illustration of coupling paths between antennas A and B at transmitting mode.

both antennas, as well as its self-impedance. In order to simply represent the aforementioned analysis, this impedance relation can be expressed as:

$$Z_{\text{in},i} = Z_{ii} + Z_{ji}\frac{I_j}{I_i} \tag{9.1}$$

where Z_{ii} and Z_{ji} are the self- and mutual impedances, respectively. I_j and I_i refer to the internal current.

The mutual impedance between two dipole antennas with two configurations (parallel and collinear) in Figure 8.21 *(Chapter 8 in "Antenna Theory Anaysis and Design,* 3rd edition," *written by Constantine A. Balanis)* is illustrated as calculating the normalized spacing between antenna elements, d/λ, where λ is the operating wavelength of the radiation elements [10]. The dipole antennas mounted by parallel configurations can be usually utilized for the planar MIMO antenna. On the other hand, the dipole antennas mounted by collinear configurations can be usually utilized for the phased array antenna. As the spacing between antenna elements in both configurations increases, the mutual impedance in all cases decreases. Therefore, the performance of the antenna array remains rarely unaffected due to the reduction of mutual coupling when the spacing between antenna elements is spatially spaced. However, in the worst case (e.g. $d/\lambda < 0.2$ in MIMO antennas and $d/\lambda < 0.5$ in array antennas) of compact multiple antennas, the deteriorated antenna performances relating to the total antenna efficiency can result in a large reduction of the realized gain. In general mobile communication devices, the antenna elements are closely spaced due to a lack of real estate, and as such, it is of critical issue to reduce the mutual coupling to improve the antenna performances.

In order to analyze the fundamental generation reason for mutual coupling, Maxwell's equations for a non-dispersive and isotropic media can be described as:

$$\nabla \cdot \bar{\mathcal{E}} = \frac{\rho}{\varepsilon} \tag{9.2}$$

$$\nabla \cdot \bar{\mathcal{H}} = 0 \tag{9.3}$$

$$\nabla \times \bar{\mathcal{E}} = -\bar{\mathcal{M}} - \mu \frac{\partial \bar{\mathcal{H}}}{\partial t} \tag{9.4}$$

$$\nabla \times \bar{\mathcal{H}} = \bar{\mathcal{J}} + \varepsilon \frac{\partial \bar{\mathcal{E}}}{\partial t} \tag{9.5}$$

where $\bar{\mathcal{E}}$, $\bar{\mathcal{H}}$, $\bar{\mathcal{J}}$, $\bar{\mathcal{M}}$, ρ, ε, μ, $\varepsilon \frac{\partial \bar{\mathcal{E}}}{\partial t}$ are the electric field, the magnetic field, the electric current density, the magnetic current density, the electric charge density, the permittivity of the media, the permeability of the media, and the displacement current, respectively. This displacement current is another current source of strong mutual coupling in multiple antenna systems [10–14]. To discuss more physical meaning in that kind of current, it should be considered utilizing two tightly coupled antennas and spatially spaced in free space. In order to account for the mutual coupling between the antennas and the current path through the displacement current, the coupling path term of closely coupled antennas about each case of MIMO antenna and phased array antenna is in detail depicted in Figure 9.4. The current path strongly coupled each antenna element, and within which EM fields exist around the path between antennas. Thus, for mitigating coupling paths, an appropriate synthesis technique or decoupling network in and out of antenna. Two main paths causing mutual coupling are **surface-wave (or current) coupling** and **space-wave coupling** [12–14]. In this chapter, a decoupling layer will be introduced to reduce mutual coupling in these paths.

Patch antennas and linear antennas with the common ground plane are widely known to occur in surface-wave modes as dielectric slabs do [12–14]. As depicted in Figure 9.5, these surface-waves are guided between the substrate and ground plane, and propagate along the dielectric-air interface with exponential decay. The power launched into surface-waves will be finally lost, thus deteriorating the radiation efficiency of patch antennas. In the case of a planar antenna, the surface-wave between antenna elements is transformed into the flowing surface current into the edge of a common ground plane. As especially illustrated in linear array antennas of Figure 9.4b, the HIS condition at the edge of the ground plane in a single antenna element could be converted with the low-impedance surface condition [4]. This converted impedance surface condition can result from coupling path through surface current flowing on the common ground plane of closely packed array elements.

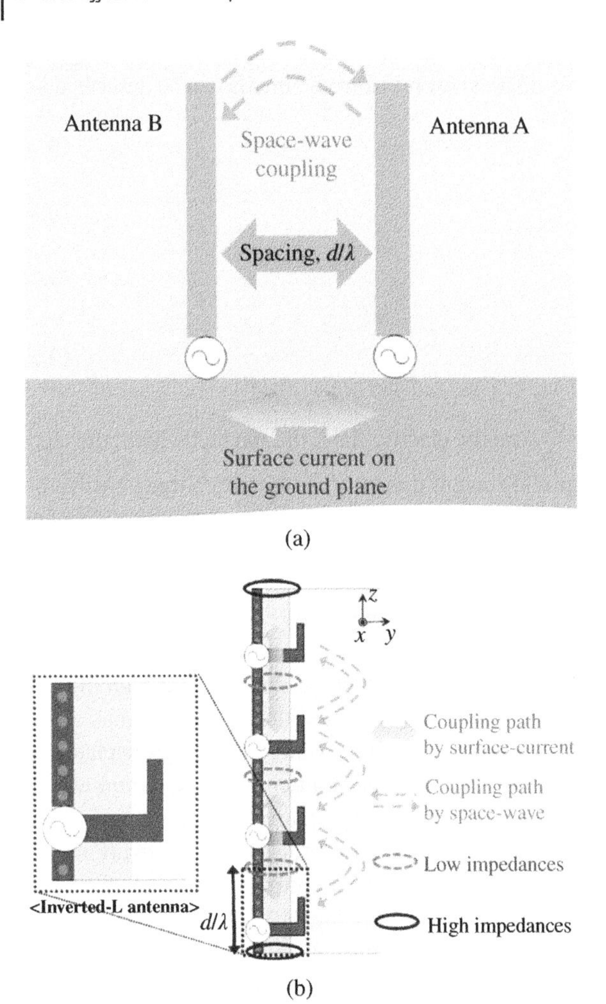

Figure 9.4 Representative modeling for sources of mutual coupling. (a) Planar MIMO antennas. (b) Linear array antennas. *Source:* From [4]. ©2021 IEEE. Reproduced with permission.

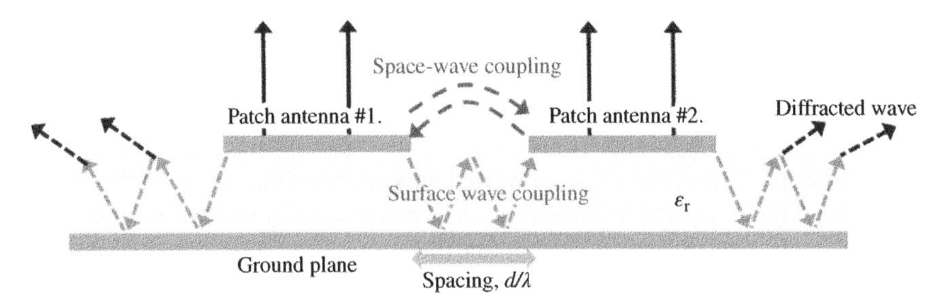

Figure 9.5 Representative modeling for sources of mutual coupling in low-profile microstrip patch antenna.

Meanwhile, the space-wave coupling path can exist not only in many types of matter but also in free space. Especially, when the spacing between antenna elements is within the near-field zone, space-wave coupling increases, wherein the fringing fields couple among each planar antenna (e.g. patch, monopole, and dipole) element and can create dominant coupling paths.

As HIS properties for coupling reduction in the surface-wave coupling path, the mushroom-like 2-D EBG, cavity walls and defected ground structures have been embedded on a common ground plane [26–35]. By those HIS structures, the mutual coupling due to surface current on a common ground plane within compact elements is suppressed and then wide-beam coverage can be secured [30, 31]. Meanwhile, in order to reduce the space-wave coupling path, the decoupling spatial resonators, such as parasitic elements and metamaterials, have been inserted between antenna elements [12, 36, 37]. Those decoupling spatial resonators provide cancelation effect of the coupling field by creating an additional coupling path between closely packed antenna elements within a compact form factor.

The 2-D EBG structures consisting of periodic unit cells have been frequently reported in EM research groups for realizing high performance within compact antenna sizes [32–35]. The 2-D EBG structures exhibit two distinctive EM properties in multiple antennae [32, 35]. One is the HIS properties reducing the surface current on the common ground plane and the other is reflection phase properties radiating energies toward the desired directions. In particular, the EBG structures in the surface-wave bandgaps operate as HIS with slow-wave behavior controlling both space-wave and surface-wave [32, 35]. Although the antenna can be placed adjacent to the EBG structures, which have in-phase reflection properties, it is difficult to achieve good impedance matching at the resonance frequency of EBG structures. At a quadratic reflection phase ($90° \pm 45°$), the antenna with EBG structures can exhibit impedance matching and high radiation efficiency by utilizing both EM properties between surface-wave bandgaps and input-match frequency band [33].

Recently, by utilizing identical operating properties of 2-D EBG structures, the antenna applications with 1-D EBG structures have been presented to exhibit good impedance matching and high radiation performance in further compact form factors rather than 2-D EBG structures [38–40]. In the initial research of [38], the interdigitated patches and meandered lines in the 1-D EBG unit cells have been designed to make the unit cell featuring sub-wavelength structures. The low-profile dipole antennas integrated with balun and 1-D EBG ground structures can achieve directive radiation patterns as a reflector at quadratic reflection phase. And, in [39], by utilizing the reflection properties of 1-D EBG ground structures, multiple antennas fabricated on handset configurations with flexible PCB (FPCB) have been investigated for the reduction of mutual coupling. Especially, for mitigating the correlation between antennas, the 1-D EBG structures have been utilized by reducing the surface current on the common ground plane or mitigating radiation energies in the opposite directions, respectively. Lastly, low-profile planar MIMO antennas featuring reduction of mutual coupling are presented using 1-D EBG and split ring resonator (SRR) structures [40]. By the simulation and measurement, it has been ascertained that the 1-D EBG and SRR structures in the MIMO antenna serve as a reflector and wave trap, respectively.

However, in the initial researches on the antenna containing 1-D EBG structures, there are some limitations to adapting the mmWave antenna. In [38], since the dipole antenna with microstrip balun for a balanced feeding is fabricated to the radiator, the overall physical length of the antenna is greater than $0.8\lambda_0$. So, the antenna impedance bandwidth is dependent on the operating bandwidth of the balun. In [40], the 1-D EBG and SRR structures are inserted between two closely located straight monopole antennas. The overall length of one antenna element is $0.47\lambda_0$ and the antenna impedance bandwidth is limited by the characteristic mode of ground plane. In addition, to achieve scalable array performance with the reduction of mutual coupling in element shapes, inter-element spacing greater than $0.5\lambda_0$ is required [38, 40]. Although auxiliary circuits limiting antenna impedance bandwidth are utilized, further miniaturization of antenna elements for robust arrays featuring broad bandwidth, high gain, wide-angle scanning, and universal antenna design should be realized despite extreme propagation environments.

9.1.3 Previous Researches on Low-Loss Feeding Networks

Traditionally, the feeding network in mmWave spectrum is realized by utilizing transmission lines (TL), power dividers, couplers, butler matrix [41]. In order to achieve high-precision design and fast analysis of mmWave feeding network, there are some considerations related to the design, fabrication and measurement techniques. First, due to a lack of available information on mmWave material properties beyond 10 GHz, unwanted discrepancies between the design and measurement of feeding networks can often occur. Second, due to the absence of calibration kits in measuring instruments supporting any desired accessory configurations, the alternative measurement techniques featuring reliable data and fast measuring time cannot be often secured. Particularly, the discrepancy between the simulated and measured insertion loss of some feeding networks featuring high mutual coupling between ports cannot be cannot be minimized.

In order to design and characterize mmWave feeding networks with a power divider fabricated on FR-4 PCB process, the multi-port measurement procedure in two cases has been discussed [42]. First, the fabricated feeding network for 4 array elements is measured and characterized, as illustrated in Figure 9.6 [42]. The feeding network consists of the island-shaped coplanar waveguide ground

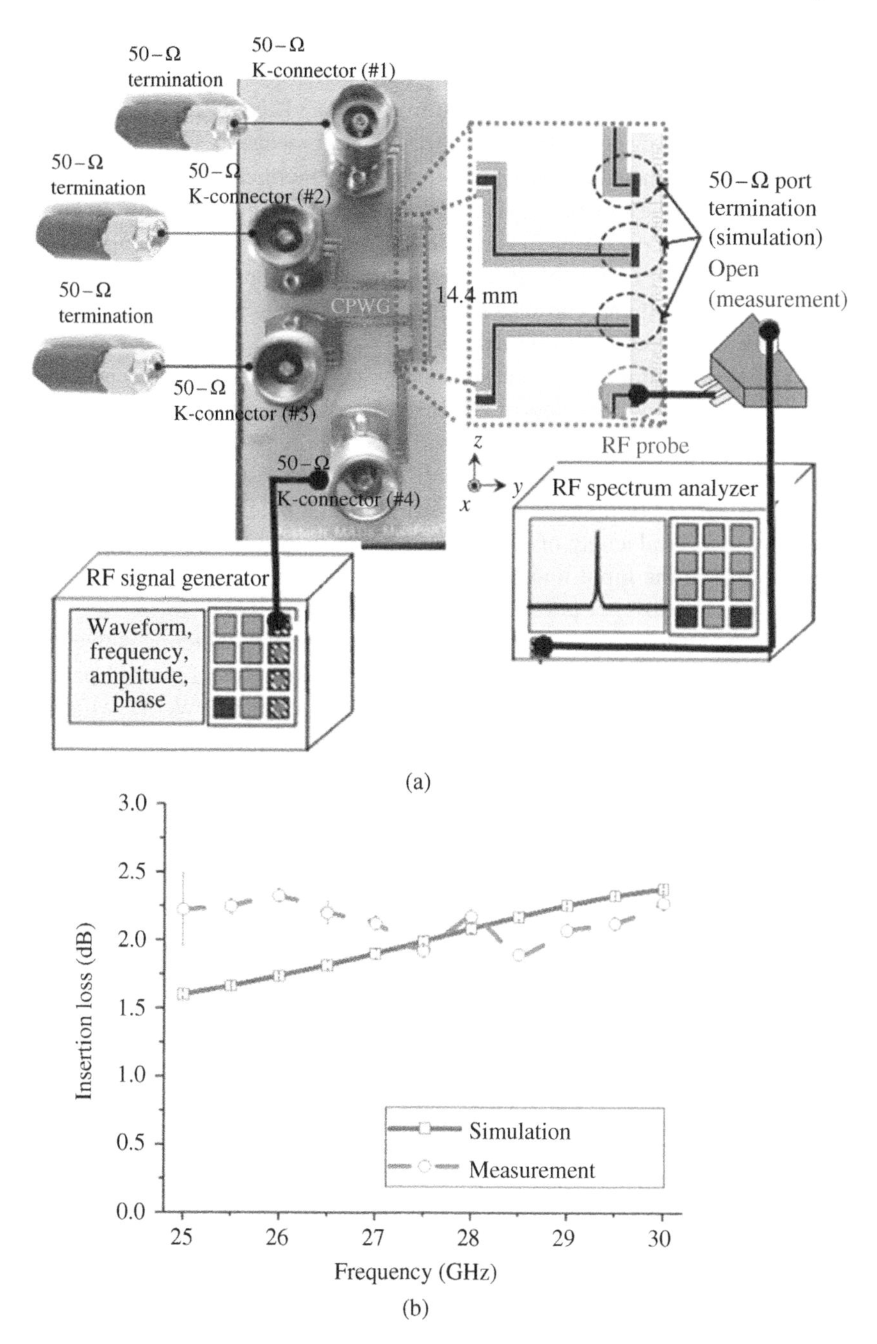

Figure 9.6 Characterization of feeding network for 4 array elements. (a) 4by4 port CPWG TLs and its measurement setup to estimate insertion loss. (b) simulated and measured insertion loss (line: average value between ports, error bar: standard deviation between ports). *Source:* From [42]. ©2019 IEEE. Reproduced with permission.

(CPWG) TL with lots of via wall structures serving as HIS for reduction of surface-wave. For both impedance matching and good radiation performance, the characteristic impedances of transmission lines among K-connector, CPWG and antenna element are designed from 50-Ω to 65-Ω that meets the PCB design rule with compact size. Since the landing pad size of the K-connectors is 10.5 mm and the entire length of the 4-elements ILA is 14.4 mm, each in-phase feeding network connected to the antenna element is required to be designed by using long TL.

Due to the absence of a conventional calibration kit supporting connector-to-probe configuration, there is no available method to measure the insertion loss by using conventional vector network analyzers (VNAs). As an alternative, by utilizing the signal generator (SG), spectrum analyzer (SA), connector, adaptor, and GGB picoprobe, the insertion loss of feeding network is measured, with the compensated loss of additionally used accessories. Since the measured mutual coupling among the K-connector is below −30 dB by using VNA, any power imbalance or oscillation derived from the reflected power through the open in other ports is not detected. The total length of each feeding network channel from the connector contact pad to the input node of each antenna element is 15 mm. Thus, the measured insertion loss per physical length is from 0.14 to 0.17 dB/mm.

Second, to achieve high gain and power combining for array antenna with eight elements, the Figure 9.7 illustrates feeding network including T-shaped power divider [42]. The fundamental TL consisting of island-shaped CPWG TL with via wall structures fabricated on FR-4 PCB are identical in Figure 9.6.

Although the design and characterization method remains identical to that discussed in Figure 9.6, power imbalance between ports and large oscillation phenomenon is detected. Since the T-shaped power divider inherently exhibits high mutual coupling between adjacent output ports beyond −2 dB, power imbalance and oscillation result from the reflected power derived from the open port. To estimate the insertion loss without any power imbalance and oscillation, the back-to-back transition structures and low mutual coupling power divider should be required.

9.2 Compact Inverted-L Antenna Element with 1-D EBG Structures

In order to implement the first example of a cost-effective antenna element in this chapter, a compact ILA topology integrating with 1-D EBG ground structures fabricated on FR-4 PCB are presented and characterized at a microwave regime. In this Section, by simulation and measurement results, it can be verified that the 1-D EBG ground structures in compact ILA (antenna size: $0.21\lambda_0 \times 0.32\lambda_0$) topology serve as HIS with slow-wave behavior controlling both space-wave and

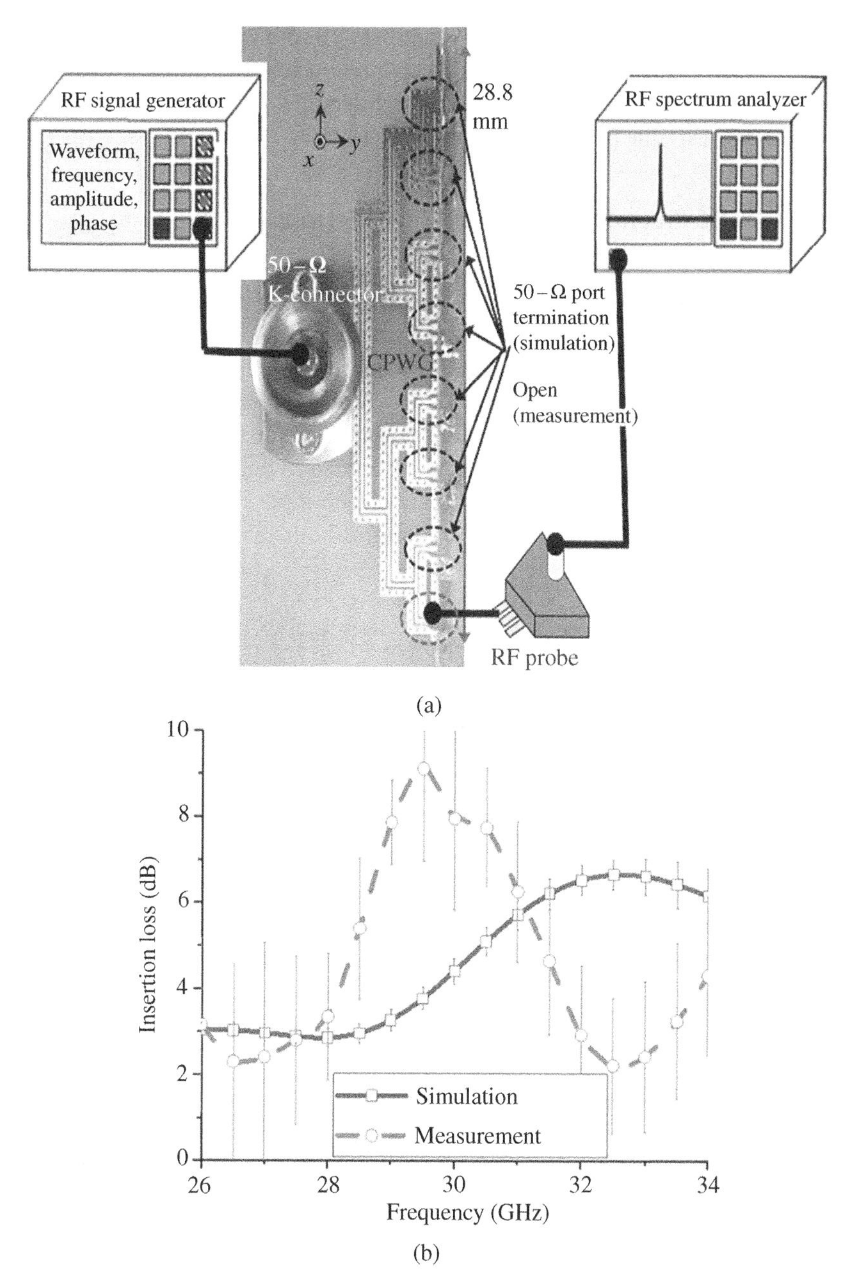

Figure 9.7 Characterization of feeding network including power divider. (a) 1by8 port CPWG TLs and its measurement setup to estimate insertion loss. (b) simulated and measured insertion loss (line: average value between ports, error bar: standard deviation between ports). *Source:* From [42]. ©2019 IEEE. Reproduced with permission.

surface-wave. In addition, without any auxiliary circuits limiting the antenna performance, the antenna with 1-D EBG ground structures featuring the HIS properties can be a promising array element, which provide high antenna performance, despite extreme propagation environments.

9.2.1 1-D EBG Ground Structures and Their Electromagnetic Characteristics

The antenna is connected through the microstrip line with the coaxial connector for a feeding accessory, and the 1-D EBG structures are embedded on the edge of the partially grounded front side of an FR-4 substrate which features a size of 70 mm × 40 mm × 1 mm and a relative permittivity of 4.4, as depicted in Figure 9.8. On the front side edge of the ground planes that are closer to the planar ILA, a 1-D EBG structure with seven periodic cells is embedded in one column to serve as HIS with slow-wave behavior. The designed technique of 1-D EBG ground structures is simultaneously intended to control both the reflection phase in

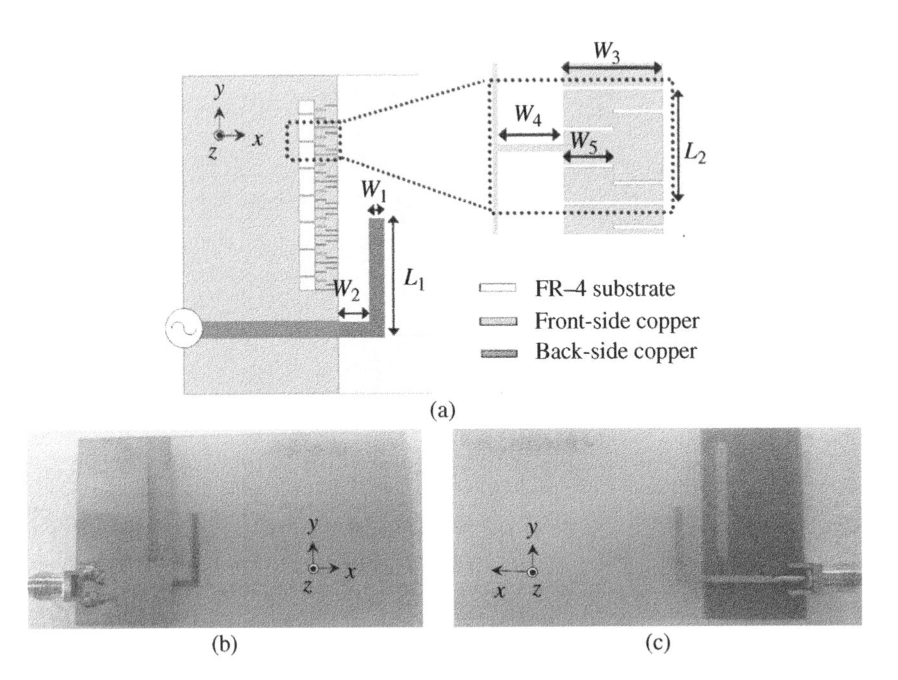

Figure 9.8 Configuration of planar ILA with 1-D EBG ground structures [15]. (a) Entire view of the fabricated antenna including the unit cell of 1-D EBG structures. (b) Photograph of front view and (c) back view of planar ILA with 1-D EBG ground structures (w_1 = 1.5 mm, w_2 = 4 mm, w_3 = 3 mm, w_4 = 2 mm, w_5 = 1.5 mm, L_1 = 15 mm, L_2 = 3.3 mm, g [slit width & edge-to-edge spacing between EBG cells] = 0.1 mm).

plane-wave illumination and surface-wave on the horizontal direction of the ground plane edge. The 1-D EBG unit cell consists of a metal patch with narrow-slitted patterns, a narrow gap capacitor between the EBG cells, and a stripline. The slit width and gap between EBG cells are designed to a minimum value of 0.1 mm due to PCB manufacturing limitations. The size of the unit cell and the number of 1-D EBG structures are the most essential factor to exhibit the unique HIS characteristics of the 1-D EBG ground structures optimized at the operating frequency, 2.4 GHz.

In order to investigate the properties of the 1-D EBG structures, the reflection phase of the partially grounded FR-4 substrates is computed using the 1-D imaginary line of the simulation manual of high-frequency structure simulation (HFSS) package in [38, 40]. As illustrated in Figure 9.9, the substrate containing 1-D EBG structures provides a quadratic reflection phase at 2.4 GHz. The substrate without the 1-D EBG structures exhibits a quadratic reflection phase above 3 GHz. By comparing the substrate with and without 1-D EBG structures, the calculated reflection phase exhibits relatively steep phase changes due to HIS properties of the 1-D EBG structures. Within the same dimensions (20 mm × 40 mm), the EBG ground structures can manipulate the reflection phase as HIS with slow-wave behavior in plane-wave illumination.

As depicted in Figure 9.10, the slow-wave behavior toward horizontal direction of the edge of 1-D EBG structures can be demonstrated not by the dispersion analysis, but by the calculated group delay. Because this simulation method is not easy

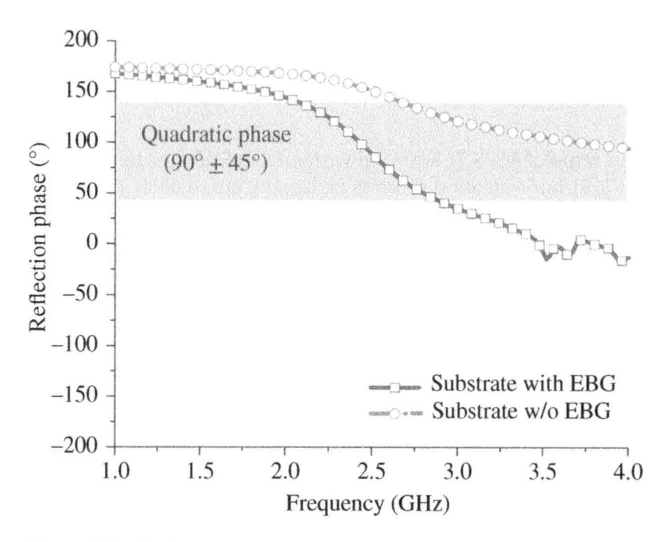

Figure 9.9 Reflection phase of partially grounded FR-4 substrates with and without 1-D EBG structures [15].

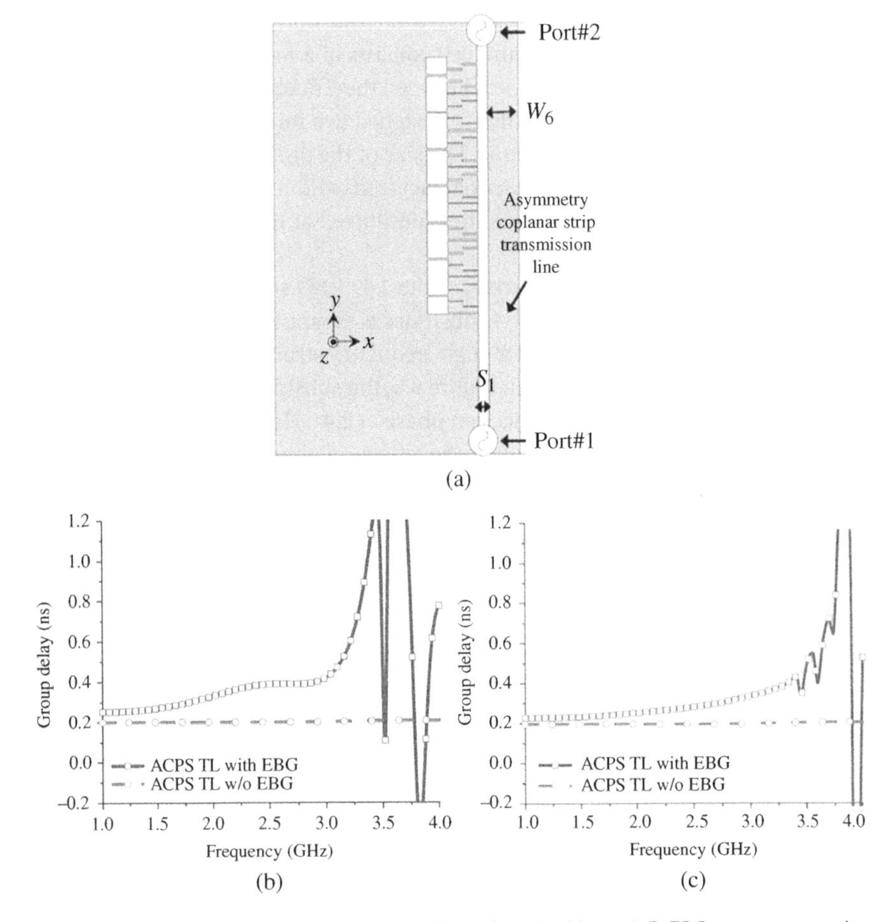

Figure 9.10 Configuration setup of ACPS TL with and without 1-D EBG structures, and the calculated group delay [15]. (a) Simulation setup to obtain the group delay. (b) 50-Ω characteristic impedance TL. (c) 100-Ω characteristic impedance TL.

to obtain dispersion analysis due to finite 1-D array structures in the ground plane, the asymmetry coplanar strip (ACPS) TL for surface-wave modeling is placed on the right side edge of the ground plane. By using two-port S-parameters in this simulation scenario, the group delay can be computed according to (9.6). The slow-wave behavior controlling the surface-wave on the ground plane edge can be verified by comparing the group delay with and without 1-D EBG structures. The group delay τ_g is defined as:

$$\tau_g = -\frac{\partial \phi}{2\pi \cdot \partial f} \tag{9.6}$$

where ϕ is the phase of S_{21} (or S_{12}) and f is frequency.

In order to investigate the unique properties of 1-D EBG structures at various surface impedance, the characteristic impedances of the ACPS TL are designed by 50-Ω and 100-Ω. In the case of 50-Ω characteristic impedance of the ACPS TL, the entire ACPS TL structure is designed to the w_6, of 4 mm and S_1 of 0.2 mm, respectively. At 100-Ω characteristic impedance, the W_6 and S_1 are 3 and 1 mm, respectively.

For both 50-Ω and 100-Ω characteristic impedances, the group delay of ACPS TL without 1-D EBG ground structures is equal to 0.2 ns in the frequency range from 1 to 4 GHz. The group delay of ACPS TL with 1-D EBG ground structures at this characteristic impedance is different despite the intrinsic surface impedance of 1-D EBG structures. Nevertheless, the slow-wave behavior due to the 1-D EBG structures can be regarded as a large group delay.

9.2.2 Compact ILA with 1-D EBG Ground Structures

In Figure 9.11, the ILA with the 1-D EBG structures operates at 2.4 GHz. The ILA without the 1-D EBG structures and the planar inverted-F antenna (PIFA) without the 1-D EBG structures are fabricated with the same lengths (L_1) in order to study the properties of the 1-D EBG structures, as depicted in the inset of Figure 9.10.

Because the ground planes without the 1-D EBG structures exhibit a quadratic reflection phase over 3 GHz, the ILA without the 1-D EBG structures operates

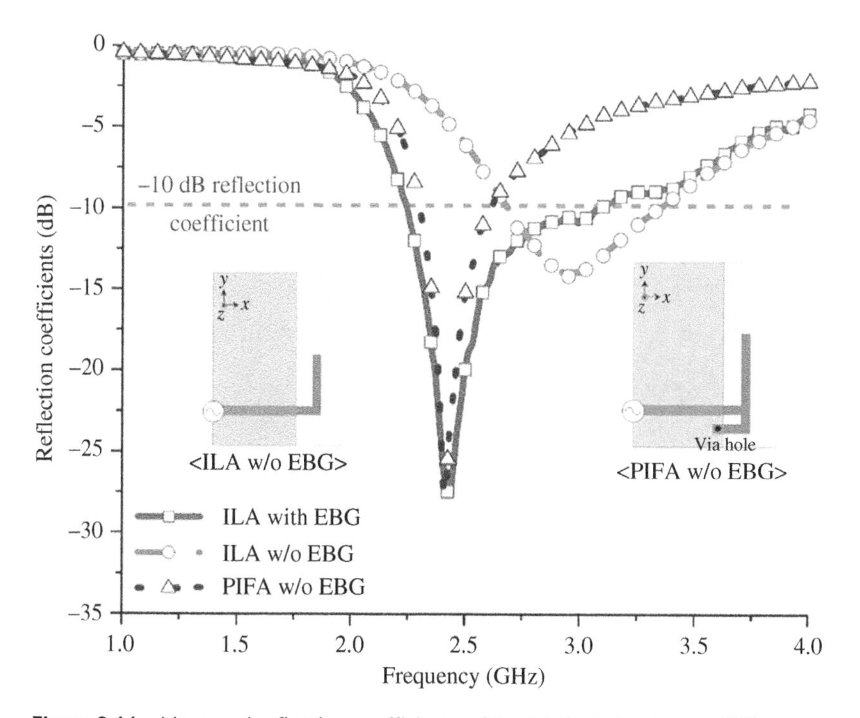

Figure 9.11 Measured reflection coefficients of the fabricated antennas [15].

near 3 GHz. In the PIFA without the 1-D EBG structures, the antenna can operate at the same frequency as that of the ILA with 1-D EBG structures due to identical monopole lengths. The ILA with the 1-D EBG structures exhibits a − 10 dB impedance bandwidth from 2.24 to 3.11 GHz. Despite the relatively steep phase changes presented in Figure 9.9, the impedance bandwidth of the ILA with 1-D EBG structures is wider than that of other antennas in Figure 9.11. Thus, the ILA topology with 1-D EBG structures can become a promising element for broadband array antennas due to the absence of auxiliary circuits that can limit the antenna impedance bandwidth, such as via holes of PIFA.

As illustrated in Figure 9.12a,b, the radiation patterns in the ILA with the 1-D EBG structures indicate a maximum gain of 2.0 dBi at the Co-polarization (E_{phi}) considered as horizontal polarization. The designed antenna at the horizontal polarization due to the quadratic reflection phase of the EBG structures can achieve directive radiation patterns toward end-fire direction (+x-direction). The radiation patterns of ILA with the 1-D EBG structures in the horizontal polarization toward end-fire direction exhibit highly directive more than those of the other antennas. Meanwhile, as depicted in Figure 9.12c,d for cross-polarization (E_{theta}), the radiation patterns of the ILA with the 1-D EBG structures are reduced more than those of the other antennas. The 1-D EBG ground structures can enhance the radiation patterns of ILA without any auxiliary circuits.

The slow-wave behavior by compact size of the ground plane with 1-D EBG structures can also be examined in the near-field distributions using HFSS. Three

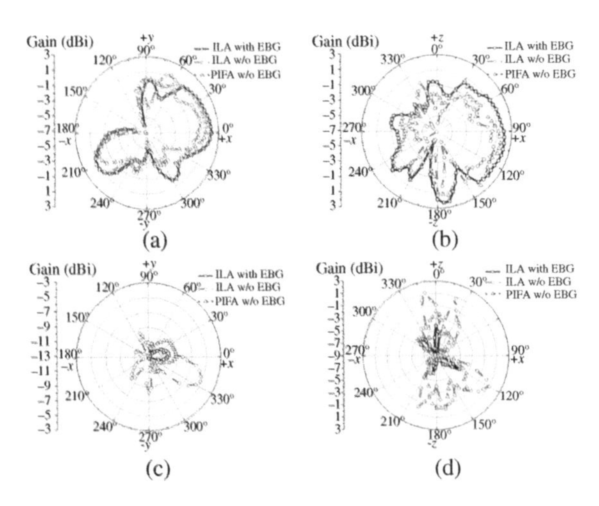

Figure 9.12 Measured radiation patterns for the fabricated antennas at operating frequency [15]. (a) xy-cut ($\theta = 90°$) @ E_{phi}-polarization. (a) zx-cut ($\phi = 0°$) @ E_{phi}-polarization. (c) xy-cut ($\theta = 90°$) (b) @ E_{theta}-polarization. (d) zx-cut ($\phi = 0°$) (c) @ E_{theta}-polarization.

types of antennas are activated via a 50-Ω coaxial connector to obtain the near-field distributions. To study the near-field in a transverse plane, a contour plot (*XY*-plane) of the total electric field distribution located at 0.5 mm higher than the patterned ground structures is computed [15, 40].

As presented in Figure 9.13a,b, at the operating frequency, the ILA with the 1-D EBG structures exhibits more intensive contour lines at the imaginary lines of AB than that of other antennas (lines CD and EF). It can be ascertained that the wave impedance variation of the ground plane edge is being made abruptly due to the HIS of the 1-D EBG structures within the small spaces. Thus, the 1-D EBG ground

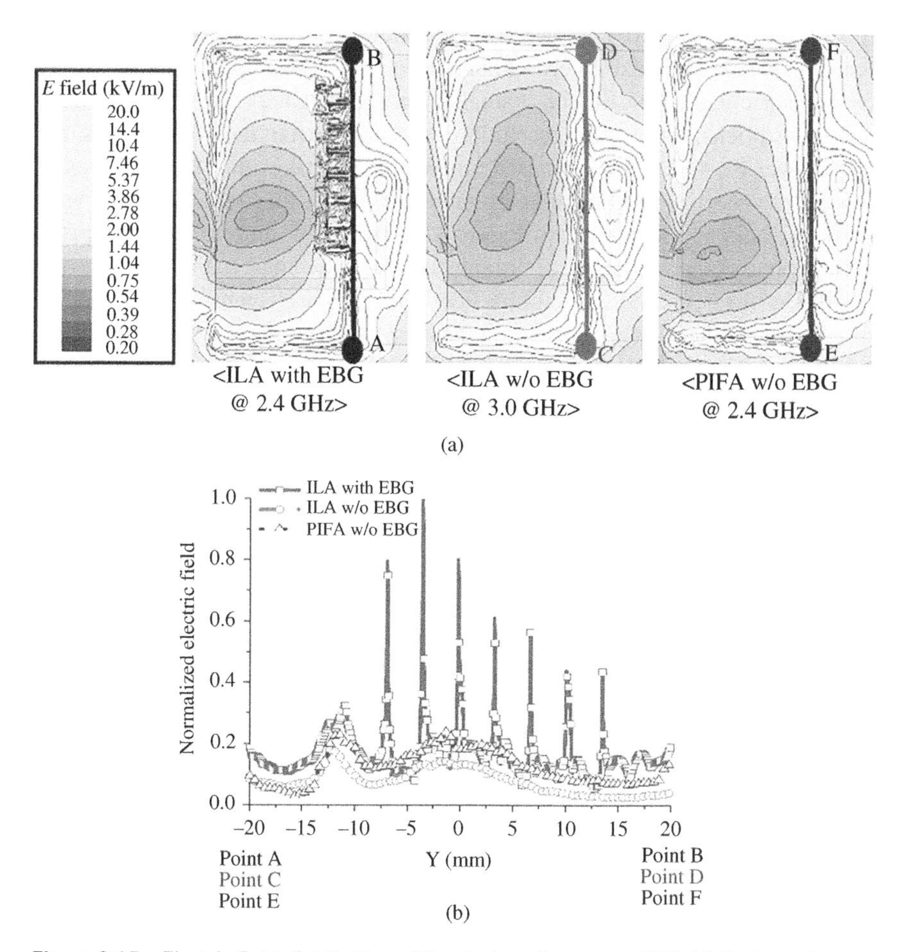

Figure 9.13 Electric field distribution of the designed antennas [15]. (a) Contour plots of the designed antennas. (b) Normalized field distributions crossing the lines AB, CD, and CD in (a).

structures in the near-field distributions can be considered as HIS with slow-wave behavior controlling surface-wave.

9.3 Low-Coupled mmWave Phased-Arrays Fabricated on FR-4 PCB

In this section, a compact phased array incorporating ILA with HIS based on single-layer FR-4 PCB will be proposed and described in detail [4]. Despite FR-4 PCB exhibiting relatively higher dielectric lossy material (dielectric loss tangent tanδ: 0.032 at 28 GHz) and insufficient fabrication resolution at mmWave spectrum, in order to achieve high performance of phased array, there are three main issues [4]. First, it is extremely difficult to obtain HIS properties of 1-D EBG ground structures fabricated on a single-layer FR-4 PCB at mmWave spectrum due to insufficient fabrication resolution. Although having those limitations, the single-layer FR-4 PCB process-based mmWave antennas without any packaging process have been spotlighted for extremely low cost, as illustrated in Figure 9.2a. Second, as already illustrated in Figure 9.4b, two main mutual coupling paths exist in phased array antenna. In order to achieve wide-angle scanning capability of a phased array, the mutual coupling between antenna elements should be minimized, despite the closely packed antenna arrays with compact antenna elements designing $0.5\lambda_0$ inter-element spacing. Although the ILA topology with 1-D EBG structures has been utilized for broadband characteristics and compact element size, the mutual coupling between antenna elements has been still a critical issue [4]. Finally, due to FR-4 PCB exhibiting relatively higher dielectric lossy material the low-loss feeding networks should be implemented [4, 42]. Thus, despite three difficult issues, broadband and wide-angle scanning capability in the compact phased array incorporating ILA with HIS based on single-layer FR-4 PCB can be achieved. Finally, The cost-effective phased array in linear array configurations can be realized without relatively bulky and complex architecture (e.g. conformal array configurations and pattern reconfigurable elements divided into several subareas) or additional packaging process.

9.3.1 Design of Single Antenna Element with HIS Structures

As depicted in Figure 9.14a, an antenna element is designed by using ILA topology with 1-D EBG structures that can operate at mmWave 5G spectrum [4]. The feeding structures of the designed ILA consist of the CPWG-to-microstrip transition line and the 50-Ω ideal port. One column-arranged 1-D EBG structure with five periodic cells featuring HIS is embedded on the partially grounded back side of a FR-4 substrate. The partially grounded FR-4 substrate of a single element is designed to a size of 0.4 mm × 2.1 mm × 3.6 mm and a relative permittivity of 4.4,

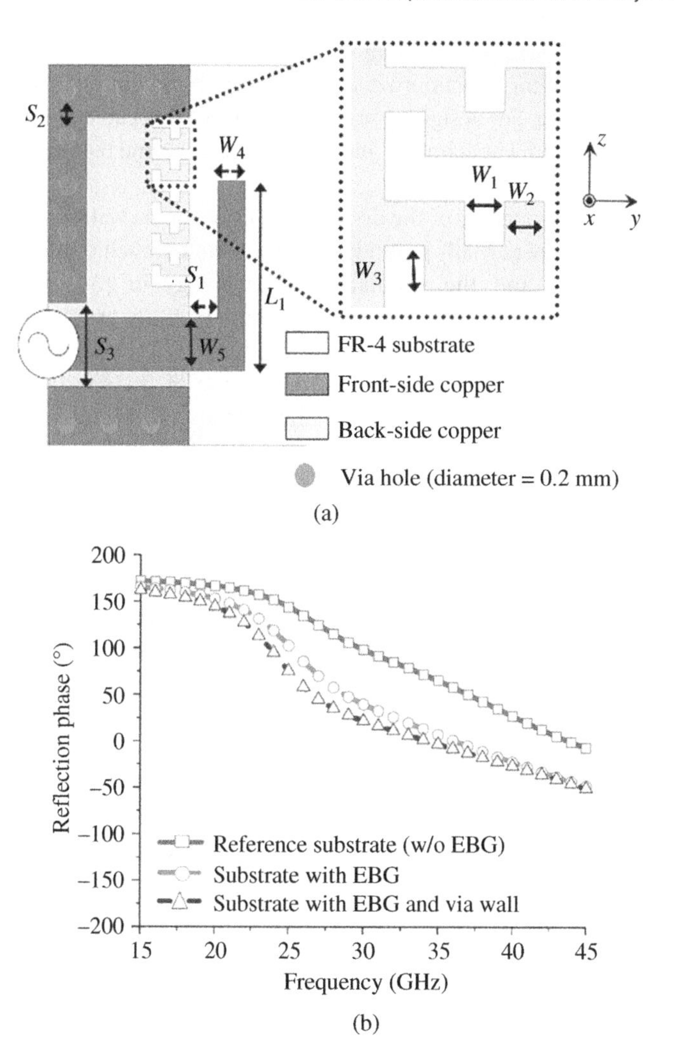

Figure 9.14 Design of single antenna element with HIS [4]. (a) Configuration of planar ILA with 1-D EBG and via wall structures. (W_1 = 0.1 mm, W_2 = 0.1 mm, W_3 = 0.1 mm, W_4 = 0.3 mm, W_5 = 0.5 mm, L_1 = 1.8 mm, S_1 = 0.3 mm, S_2 = 0.25 mm, S_3 = 0.8 mm). (b) Reflection phase of the partially grounded FR-4 substrates containing the 1-D EBG and via wall structures. *Source:* From [4]. ©2021 IEEE. Reproduced with permission.

respectively. The 1-D EBG unit cell consists of a meandered stripline with a minimum width/spacing of 0.1 mm along the PCB design rule that is dependent on manufacturing resolution limits. Moreover, to suppress the mutual coupling path through the surface current on the common ground plane, the via wall structures featuring HIS are additionally embedded onto both edges of the ground plane.

By integrating ILA with 1-D EBG and via wall structures, the antenna element can mitigate the mutual coupling and improve array performance within $0.39\lambda_0$ (3.6 mm) inter-element spacing. For designing via wall structures that can be realized into the PCB, the minimum via diameter and pitch (S_2) are 0.2 and 0.45 mm, respectively.

In order to investigate EM properties of the designed 1-D EBG or via wall structures, the reflection phase of the partially grounded FR-4 substrates is achieved by utilizing the HFSS package and the simulation methodology in [38, 40]. Figure 9.14b illustrates the computed reflection phase of the partially grounded FR-4 substrates containing the 1-D EBG and via wall structures. Within identical size (2.1 mm × 3.6 mm) of the substrates, the reflection phase of the substrate including both 1-D EBG and via wall structures exhibits relatively steep phase changes rather than that of the other substrates. When the reflection phase of the substrate including both 1-D EBG and via wall structures is from 135° to 0°, the input-match frequency band is from 23 to 34 GHz as referred to [15, 16].

As depicted in Figure 9.15a, to investigate the effects with 1-D EBG and via wall structures, the reference ILA and the ILA with 1-D EBG structures are designed with the same physical monopole lengths (L_1). Since the reflection phases of the designed three ILA prototypes are changed, each input-match frequency in reference ILA, ILA with EBG, and ILA containing both EBG and via wall can be shifted at 33, 27.5, and 24 GHz, respectively. Due to the lowest input-match frequency within identical spacing (3.6 mm), an electrical length of the ILA with 1-D EBG and via wall structures is more compressed rather than other antennas. As presented in Figure 9.15b, despite a more electrically compressed configuration, the ILA topology with 1-D EBG and via wall structures can be expected to achieve high gain at horizontal polarization (E_{theta}) in broadband.

9.3.2 Compact Phased Array Antenna in 4 Elements

In order to realize FR-4 PCB process-based mmWave phased array, feeding networks featuring low-loss and good impedance matching are required. As presented in Figure 9.16, the island-shaped coplanar waveguide ground (CPWG) TL with lots of via wall structures serving as HIS is utilized to suppress surface-wave [26, 27, 43]. For both impedance matching and good radiation performance, the characteristic impedances of transition structures among K-connector, CPWG and antenna element are designed from 50 to 65-Ω for good impedance matching. The entire length of each channel from the connector contact pad to the antenna is 15 mm. The measured insertion loss per physical length is from 0.14 to 0.17 dB/mm in [43]. By using this feeding network and three types of the designed antennas, three types of compact phased arrays in 4-elements are fabricated as illustrated in Figure 9.16.

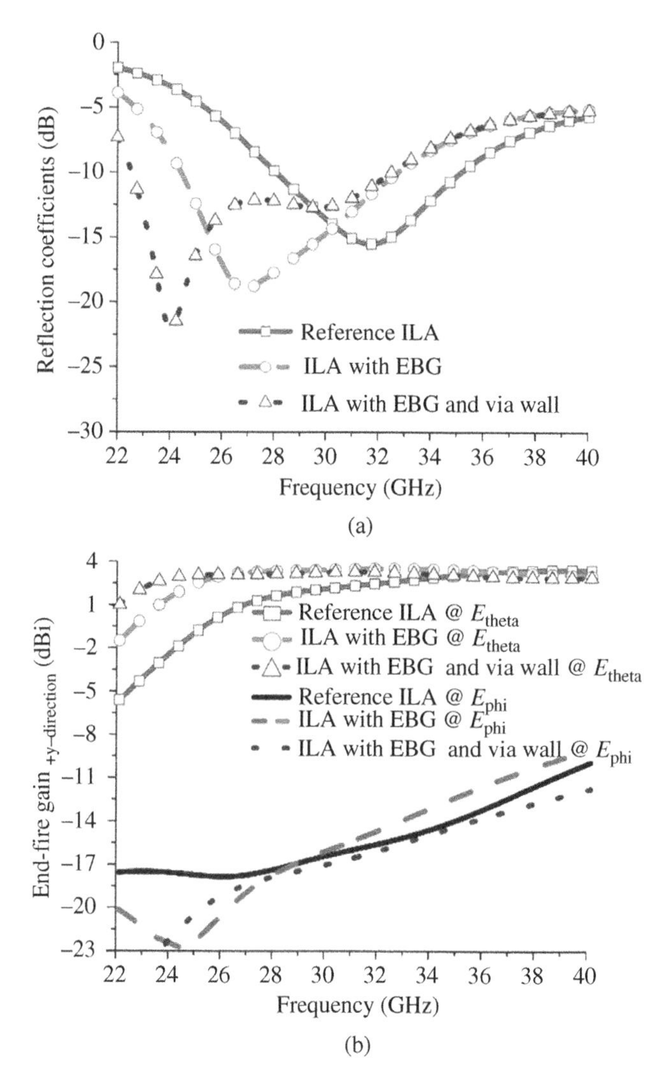

Figure 9.15 Simulated results of single antenna elements [4]. (a) Reflection coefficients. (b) End-fire gain (+*y*) versus frequency. *Source:* From [4]. ©2021 IEEE. Reproduced with permission.

Figure 9.17 illustrates the measured results of three types of fabricated array antennas to describe the design methodology that reduces mutual coupling and achieves good antenna performance. Among three types of fabricated antennas in Figure 9.17a, the antenna containing ILA integrated with 1-D EBG and via wall structures exhibits broad impedance bandwidth between 22 and 40 GHz below a −7.5 dB reflection coefficient. Figure 9.17b depicts the averaged mutual coupling

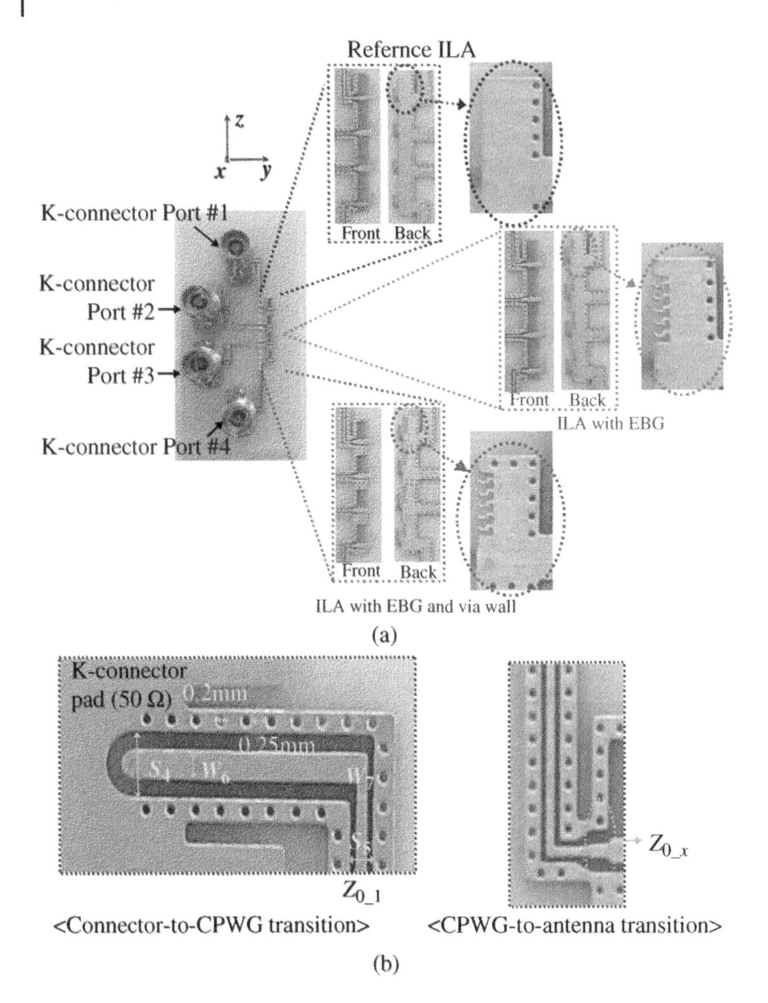

Figure 9.16 (a) Photograph of three types of fabricated array antennas with identical feeding networks. (b) Photograph of feeding structures of the fabricated array antennas (W_6 = 0.6 mm, W_7 = 0.2 mm, S_4 = 1.0 mm, S_5 = 0.4 mm, Z_{0_1} = 65 Ω, Z_{0_x} = 60 Ω). *Source:* From [4]. ©2021 IEEE. Reproduced with permission.

of the fabricated antennas by using K-connector for feeding structures. The averaged mutual coupling of the fabricated antenna incorporating ILA with 1-D EBG and via wall structures can be suppressed at beyond 30 GHz by HIS properties of via wall. Despite larger mutual coupling at 26 GHz than other antennas due to good impedance matching, the coupling level of ILA topology both with 1-D EBG and via wall can be maintained below −12 dB.

By using a commercial RFIC beamforming chipset, the radiation patterns of three types of fabricated linear array antennas are measured as illustrated in

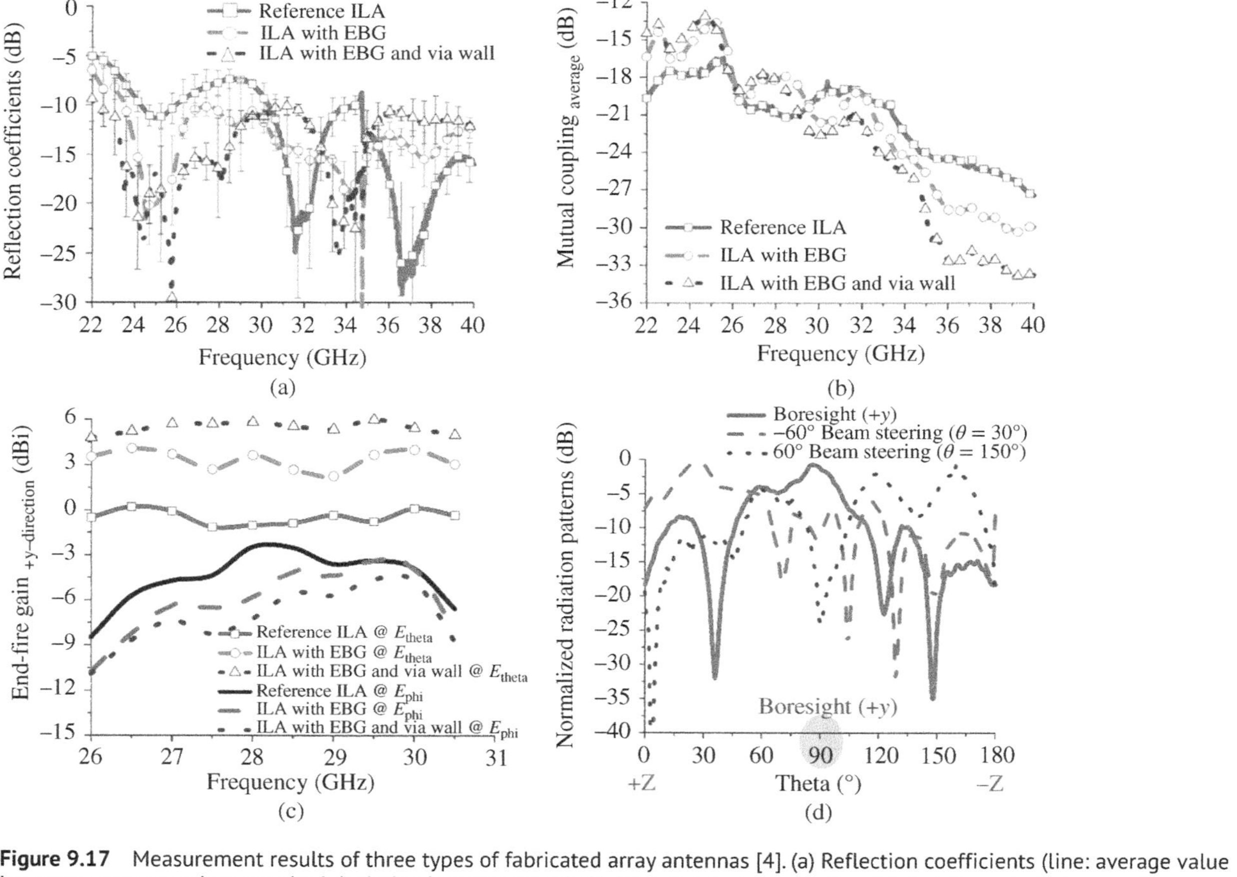

Figure 9.17 Measurement results of three types of fabricated array antennas [4]. (a) Reflection coefficients (line: average value between ports, error bar: standard deviation between ports). (b) Averaged mutual coupling. (c) End-fire gain (+*y*) versus frequency. (d) Measured radiation patterns of the antenna by using a commercial RFIC beamforming chipset @ 28 GHz (e.g. ILA with EBG and via wall structures). *Source:* From [4]. ©2021 IEEE. Reproduced with permission.

Figure 9.17c,d [4, 16]. The beamforming chipset supports 4 Tx/Rx channels and includes a 5-bit phase and a 5-bit gain control. The chipset operates in half-duplex mode from 26.5 to 29.5 GHz. At in-phase feeding, the antenna incorporating ILA with 1-D EBG and via wall structures can achieve end-fire gain of 6 dBi and a low cross-polarization level of 12 dB toward the end-fire direction at 28 GHz, respectively. By choosing ILA topology with 1-D EBG and via wall structures, the antenna containing ILA integrated with 1-D EBG and via wall structures can also exhibit 120° (±60°) beam coverage within 3 dB scan loss at 28 GHz, as depicted in Figure 9.17d. To satisfy the 120° (±60°) beam coverage within 3 dB scan loss, the required phase difference between adjacent elements is set to 120° [16]. Due to the via wall structures as HIS for reduction of mutual coupling, thus, the antenna containing ILA topology with 1-D EBG and via wall structures can achieve broader bandwidth and higher end-fire gain with wide-angle scanning capability than the other antenna.

The characteristics of an array antenna containing ILA topology integrated with 1-D EBG and via wall structures can be verified through surface current distribution on the ground plane of the antenna elements. In order to discuss specific mutual coupling paths by surface current on the common ground plane as depicted in Figure 9.18a, the averaged surface currents are detected at the imaginary lines of AB, CD, and EF between antenna elements. Due to HIS properties of via wall structures, the surface currents at the imaginary lines can be significantly reduced as depicted in Figure 9.18b–d. Thus, the mutual coupling path by surface current on the common ground plane can be reduced due to the embedded via wall structures.

The error vector magnitude (EVM), which is a representative factor in over-the-air (OTA) system performance of mmWave transceiver with antenna module, is a critical criterion to determine signal linearity in high power. Since the FR-4 substrate features high dielectric lossy properties, the practicality of the antenna should be demonstrated by EVM performance comparison with mmWave 5G NR standard specification [44]. By using an identical beamforming chipset in previous beamforming tests, the OTA system performance of the fabricated antenna is measured for three different scan angles, as illustrated in Figure 9.19a. The measured antenna is the phased array containing ILA topology integrated with 1-D EBG and via wall structures. During OTA performance tests, the signal bandwidth of 100 MHz at a 28 GHz carrier frequency and QPSK modulation scheme are utilized, respectively. In order to create the 5G NR waveform, Keysight M9505A arbitrary waveform generator (AWG) is used.

In order to demonstrate the OTA performance of wide-beam coverage of the presented antenna, three constellation diagrams are each direction (e.g. boresight direction and maximum two beam steering directions [±60°]) are measured [16]. Figure 9.19b illustrates that the maximum EVM is 16.6% at a 60° scan

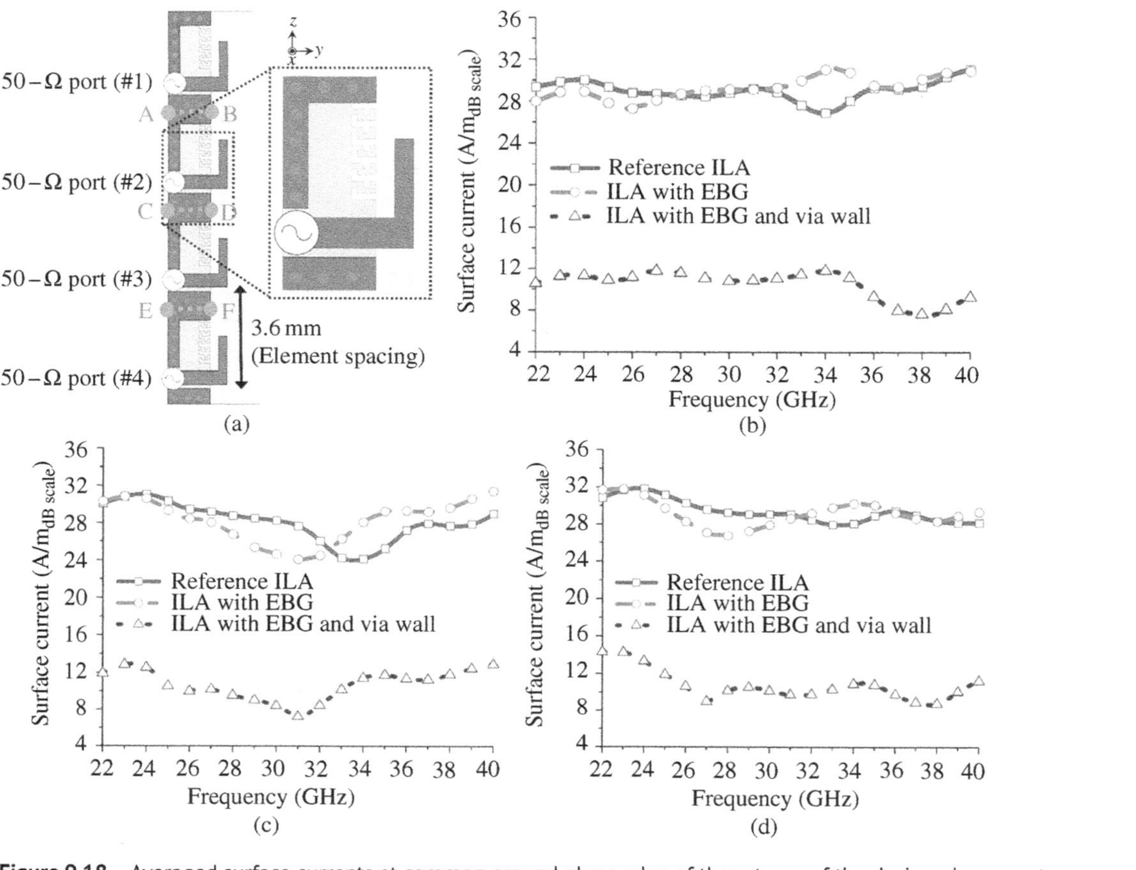

Figure 9.18 Averaged surface currents at common ground plane edge of three types of the designed array antennas. (a) Configuration of array antennas containing ILA with 1-D EBG and via wall structures. (b) Line AB. (c) Line CD. (d) Line EF. *Source:* From [4]. ©2021 IEEE. Reproduced with permission.

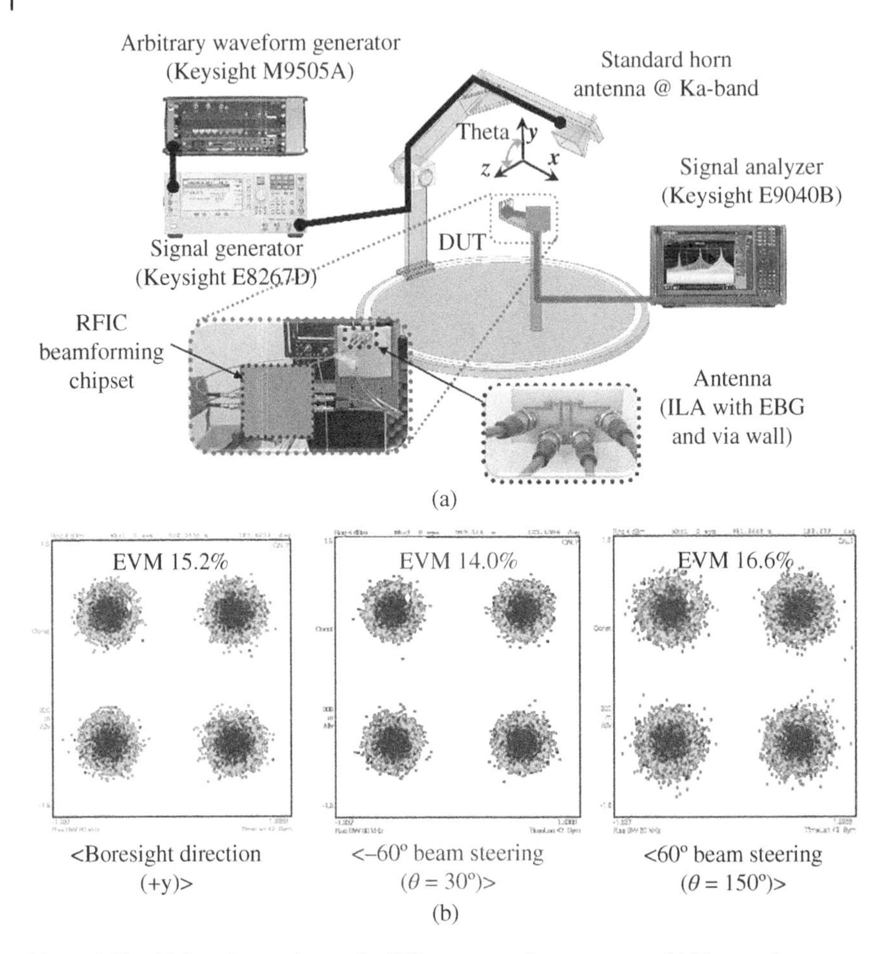

Figure 9.19 (a) Experimental setup for OTA system performance tests. (b) Measured constellations at three different scan angles (e.g. ILA with EBG and via wall structures). *Source:* From [4]. ©2021 IEEE. Reproduced with permission.

angle due to high side lobe level at $\theta = 120°$ in Figure 9.17d. The measured OTA performance results are accomplished EVM research point at QPSK modulation scheme as compared with mmWave 5G NR standard specification.

9.3.3 8-Elements Large-Scaled Linear Array Antenna

To meet the link budget in a mmWave 5G UE, 8-elements or more large-scaled linear arrays featuring increased directivity and the reduced side lobe level have been studied [4]. However, due to the absence of commercial beamforming chipset simultaneously controlling 8-element array, 1by8 power divider is required to demonstrate the performance of the proposed phased array antenna.

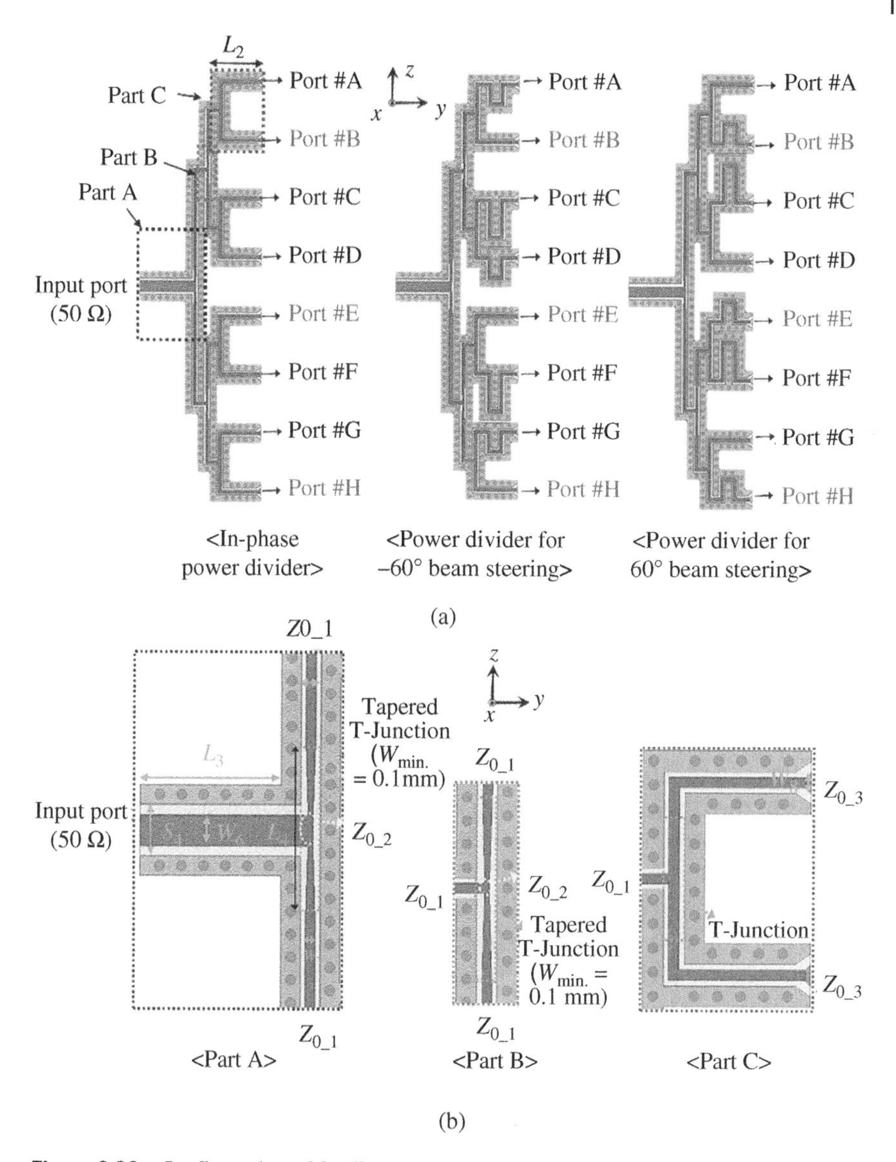

Figure 9.20 Configuration of feeding networks for array antenna with 8-elements. (a) T-junction power divider for in-phase, −60° beam steering and 60° beam steering. (b) Each specific part of T-junction power divider for in-phase. (W_8 = 0.5 mm, L_2 = 3.25 mm, L_3 = 2.9 mm, L_4 = 3.6 mm, Z_{0_2} = 91 Ω, Z_{0_3} = 52 Ω). *Source:* From [4]. ©2021 IEEE. Reproduced with permission.

By using 1 by 8 T-junction power divider featuring the predetermined phased delay, two different feeding networks for beam steering are additionally designed as shown in Figure 9.20a. The input port is defined as the connector contact pad by using K-connector, while the output ports from A to H are modeled as 50-Ω load to antenna element. In order to realize low loss feeding network within tightly inter-element spacing, the ground plane including via wall structures between each part of power divider is partially overlapped and compressively designed. The entire length of in-phase T-junction power divider from the input port to the output port is minimized by 19.1 mm. To realize broad impedance bandwidth within the PCB design rule, the tapered T-junctions with quarter-wavelength impedance transformers are inserted in part A and B, as illustrated in Figure 9.20b. For achieving 120° phased delay in T-junction power divider for ±60° beam steering at 28 GHz, the physical length difference toward each adjacent element in part C is 2 mm.

As already mentioned in Section 9.1.3, the T-junction power divider inherently exhibits high mutual coupling between adjacent output ports toward each antenna element. Due to high mutual coupling level, the T-junction power divider can often occur oscillation and power imbalance by the reflected power from open port [43]. Thereby, a comprehensive research methodology of feeding network including a power divider for multi-port antenna arrays should be required during design, measurement, and analysis step.

Figure 9.21a illustrates the symmetrical back-to-back configuration including 1 by 8 T-junction power dividers is a proper solution to estimate the accurate insertion loss of feeding network without any unstable measurement. To consider the connector impact on the experiments, the symmetrical back-to-back sample with T-junction power dividers has been simulated by containing landing pad modeling configurations of connector. As depicted in Figure 9.21b, the measured insertion loss in the fabricated T-junction power dividers with in-phase is from 3.5 to 4.5 dB below 36 GHz. It can be verified that the simulated and measured insertion loss of T-junction power divider exhibits similar results in [43]. Due to additional internal loss of T-junction power divider, the measured insertion loss power physical length is from 0.18 to 0.23 dB/mm. The additional internal loss of 1 by 8 T-junction power divider can result from the little mismatch condition of the tapered T-junction by fabrication resolution limit, more than feeding networks for 4 array elements as mentioned in Section 9.1.3. In particular, the maximum characteristic impedance of the tapered T-junction power divider expect to be 91-Ω. However, despite insufficient fabrication resolution, the fabricated symmetrical T-junction power divider still exhibits both wide impedance bandwidth and low loss from 26 to 36 GHz.

By using the 1 by 8 T-junction power divider and 8 array elements incorporating ILA topology with 1-D EBG and via wall structures, Figure 9.22a shows three samples of the fabricated array antennas. In order to characterize wide-angle scanning, three samples of antennas for each beam steering direction (boresight and ± maximum beam steering angle) are fabricated. In addition, by considering parasitic effect of the connector larger than array antenna in the experiments,

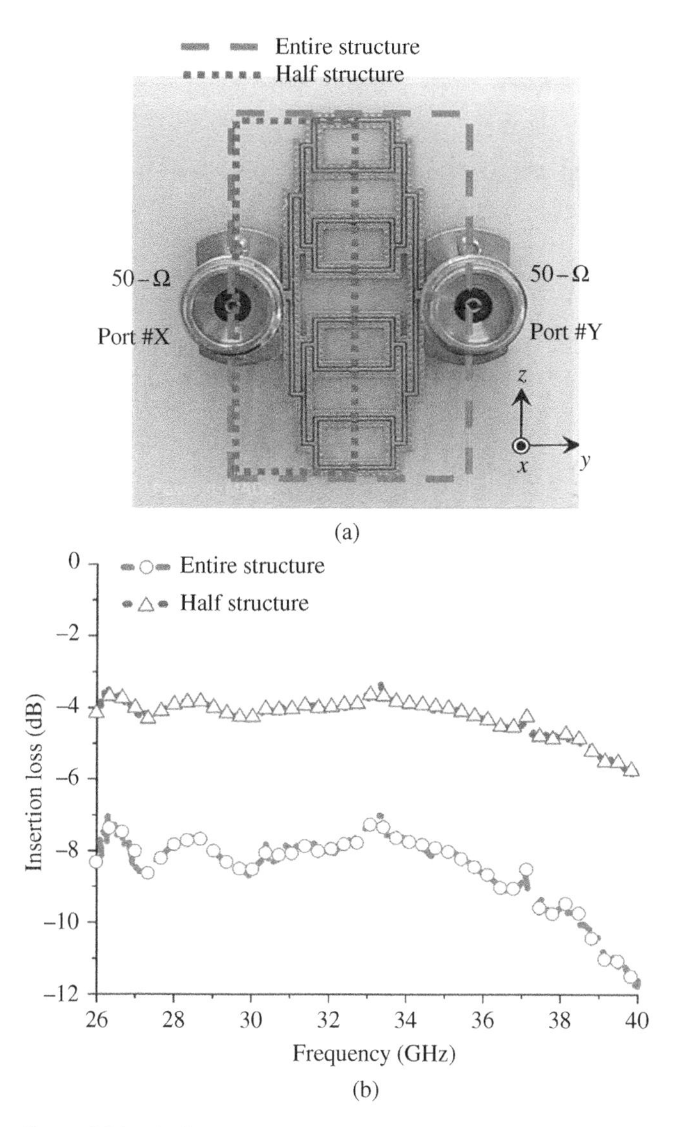

Figure 9.21 (a) Photograph of symmetrical T-Junction power divider. (b) Measured insertion loss of symmetrical T-junction power divider. *Source:* From [4]. ©2021 IEEE. Reproduced with permission.

three samples of the fabricated array antennas have been designed by containing landing pad modeling figuration of connector. Due to the designed 1 × 8 T-junction power divider including the tapered T-junction insertion, the fabricated array antennas can exhibit broad impedance bandwidth between 26 and 40 GHz below −7.5 dB reflection coefficient, as depicted in Figure 9.22b.

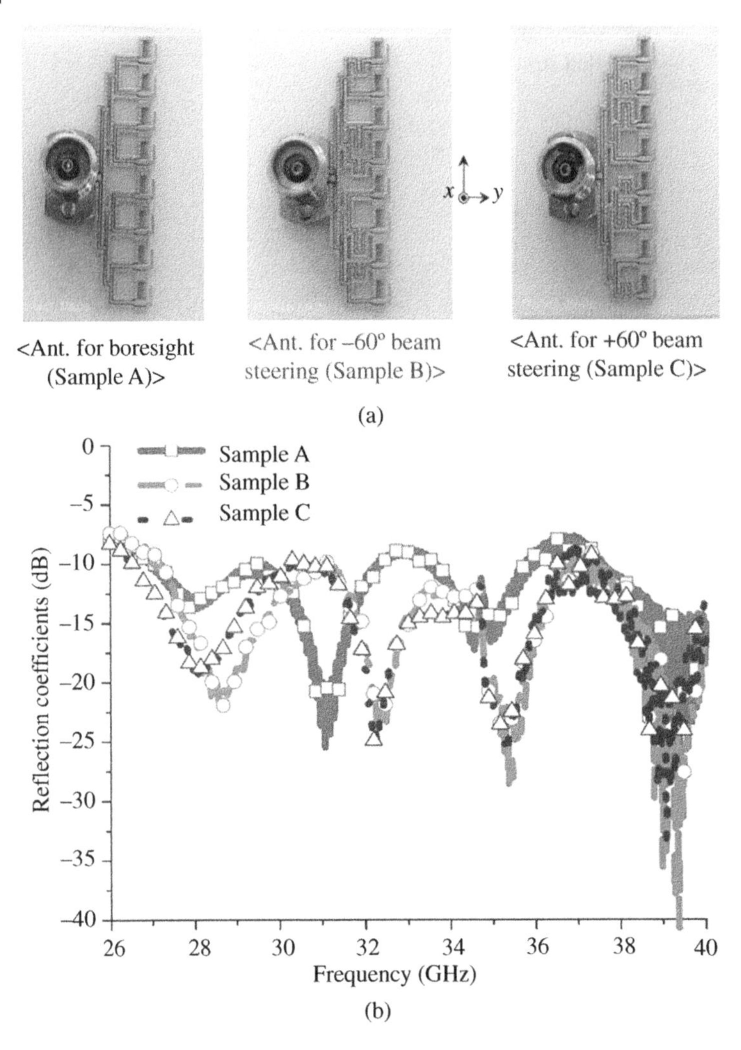

Figure 9.22 (a) Photograph of three samples of the fabricated array antennas. *Source:* From [4]. ©2021 IEEE. Reproduced with permission. (b) Measured reflection coefficients [4].

In order to investigate the performance of the proposed phased array featuring directional fan-beam, the radiation patterns of the fabricated array antennas in a horizontal plane are measured and compared. As illustrated in Figure 9.23a–f, the fabricated array antennas exhibit more than 110° (±55°) beam scanning coverage within 3 dB scan loss from 26 to 36 GHz. Due to large-scaling array, the 3-dB beamwidth in main beam becomes narrow and the side lobe level is reduced, as compared in Figure 9.17d. Thus, despite ±5° beam squint as

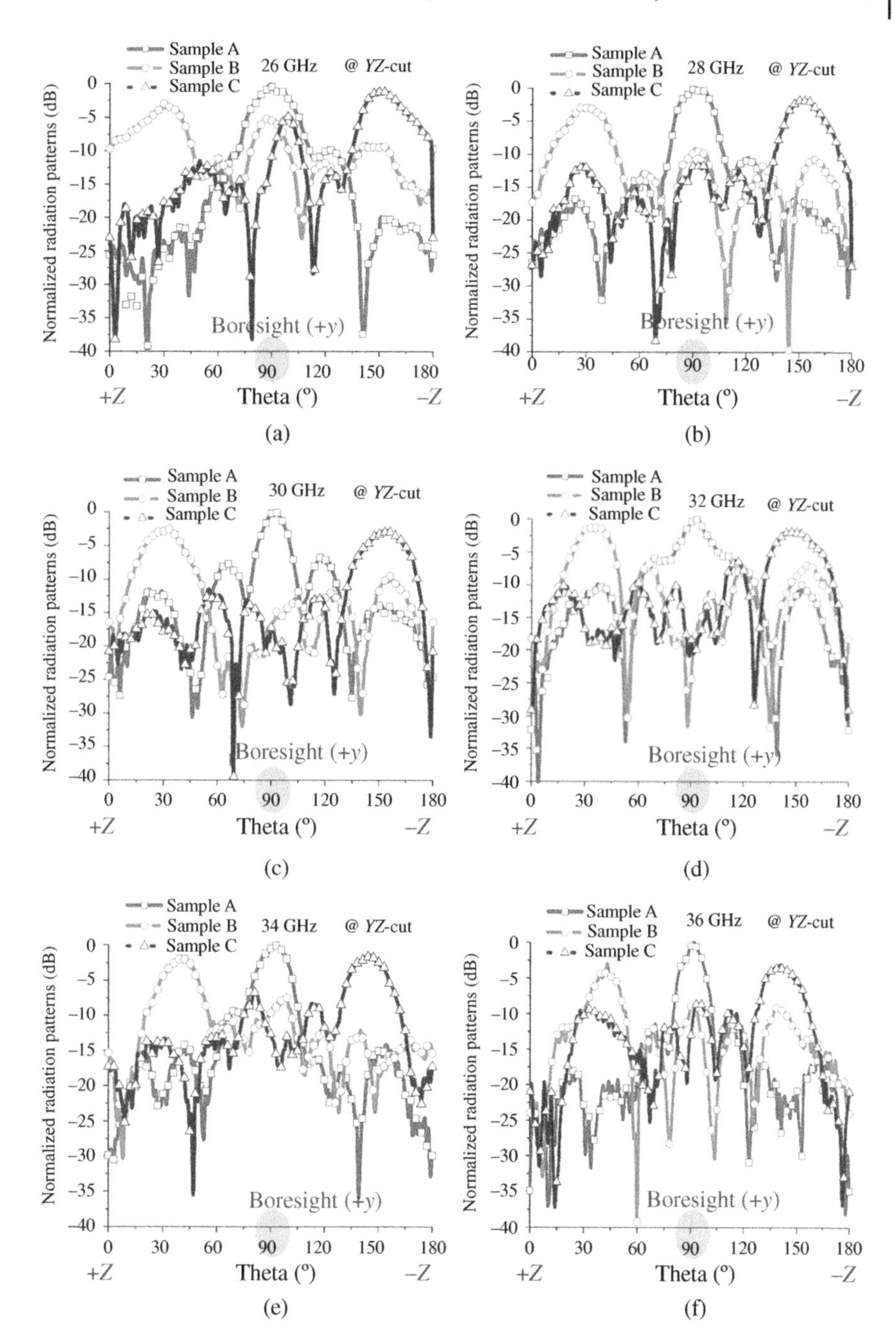

Figure 9.23 Measured radiation patterns (E_{theta}-polarization) of the fabricated array antennas for beam steering test [4]. (a) 26 GHz. (b) 28 GHz. (c) 30 GHz. (d) 32 GHz. (e) 34 GHz. (f) 36 GHz. *Source:* From [4]. ©2021 IEEE. Reproduced with permission.

illustrated in Figure 9.23a–f, wide-angle scanning capabilities 3 dB scan loss from 26 to 36 GHz are achieved.

The mmWave antennas within the mobile terminal should exhibit a quasi-isotropic spherical coverage using wide-angle beam steering condition [5–8]. Moreover, for reliable spherical coverage, the maximum realized gain should be high and the slope of coverage efficiency should be relatively steep [6, 8]. In this work, despite insufficient fabrication resolution and large loss materials, the practicality of this design methodology based on single-layer FR-4 PCB packaging technology should be evaluated by using the simulated total scan pattern (TSP) and coverage efficiency.

The TSP and coverage efficiency have been computed as illustrated in Figures 9.24a–f and 9.25. In order to illustrate the continuous scanning results, the additionally designed antennas with predetermined phase delay lines for ±15°, ±30°, and ±45° beam steering are also utilized [4]. By extracting the maximum realized gain value at every angular distribution point, the TSP is calculated from all continuous beam steering radiation patterns in Figure 9.24a–f. Due to fan-beam characteristics, wide-angle coverage beyond 110° (±55°) beam scanning coverage from 26 to 36 GHz is achieved.

The coverage efficiency has defined as the ratio between maximum solid angle and coverage solid angle [6, 8]. The coverage solid angle is computed by the TSP higher than a certain threshold realized gain; the maximum solid angle is 4π steradians. Despite the relatively lossy FR-4 substrate material properties and insufficient fabrication resolution, Figure 9.25 illustrates coverage efficiency beyond 50% for the realized gain of −1.5 dBi from 26 to 36 GHz.

9.4 Conclusion

This chapter presents elaborate design and characterization methodologies of mmWave phased array antenna incorporating ILA with HIS based on cost-effective fabrication technique. First, research motivation and strategy of single-layer FR-4 PCB-based antenna technologies with low-loss feeding networks are presented, despite insufficient fabrication resolution, high dielectric lossy media at mmWave spectrum, and high mutual coupling within $0.5\lambda_0$ inter-element spacing. Second, as an example of cost-effective antenna element, a compact ILA topology integrating with 1-D EBG ground structures has been investigated to enhance antenna performance at microwave regime [15]. Through simulation and measurement, it is confirmed that the 1-D EBG ground structures in ILA topology for small form factor can serve as HIS with slow-wave behavior controlling both space-wave and surface-wave. Afterward, by utilizing the identical properties of ILA topology with 1-D EBG structures based on single-layer FR-4 PCB,

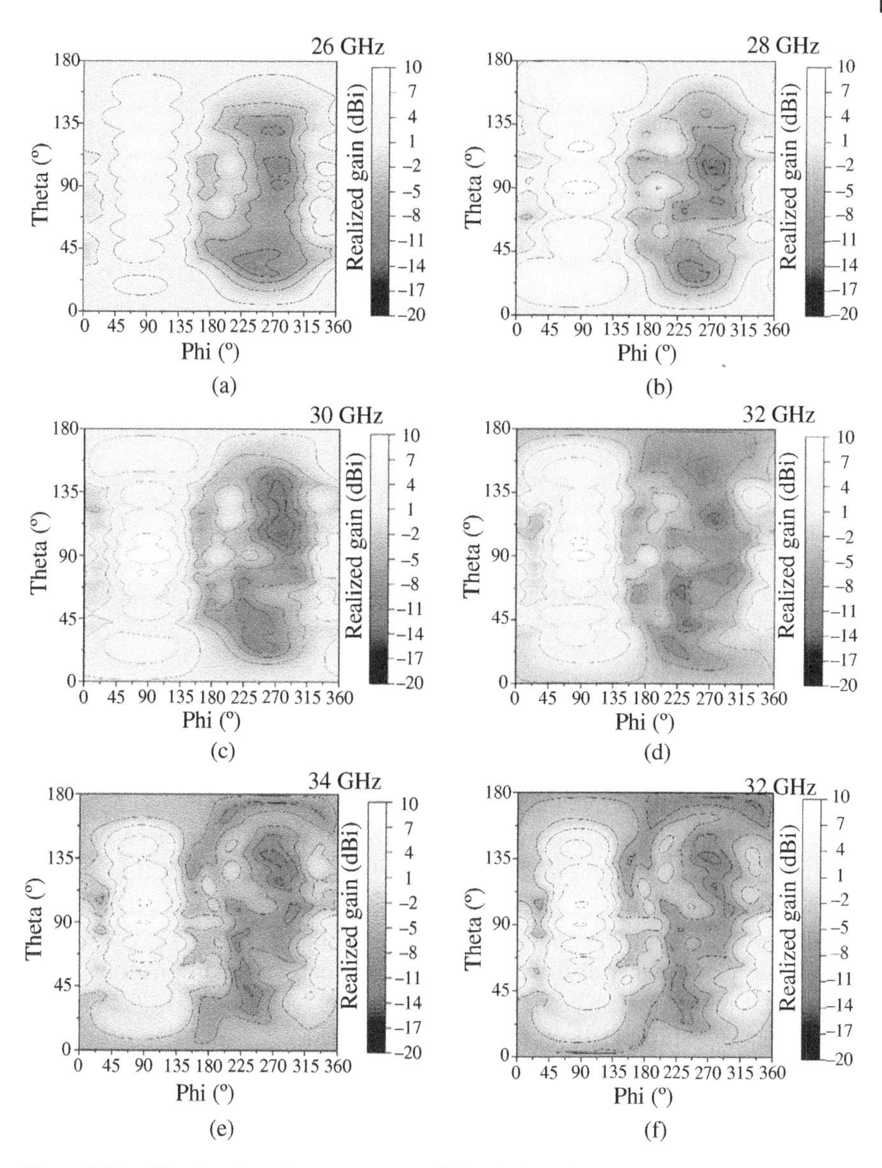

Figure 9.24 Simulated total scan patterns of the designed array antennas. (a) 26 GHz (b) 28 GHz (c) 30 GHz (d) 32 GHz (e) 34 GHz (f) 36 GHz. *Source:* From [4]. ©2021 IEEE. Reproduced with permission.

a compact mmWave phased array additionally incorporating via wall structures for HIS properties has been proposed [4]. Due to economic fabrication process and wide angle scanning capability in broadband, the proposed antenna can become a promising candidate for mmWave 5G global roaming services in mobile

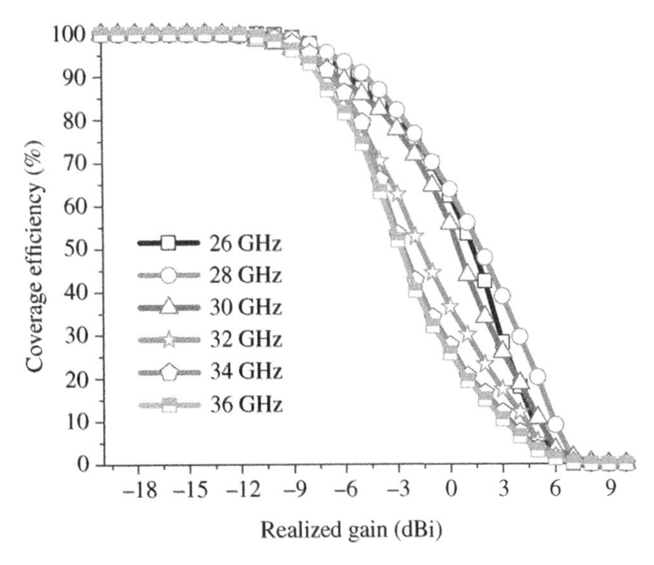

Figure 9.25 Simulated coverage efficiency of the designed array antennas. *Source:* From [4]. ©2021 IEEE. Reproduced with permission.

terminals. Moreover, by utilizing those design and analysis methodologies containing antenna element topology and low-loss feeding networks, three additional research themes could be expected as following sentences:

i) Design and analysis methodologies of feeding networks for mobile handsets:
The design and analysis methodologies of mmWave phased array containing low-loss feeding networks based on single-layer FR-4 PCB in mobile handsets should also be investigated, as discussed in [45]. In order to realize the optimized mmWave antenna for quasi-isotropic spherical coverage without any performance distortion in mobile terminals due to the proximity presence of polycarbonate carriers, additional novel research on GCPW TL with via fences featuring HIS property can be expected.

ii) Design methodologies of mmWave optically transparent antenna:
Despite absence of any auxiliary circuits, such as balun utilizing wire bonding and TGV process, the design methodology of ILA topology with 1-D EBG ground structures can be applied in mmWave optically transparent antenna based on low-cost fabrication process [46].

iii) Design and analysis methodologies of sub-THz AiP and AoC applications:
By employing via fences featuring HIS property to preserve radiation performance in GCPW TL topology, this design and analysis methodology about fundamental power flow and analysis of feeding networks can be applied in sub-THz AiP and AoC applications [18–23].

References

1 T. S. Rappaport, S. Sun, R. Mayzus, H. Zhao, Y. Azar, K. Wang, G. N. Wong, J. K. Schulz, M. Samimi, and F. Gutierrez, "Millimeter wave mobile communications for 5G cellular: it will work!," *IEEE Access*, vol. 1, pp. 335–349, 2013.

2 W. Roh, J.-Y. Seol, J. H. Park, B. Lee, J. Lee, Y. Kim, J. Cho, K. Cheun, and F. Aryanfar, "Millimeter-wave beamforming as an enabling technology for 5G cellular communications: theoretical feasibility and prototype results," *IEEE Commun. Mag.*, vol. 52, no. 2, pp. 106–113, 2014.

3 Y. Huo, X. Dong, W. Xu, and M. Yuen, "Enabling multi-functional 5G and beyond user equipment: a survey and tutorial," *IEEE Access*, vol. 7, pp. 116975–117008, 2019.

4 J.-Y. Lee, J. Choi, D. Choi, Y. Youn, and W. Hong, "Broadband and wide-angle scanning capability in low-coupled mm-wave phased-arrays incorporating ILA with HIS fabricated on FR-4 PCB," *IEEE Trans. Veh. Technol.*, vol. 70, no. 3, pp. 2076–2088, 2021.

5 W. Hong, K.-H. Baek, Y. Lee, Y. Kim, and S.-T. Ko, "Study and prototyping of practically large-scale millimeter-wave antenna systems for 5G cellular devices," *IEEE Commun. Mag.*, vol. 52, no. 9, pp. 63–89, 2014.

6 J. Helander, K. Zhao, Z. Ying, and D. Sjoberg, "Performance analysis of millimeter-wave phased array antennas in cellular handsets," *IEEE Antennas Wirel. Propag. Lett.*, vol. 15, pp. 504–507, 2016.

7 W. Hong, K.-H. Baek, and S. Ko, "Millimeter-wave 5G antennas for smartphones: overview and experimental demonstration," *IEEE Trans. Antennas Propag.*, vol. 65, no. 12, pp. 6250–6261, 2017.

8 K. Zhao, S. Zhang, Z. Ho, O. Zander, T. Bolin, Z. Ying, and G. F. Pedersen, "Spherical coverage characterization of 5G millimeter wave user equipment with 3GPP specifications," *IEEE Access*, vol. 7, pp. 4442–4452, 2019.

9 R. G. Vaughan and J. B. Andersen, "Antenna diversity in mobile communications," *IEEE Trans. Veh. Technol.*, vol. 36, no. 4, pp. 149–172, 1987.

10 C. A. Balanis, "*Antenna Theory 3rd Edition: Analysis and Design*," Wiley-Interscience, Hoboken, NJ, USA, 2005.

11 J. Mietzner, R. Schober, L. Lampe, W. H. Gerstacker, and P. A. Hoeher, "Multiple-antenna techniques for wireless communications - a comprehensive literature survey," *IEEE Commun. Surv. Tutorials*, vol. 11, no. 2, pp. 87–105, 2009.

12 M. M. Bait-Suwailam, M. S. Boybay, and O. M. Ramahi, "Electromagnetic coupling reduction in high-profile monopole antennas using single-negative magnetic metamaterials for MIMO applications," *IEEE Trans. Antennas Propag.*, vol. 58, no. 9, pp. 2894–2902, 2010.

13 M. M. Bait-Suwailam, "*Metamaterials for Decoupling Antenna and Electromagnetic Systems*," Doctor of philosophy, University of Waterloo, Ontario, Canada, 2011.

14 C. Wang, E. Li, and D. F. Sievenpiper, "Surface-wave coupling and antenna properties in two dimensions," *IEEE Trans. Antennas Propag.*, vol. 65, no. 10, pp. 5052–5060, 2017.

15 J.-Y. Lee, J. Choi, J.-H. Jang, and W. Hong, "Performance enhancement in compact inverted-L antenna by using 1-D EBG ground structures and beam directors," *IEEE Access*, vol. 7, pp. 93264–93274, 2019.

16 J.-Y. Lee, J. Choi, D. Choi, Y. Youn, J. Park, and W. Hong, "FR-4 PCB process-based mm-wave phased array antenna using planar high-impedance surfaces," *The 50th European Microwave Conference (EuMW-EuMC 2020)*, Utrecht, pp. 583–586, 2021.

17 G. Hu, J. Kang, L. W. T. Ng, X. Zhu, R. C. T. Howe, C. G. Jones, M. C. Harsam, and T. Hasan, "Functional inks and printing of two-dimensional materials," *RSC Chem. Soc. Rev.*, vol. 47, pp. 3265–3300, 2018.

18 Y. Zhang and J. Mao, "An overview of the development of antenna-in-package technology for highly integrated wireless devices," *Proc. IEEE*, vol. 107, no. 11, pp. 2265–2280, 2019.

19 J. Park, H. Seong, Y. N. Whang, and W. Hong, "Energy-efficient 5G phased arrays incorporating vertically polarized endfire planar folded slot antenna for mmWave mobile terminals," *IEEE Trans. Antennas Propag.*, vol. 68, no. 1, pp. 230–241, 2020.

20 A. Hagelauer, M. Wojnowski, K. Pressel, R. Weigel, and D. Kissinger, "Integrated systems-in-package: heterogeneous integration of millimeter-wave active circuits and passives in fan-out wafer-level packaging technologies," *IEEE Microwave Mag.*, vol. 19, no. 1, pp. 48–56, 2018.

21 H. M. Cheema and A. Shamim, "The last barrier: on-chip antennas," *IEEE Microwave Mag.*, vol. 14, no. 1, pp. 79–91, 2013.

22 S. Pan, F. Caster, P. Heydari, and F. Capolino, "A 94-GHz extremely thin metasurface-based BiCMOS on-chip antenna," *IEEE Trans. Antennas Propag.*, vol. 62, no. 9, pp. 4439–4451, 2014.

23 M. R. Karim, X. Yang, and M. F. Shafique, "On chip antenna measurement: a survey of challenges and recent trends," *IEEE Access*, vol. 6, pp. 20320–20333, 2018.

24 R. A. Alhalabi and G. M. Rebeiz, "High-efficiency angled-dipole antennas for millimeter-wave phased array applications," *IEEE Trans. Antennas Propag.*, vol. 56, no. 10, pp. 3136–3142, 2008.

25 S. X. Ta, H. Choo, and I. Park, "Broadband printed-dipole antenna and its arrays for 5G applications," *IEEE Antennas Wirel. Propag. Lett.*, vol. 16, pp. 2183–2186, 2017.

26 W. Hong, K.-H. Baek, and A. Goudelev, "Multilayer antenna package for IEEE 802.11ad employing ultralow-cost FR4," *IEEE Trans. Antennas Propag.*, vol. 60, no. 12, pp. 5932–5938, 2012.

27 W. Hong, K.-H. Baek, and A. Goudelev, "Grid assemble-free 60-GHz antenna module embedded in FR-4 transceiver carrier board," *IEEE Trans. Antennas Propag.*, vol. 61, no. 4, pp. 1573–1580, 2013.

28 K.-L. Wu, C. Wei, X. Mei, and Z.-Y. Zhang, "Array-antenna decoupling surface," *IEEE Trans. Antennas Propag.*, vol. 65, no. 12, pp. 6728–6738, 2017.

29 S. Zhang, X. Chen, and G. F. Pedersen, "Mutual coupling suppression with decoupling ground for massive MIMO antenna arrays," *IEEE Trans. Veh. Technol.*, vol. 68, no. 8, pp. 7273–7282, 2019.

30 Y.-F. Cheng, X. Ding, W. Shao, and B.-Z. Wang, "Planar wide-angle scanning phased array with pattern-reconfigurable windmill-shaped loop elements," *IEEE Trans. Antennas Propag.*, vol. 65, no. 2, pp. 932–936, 2017.

31 L. Gu, Y.-W. Zhao, Q.-M. Cai, Z.-P. Zhang, B.-H. Xu, and Z.-P. Nie, "Scanning enhanced low-profile broadband phased array with radiator-sharing approach and defected ground structures," *IEEE Trans. Antennas Propag.*, vol. 65, no. 11, pp. 5846–5854, 2017.

32 D. Sievenpiper, L. Zhang, R. F. J. Broas, N. G. Alexopolous, and E. Yablonovitch, "High-impedance electromagnetic surfaces with a forbidden frequency band," *IEEE Trans. Microwave Theory Tech.*, vol. 47, no. 11, pp. 2059–2074, 1999.

33 F. Yang and Y. Rahmat-Samii, "Reflection phase characterizations of the EBG ground plane for low profile wire antenna applications," *IEEE Trans. Antennas Propag.*, vol. 51, no. 10, pp. 2691–2703, 2003.

34 J. Liang and H.-Y. D. Yang, "Radiation characteristics of a microstrip patch over an electromagnetic bandgap surface," *IEEE Trans. Antennas Propag.*, vol. 55, no. 6, pp. 1691–1697, 2007.

35 S. Ghosh, T.-N. Tran, and T. Le-Ngoc, "Dual-layer EBG-based miniaturized multi-element antenna for MIMO systems," *IEEE Trans. Antennas Propag.*, vol. 62, no. 8, pp. 3985–3997, 2014.

36 A. C. K. Mak and C. R. Rowell, and R. D. Murch, "Isolation enhancement between two closely packed antennas," *IEEE Trans. Antennas Propag.*, vol. 56, no. 11, pp. 3411–3419, 2008.

37 D. A. Ketzaki and T. V. Yioultsis, "Metamaterial-based design of planar compact MIMO monopoles," *IEEE Trans. Antennas Propag.*, vol. 61, no. 5, pp. 2758–2766, 2013.

38 S.-H. Kim, T. T. Nguyen, and J.-H. Jang, "Reflection characteristics of 1-D EBG ground plane and its application to a planar dipole antenna," *Prog. Electromagn. Res.*, vol. 120, pp. 51–66, 2011.

39 S.-H. Kim, J.-Y. Lee, T. T. Nguyen, and J.-H. Jang, "High-performance MIMO antenna with 1-D EBG ground structures for handset applications," *IEEE Antennas Wirel. Propag. Lett.*, vol. 12, pp. 1468–1471, 2013.

40 J.-Y. Lee, S.-H. Kim, and J.-H. Jang, "Reduction of mutual coupling in planar multiple antenna by using 1-D EBG and SRR structures," *IEEE Trans. Antennas Propag.*, vol. 63, no. 9, pp. 4194–4198, 2015.

41 D. M. Pozar, "*Microwave Engineering*," 4th Edition, Wiley-John Wiley & Sons, Inc., Hoboken, NJ, USA, 2012.

42 J.-Y. Lee, D. Choi, K. Kong, J. Yun, Y. Youn, W. Kwon, and W. Hong, "Design and measurement considerations of feeding network including power divider for multi-port antenna arrays," *IEEE International Conference on Antenna Measurements and Applications (CAMA),* Bali, pp. 279–281, 2019.

43 A. Sain and K. L. Melde, "Impact of ground via placement in grounded coplanar waveguide interconnects," *IEEE Trans. Compon. Packag. Manuf. Technol.*, vol. 6, no. 1, pp. 136–144, 2016.

44 "Technical Specification Group Radio Access Network, document TR 38.817-01 V 16.1.0", 3GPP, 2019. (Online). Available: https://portal.3gpp.org/desktopmodules/Specifications/SpecificationDetails.aspx?specificationId=3359.

45 J.-Y. Lee, J. Choi, B. Kim, Y. Oh, and W. Hong, "Performance enhancement of mm-wave phased arrays for mobile terminals through grounded coplanar waveguide feeding networks with via fences," *Front. Commun. Networks*, vol. 2, no. 741533, pp. 1–11, 2021.

46 J.-Y. Lee, S. Jung, Y. Youn, J. Park, W. Kwon, and W. Hong, "Optically transparent 1-D EBG antenna using sub-skin depth thin-film alloy in the Ka-band," *13th European Conference on Antennas and Propagation (EuCAP 2019)*, Krakow, pp. 1–3, 2019.

10

Millimeter-Wave Antenna-on-Display for 5G Mobile Devices

The advent of widescreen mobile devices (e.g. smartphones) featuring wireless connectivity has significantly changed our lives and revolutionized the electronics industry over the past few decades [1]. Widescreens with touch input capabilities not only provide information in visual form but also enable users to directly interact with devices. Moreover, this information from users and devices can be shared via wireless communication protocols, such as Bluetooth, Wi-Fi, 3G, 4G, and 5G standards [2–6]. Naturally, every device should be equipped with radio frequency (RF) transceiver modules, which include antennas and monolithic microwave integrated circuits (MMICs). In addition, it is expected that a plural number of antennas and MMICs will be required for massive multi-input multi-output (MIMO) antenna technologies in the upcoming beyond 5G wireless standards to further enhance channel capacities of future wireless links [7–10].

Electromagnetic waves cannot propagate through metallic boundaries. Naturally, antennas and MMICs have been placed in the areas outside the display components containing metallic parts for all wireless devices such as laptops, TVs, and portable devices [11–16]. For instance, for 4G long-term evolution (LTE) RF front-end modules, the metallic rim of mobile devices is utilized as the antenna to minimize the required real estate [11–13]. In the case of 5G mmWave mobile devices, multiple antenna-integrated modules are placed within the bezel area [14] and the back cover [15, 16] region of mobile devices. In addition, the recent proliferation of cloud computing and ultrafast wireless communication has introduced new display-driven designs featuring full-screen, higher resolution, and new form factors, such as wearable, curved, rollable, and foldable displays [17–20]. Collectively, it is clear that there will be less space available for wireless circuits while the wireless technological requirements become increasingly stringent. This introduces an unprecedented, critical tradeoff between the design and

Microwave and Millimeter-Wave Antenna Design for 5G Smartphone Applications, First Edition. Wonbin Hong and Chow-Yen-Desmond Sim.

Published 2023 by John Wiley & Sons, Inc.

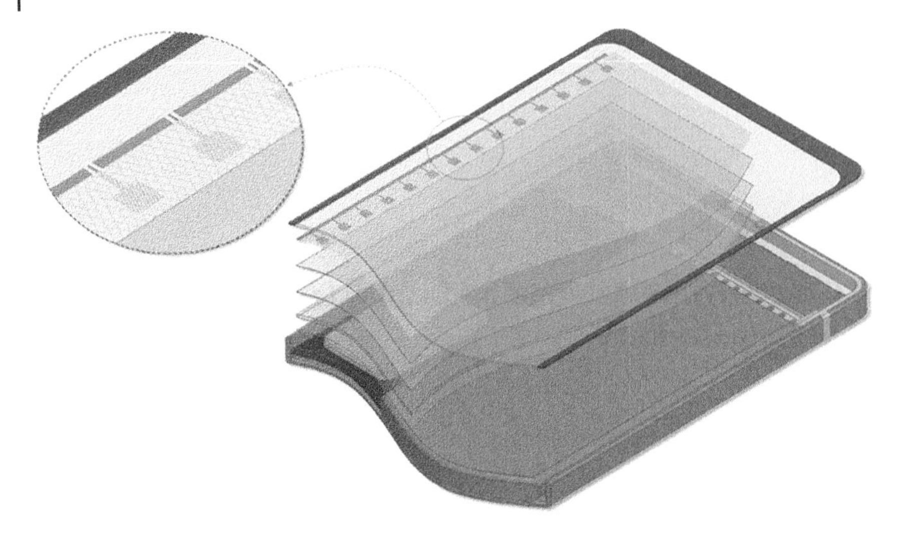

Figure 10.1 Illustration of the AoD concept for future UE. *Source:* From [21]. ©2019 IEEE. Reproduced with permission.

wireless capability of upcoming wireless devices and requires further investigation into the implementation of mmWave and THz antenna/RF systems dedicated to beyond 5G mobile devices.

In contrast to antenna-in-package strategies, which have been covered in Chapters 6–8, the antenna-on-display (AoD) (see Figure 10.1) concept enables antennas to be integrated within the view area of high-resolution OLED or LCD panels of cellular handsets while remaining unnoticeable to the human eye through the use of nanoscale patterns [21–23]. This chapter focuses on three specific challenges in order to investigate potential applications of AoD technology.

1) We adhere to the conventional definition of a transparent antenna, which is defined as a transparent electrode exhibiting optical transparency[1] of 80% or more [24]. However, it should be noted that optical invisibility, which is the main focus of Section 10.2, conceptually differs from optical transparency in that the trace lines of transparent electrodes can be observed by the human eye. Therefore, mmWave AoDs must be devised using optical invisibility techniques in order to be applied to OLED and LCD, similar to conventional touch-screen sensors.

---

1 Optical transparency is a physical property that defines the ratio of the incident light passing through a material without scattering. The optical transparency in this thesis paper is measured using a Murakami Color Research Laboratory HM-150 haze meter or a Konica Minolta CM-3700A spectrophotometer.

2) The formulation of a phased-array antenna configuration is essential in achieving high directivity and sufficient beamsteering coverage. Section 10.3 aims to demonstrate the mmWave 5G AoD radio topology is fully integrated within a 537 ppi density (pixels-per-inch) OLED panel while achieving full optical invisibility to the human eye.
3) During the previous two research phases, it has been demonstrated that AoD technology enables the realization of an optically transparent antenna fully integrated within the display component of UE. However, in previous studies, fully integrated mmWave RF front-end and touch sensor components are thoroughly excluded from the stage of design and implementation due to technical challenges. Therefore, the final goal of Section 10.4 is to advance the AoD concept to a flexible, invisible hybrid electromagnetic sensor (HEMS) that dually functions as a radio-frequency transceiver and a high-resolution touchscreen display.

10.1 Performance Metrics of mmWave 5G Mobile Antenna Systems

The performance metrics of the mmWave antenna system featuring beamforming and beamsteering capabilities are reviewed in this section for a better understanding of the potential of AoD technology.

10.1.1 Spherical Coverage Requirements

The 3rd Generation Partnership Project (3GPP) has standardized the uplink and downlink spherical coverage requirements for mmWave 5G UE using two benchmarks: equivalent isotropically radiated power (EIRP) for uplink and effective isotropic sensitivity (EIS) for downlink values at certain percentile on the cumulative distribution function (CDF) curve [25]. After measuring the EIRP or EIS values on the test point in the entire sphere, the CDF of mmWave UE in uplink and downlink modes can be calculated using Eqs. (10.1) and (10.2), respectively. The right-hand side of Eqs. (10.1) and (10.2) represents the probability that the measured EIRP/EIS values of the UE will take a value less than or equal to a threshold EIRP/EIS value.

$$\mathrm{CDF}(\mathrm{EIRP}) = P\left(\mathrm{EIRP}_{\mathrm{UE}}\left[\theta,\varphi\right] \leq \mathrm{EIRP}\right) \tag{10.1}$$

$$\mathrm{CDF}(\mathrm{EIS}) = P\left(\mathrm{EIS}_{\mathrm{UE}}\left[\theta,\varphi\right] \leq \mathrm{EIS}\right) \tag{10.2}$$

Table 10.1 UE minimum peak EIRP and spherical coverage EIRP for power class 3 [22].

Operating band	Min peak EIRP (dBm) at 100% tile CDF	Min spherical coverage EIRP (dBm) at 50%-tile CDF
n257 (26.50–29.50 GHz)	22.4	11.5
n258 (24.25–27.50 GHz)	22.4	11.5
n259 (39.50–43.50 GHz)	18.7	5.8
n260 (37.00–40.00 GHz)	20.6	8
n261 (27.50–28.35 GHz)	22.4	11.5

For example, Table 10.1 shows the minimum peak EIRP at 100%-tile CDF and the minimum peak EIRP at 50%-tile CDF, defined by 3GPP for uplink of power class 3 UE (e.g. mobile phones). The minimum peak EIRP at 100%-tile CDF evaluates the peak EIRP value in the main beam direction and the minimum peak EIRP at 50%-tile CDF determines whether sufficient EIRP can be generated for more than half of the entire spherical coverage. Therefore, the commercial mmWave UE has to ensure the threshold EIRP and EIS values at predefined percentile CDF according to the 3GPP's measurement standard. Again, it becomes imperative to maximize the beamsteering angles of mmWave 5G phased-array antennas within cellular handsets to enhance the CDF of EIRP/EIS.

10.1.2 Error Vector Magnitude (EVM) Requirement

The error vector magnitude (EVM) is a performance metric used to quantify the performance of a digital radio transmitter or receiver and is degraded by noise, distortion, spurious signals, and phase noise. Therefore, EVM is considered a system-level performance metric that comprehensively and quantitatively holds the quality of a wireless receiver or transmitter for digital wireless communications.

The EVM is calculated using two following values: (i) The error vector is the vector between the ideal constellation point and the actual received point in the I–Q plane. The magnitude of this error vector is denoted and quantified as P_{error}. (ii) As a reference, the vector between the ideal constellation and the origin is defined in the I–Q plane. The magnitude of this reference vector is denoted and quantified as $P_{\text{reference}}$. The EVM is the root-mean-square (RMS) mean amplitude of the error vector (P_{error}) normalized to the ideal signal amplitude ($P_{\text{reference}}$). The EVM is typically expressed in percent using Eq. (10.3)

$$\text{EVM}(\%) = \sqrt{\frac{P_{\text{error}}}{P_{\text{reference}}}} \times 100\% \tag{10.3}$$

Table 10.2 5G NR sub-6 GHz Requirements for EVM [27].

Parameter	Unit	Average EVM level
Pi/2-BPSK	%	30
QPSK	%	17.5
16 QAM	%	12.5
64 QAM	%	8
256 QAM	%	3.5

Table 10.3 5G NR mmWave minimum requirements for EVM [25].

Parameter	Unit	Average EVM level	Reference signal EVM level
Pi/2 BPSK	%	30.0	30.0
QPSK	%	17.5	17.5
16 QAM	%	12.5	12.5
64 QAM	%	8.0	8.0

The definition of EVM is highly dependent on the standard being used. For example, the 3GPP defines exactly how to measure EVM in sub-6 GHz [26, 27], and mmWave NR applications [25].

In Sub-6 GHz systems, the EVM is only measured at the highest transmission (TX) power level [26]. On the other hand, in the beamforming mmWave system, the system EVM, which is defined as the combination of the transmitter (Tx) and receiver (Rx) EVMs, is used as a system-level performance metrics according to the latest version of the 3GPP 5G NR specification [25, 27]. EVM values in sub-6 GHz and mmWave bands of 5G NR according to modulation schemes are summarized in Tables 10.2 and 10.3.

10.2 Optically Invisible Antenna-on-Display Concept

This section reviews the transparent conductive films (TCFs) and display-related materials/processes to realize the AoD concept. In particular, the design process of the unit cell of transparent electrodes and the optically invisible antennas is explained in detail from the viewpoint of compatibility with display and RF performance.

10.2.1 Material and Process

TCFs for transparent antennas can be classified as transparent conductive oxides [24, 28–30], nano carbons (graphene [31–33], carbon nanotube [34, 35]), conductive polymers [36–38], or metallic nano structures (nanowire [39–41], ultra-thin metallic film [42, 43]). As an alternative to ultra-thin metallic films, metal mesh grids with grid widths of tens of microns have recently been widely investigated in order to realize optically transparent antennas and related applications [30, 44–49]. However, when placed above active displays, the tens of micron-grid widths cause the Moiré phenomenon [50], severely impairing the visual quality of the display panel. In order to suppress undesired Moiré patterns, photolithography is used in this paper to formulate diamond-shaped metal grids.

A photolithography process is used to implement diamond-grid unit cells. As shown in Figure 10.2, the Ag-alloy pattern (L4) is chemically etched into a glass. Afterward, this patterned Ag-alloy is deposited on the 40 μm-thick transparent dielectric substrate (L5) using a transfer process. A secondary Ag-alloy layer (L8) is then formulated on the bottom of the glass substrate (OLED display panel region) (L7) to function as the electrical ground for the AoD topology using an identical procedure. Finally, 100 μm-thick optically clear adhesive (OCA) (L6) is added between the transparent dielectric substrate and 700 μm-thick glass (OLED display panel region). A glass substrate featuring a relative dielectric constant of 5.5 and thickness of 700 μm is used throughout Section 10.2. In real-life products, a front glass will encompass the entire AoD for protection from scratches as well as chemical and water damage. During this research, an insulator layer is applied above the antenna layer to prevent mechanical damage and oxidization.

It should be noted that Ag-alloy, copper, and aluminum are eligible candidates for electrodes for this particular photolithography process. In this work, 2000 Å-thick Ag-alloy is selected due to its low heat, corrosion resistance, and high conductivity.

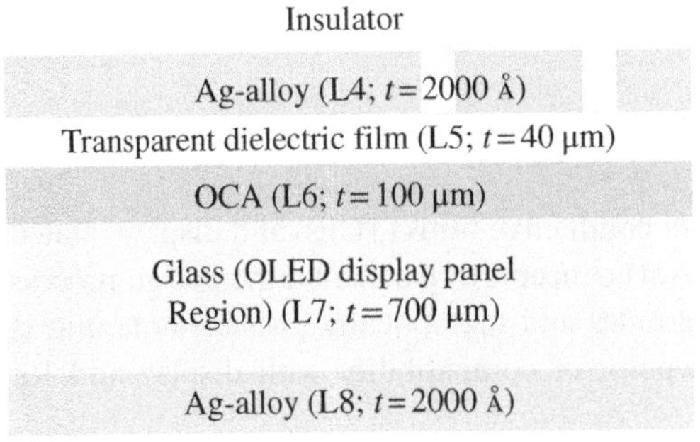

Figure 10.2 Stack-up of the mmWave AoD. *Source:* From [21]. ©2019 IEEE. Reproduced with permission.

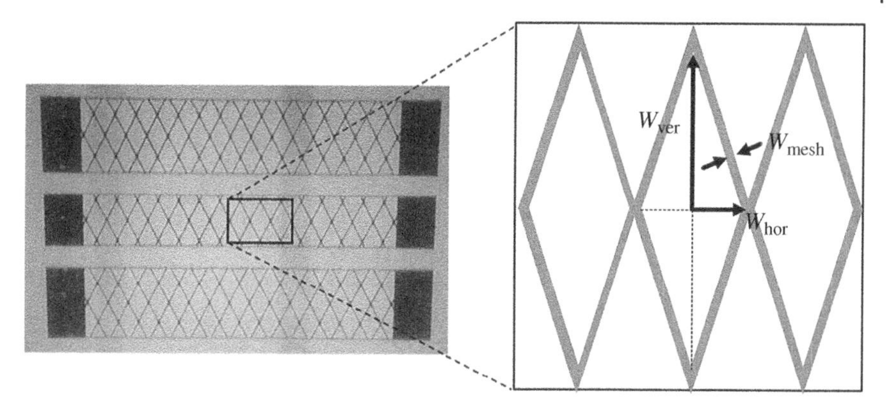

Figure 10.3 Photograph of transparent diamond-grid CPW. *Source:* From [21]. ©2019 IEEE. Reproduced with permission.

The coplanar waveguide (CPW) test samples are fabricated using the aforementioned materials as electrodes, and their electrical characteristics are first examined. A photograph of the fabricated CPW test sample is presented in Figure 10.3. The width and height of the diamond-grid unit cell are set to be $w_{ver} = 100\,\mu m$ and $w_{hor} = 50\,\mu m$, featuring more than 88% optical transparency. The width of the signal line, the ground, and the pitch between the signal line and the ground are configured to be 250, 350, and 100 μm, respectively.

10.2.2 Parametric Studies on Material Thickness and Width

In order to ensure optical transparency, the thickness of Ag-alloys must be confined to several hundreds of nanometers. Since this is sufficiently less than the skin depth at 28 GHz, the insertion loss characteristics are further examined using a series of fabricated Ag-alloy CPW test samples with identical physical properties. The thickness of the Ag-alloy for each test sample was configured to be 700, 1000, and 2000 Å, respectively. Each sample was fabricated using the previously discussed process. At the time of this study, the S-parameters are measured to range from 100 MHz to 18 GHz using a KEYSIGHT N5230A PNA-L. The measured and simulated results using ANSYS indicate a nonlinear deterioration of the insertion loss as a function of Ag-alloy thickness, t, as shown in Figure 10.4. For instance, the 2000 Å-thick CPW features a 0.4 dB/mm insertion loss at 28 GHz, while this is worsened to 1.45 dB/mm in the case of the 700 Å-thick test sample. Figure 10.5 presents the effect of the line width, w_{mesh}, of the diamond-grid unit cell on the insertion loss properties at mmWave frequencies. The thickness of the Ag-alloy is fixed at 2000 Å. By increasing w_{mesh} from 3 to 5 μm, the optical transparency decreases from 88% to 86%, while the insertion loss is improved by 0.3 dB/mm at 28 GHz.

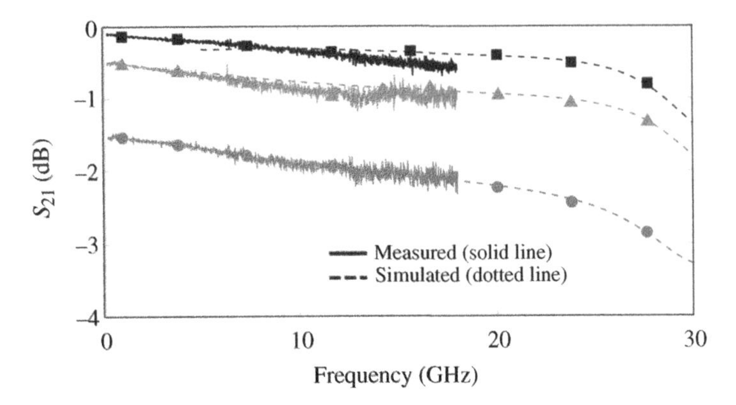

Figure 10.4 S_{21} of the 2 mm-long CPW as a function of the thickness of Ag-alloy. Square: t = 2000 Å; Triangular: t = 1000 Å; Circle: t = 700 Å. *Source:* From [21]. ©2019 IEEE. Reproduced with permission.

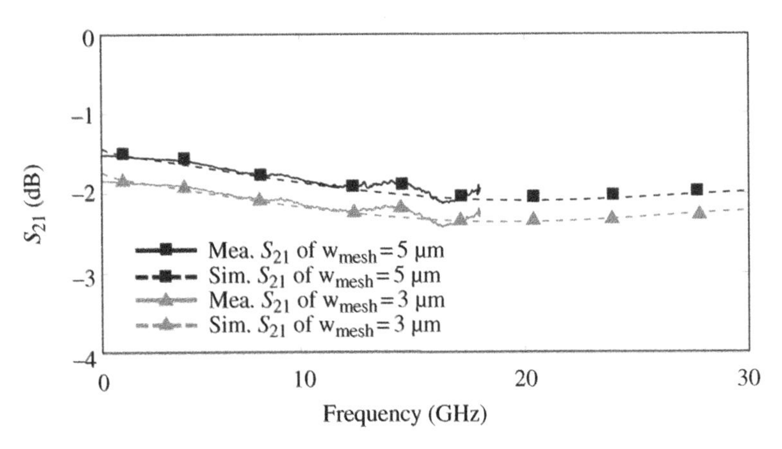

Figure 10.5 S_{21} of the transparent 10 mm-long CPW as a function of w_{mesh}. *Source:* From [21]. ©2019 IEEE. Reproduced with permission.

10.2.3 Optically Transparent Antenna with Corrugated Edges

Transparent antennas are susceptible to Moiré[2] and haze [51] phenomena. Moiré is induced by the interference between the RGB (red, green, blue) pixels of active displays and the periodic unit cells (rectangular or diamond) that constitute the antenna topologies. Haze is defined as the ratio of the amount of light deviated

2 The Moiré phenomenon is caused by the interaction of light through two or more periodic layers. This Moiré phenomenon appears as a wave pattern and degrades the image quality of the display.

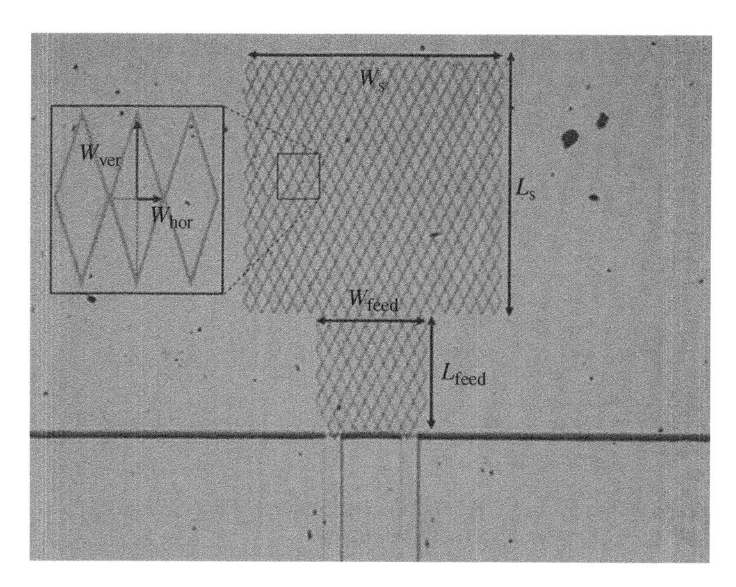

Figure 10.6 Photograph of the fabricated transparent diamond-grid antenna featuring corrugated edges along the outer boundaries. *Source:* From [21]. ©2019 IEEE. Reproduced with permission.

by 2.5° to the incident angle in total light. In this section, a 28 GHz transparent diamond-grid antenna element topology that can suppress Moiré in the presence of active displays is first investigated and subsequently converted into an optically invisible diamond-grid pattern configuration.

A transparent TM_{010} patch antenna element consisting of diamond-grid unit cells is designed, fabricated, and illustrated in Figure 10.6. The resonant frequency is determined by adjusting the length of the patch ($L_S = 2.06$ mm) and the quality factor is further optimized by controlling the width of the patch ($W_S = 2.17$ mm). An impedance transformer with a width of $W_{feed} = 0.93$ mm and a length of $L_{feed} = 0.98$ mm is added to the feed line. The proposed transparent diamond-grid antenna features corrugated edges along the outer boundaries. The corrugated edges are devised so as to disguise the antenna region and avoid Moiré and optical haze (i.e. fuzziness).

The prototype is visually inspected at Dongwoo Fine-Chem, Pyeongtaek, Korea. The optical transparency and haze are measured using the Murakami Color Research Lab HM-150 haze meter and determined to be 88% and 2.8%, respectively. The devised transparent diamond-grid antenna achieves 0.63 GHz (26.65–27.28 GHz) measured and 1.2 GHz (27.9–29.1 GHz) simulated impedance bandwidths, as shown in Figure 10.7. The discrepancy between the measured and simulated results can primarily be attributed to the fabrication tolerances and deviation in the electrical properties of the substrate.

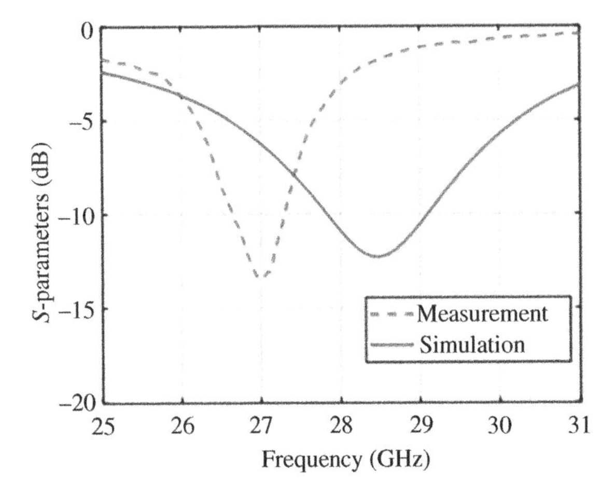

Figure 10.7 Measured and simulated S_{11} of transparent diamond-grid antenna. *Source:* From [21]. ©2019 IEEE. Reproduced with permission.

10.2.4 Optically Invisible Antenna with Dummy-Grids

Despite the high optical transparency, the outer trace lines of the aforementioned transparent diamond-grid antenna can be detected by the human eye when placed above light-emitting displays. In order to render complete optical invisibility of the designed transparent diamond-grid antenna, the antenna region is encompassed with dummy grids so as to eliminate any difference in optical transparency. The topologies of the dummy grids are identical to the diamond grids comprising the transparent diamond-grid antenna. A microscopic view of the fabricated optically invisible diamond-grid antenna is shown in Figure 10.8. The dummy grids are spaced at fixed intervals, w_{gap}, from the corrugated edges of the original transparent diamond-grid antenna region, as demonstrated in the red circle in Figure 10.8. The optimized w_{gap} for maintaining the optical transparency and the antenna efficiency is determined to be 3 μm. It should be noted that minimizing w_{gap} is advantageous in achieving optical invisibility for antennas integrated within display panels with higher resolution. However, this exacerbates an open-circuit effect between the antenna region and the dummy grids.

The open-circuit boundary degrades radiation efficiency due to the near-field energy stored between the antenna and dummy-grid patterns. However, the resonant frequency is predominantly determined by the electrical length of the radiator and is independent of the dummy grids. The current distributions of the outermost frames of the transparent and optically invisible diamond-grid antennas are compared in Figure 10.9. It is noted that regardless of the dummy-grid patterns, the current components along the outermost frames have the highest

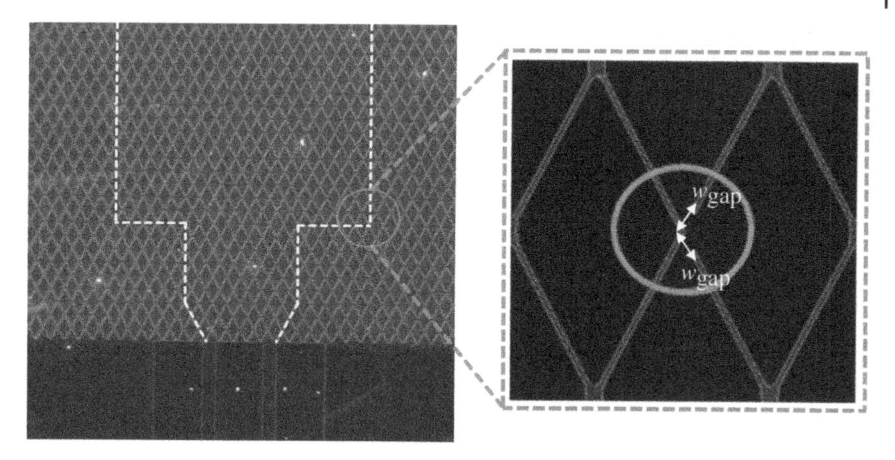

Figure 10.8 Photograph of the optically invisible diamond-grid antenna featuring dummy grids around the corrugate edges. The dotted line indicates the boundary between the electrically continuous antenna region and the dummy-grid region. *Source:* From [21]. ©2019 IEEE. Reproduced with permission.

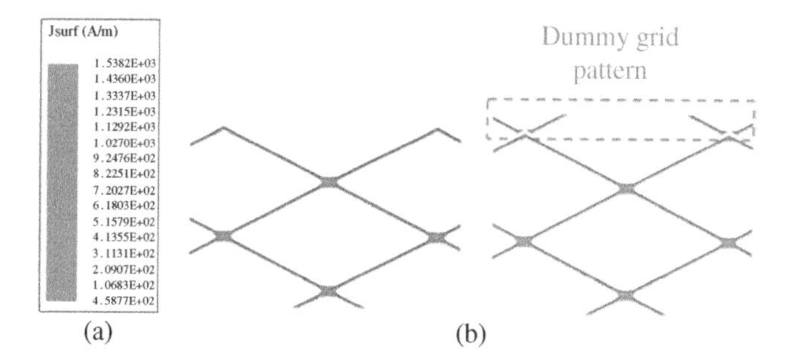

Figure 10.9 Current distributions of the transparent diamond-grid antenna (a) and the optically invisible diamond-grid antenna (b). *Source:* From [21]. ©2019 IEEE. Reproduced with permission.

amplitude and remain the dominant factor at the desired TM_{010} mode. The effect of the separation, w_{gap}, between the dummy grids and the antenna region is studied in Table 10.4. As predicted, reducing w_{gap} adversely affects the radiation resistance of the antenna, and the antenna efficiency and antenna gain are reduced accordingly. Further methods, such as devising novel grid unit cells, which are less susceptible to radiation degradation, should be researched in future works.

The input reflection coefficient $|S_{11}|$ of the optically invisible diamond-grid antenna is illustrated in Figure 10.10 and is ascertained to respectively feature 3.13 GHz (26–29.13 GHz) measured and 2.6 GHz (27.1–29.7 GHz) simulated

Table 10.4 Comparison of bandwidth and simulated gain as a function of w_{gap} for the optically invisible diamond-grid antenna.

w_{gap} (μm)	Radiation gain (dBi)	Efficiency (%)	Radiation gain (dBi)
3	1.18	31.3%	1.18
5	1.36	32.6%	1.36
7	1.40	32.9%	1.40
10	1.43	33.1%	1.43

Source: From [21]. ©2019 IEEE. Reproduced with permission.

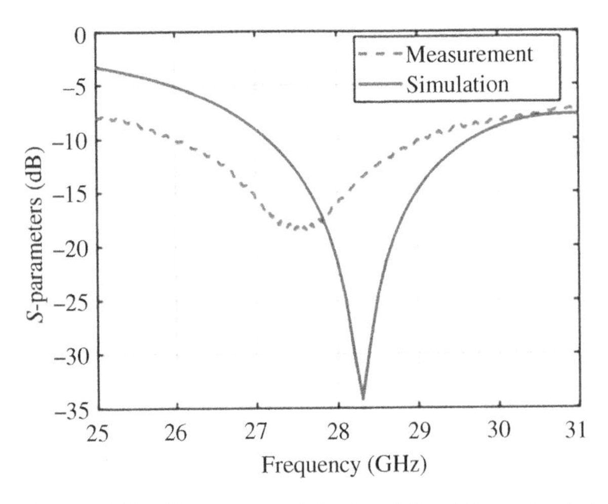

Figure 10.10 Measured and simulated S_{11} of the optically invisible diamond-grid antenna. *Source:* From [21]. ©2019 IEEE. Reproduced with permission.

impedance bandwidths. The resonance point was shifted by approximately 0.7 GHz due to the fabrication error.

The far-field radiation patterns of the fabricated antennas are measured using a millimeter-wave anechoic chamber at POSTECH, Pohang, Korea. The antenna under test (AUT) is fixed to the foam on the probe stage in order to minimize measurement error due to the dielectric effect as shown in Figure 10.11. Then, the revolving arm with the standard gain horn antenna (A-INFO 26.5-40) is rotated around the AUT. The range of the measured H-plane (x–z plane) and E-plane (y–z plane) of the proposed antenna is limited by the blockage and scattering incurred by nearby metallic objects and components such as the wafer probe, probe stage,

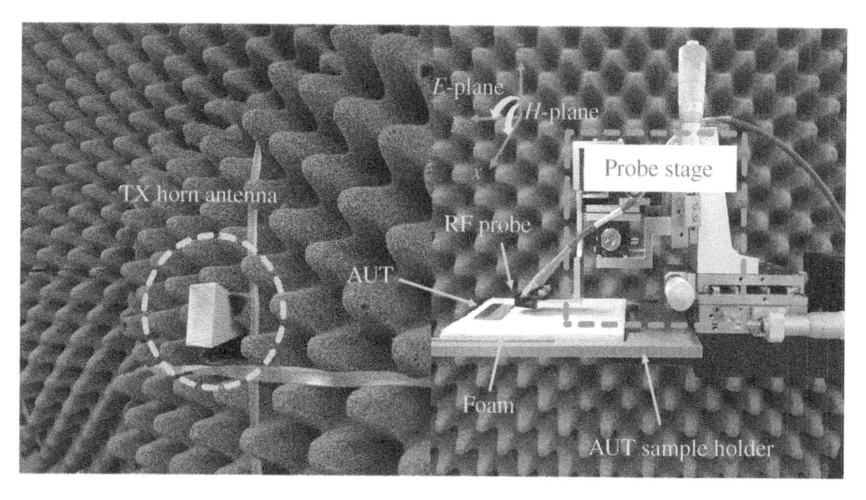

Figure 10.11 The mmWave far-field antenna chamber setup. *Source:* From [21]. ©2019 IEEE. Reproduced with permission.

chuck, and microscope. The measured normalized radiation pattern of the transparent diamond-grid antenna exhibits a measured gain of 4.05 dBi at 27 GHz, as shown in Figure 10.12a. The co- and cross-polarized radiation intensity differences are more than 15 dB in all directions of interest in the E- and H-planes. The radiation patterns of the optically invisible diamond-grid antenna are presented in Figure 10.12b. The measured gain is 0.98 dBi at 27.5 GHz. The measured gain decreased by 3.07 dB as compared to the transparent diamond-grid antenna, which is attributed to the open-circuit effect of the adjacent dummy grids. The radiation efficiency is calculated to be 52.37% for the transparent diamond-grid antenna and 29.83% for the optically invisible diamond-grid antenna, respectively as summarized in Table 10.5.

10.3 OLED Display-Integrated Optically Invisible Phased-Arrays

In this section, the phased-array AoD is demonstrated using the previously devised AoD topology. It should be noted that the presence of the integrated phased-array AoD should neither comprise nor degrade the visual quality of high-resolution display panel nor the functionality of the cellular handset in any situation. Concurrently, it is equally important to sufficiently function as a beamforming antenna at the 28 GHz spectrum.

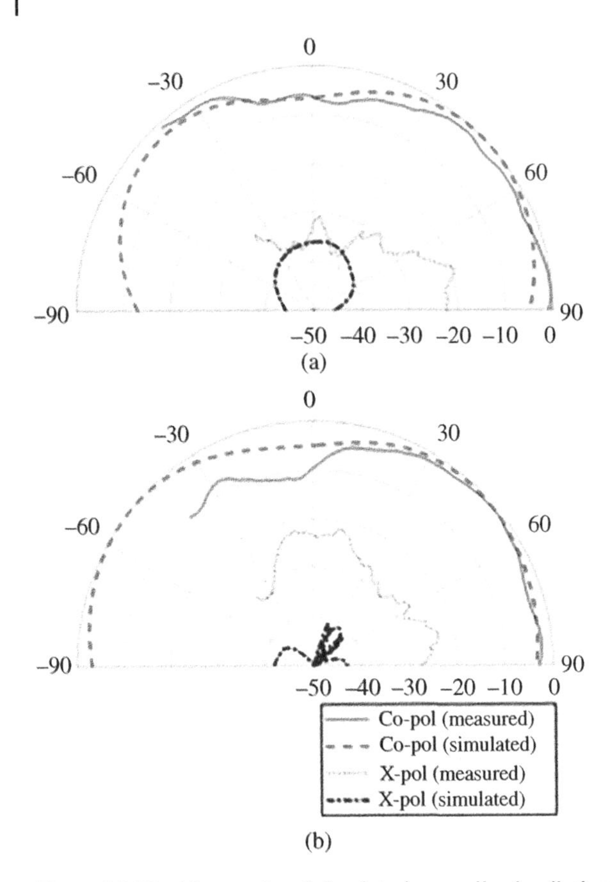

Figure 10.12 Measured and simulated normalized radiation pattern (E-plane) of the (a) transparent diamond-grid antenna and (b) optically invisible diamond-grid antenna array. *Source:* From [21]. ©2019 IEEE. Reproduced with permission.

Table 10.5 Calculated efficiency of the transparent and optically invisible diamond-grid antenna elements.

	Simulated Dir (dBi)	Measured gain (dBi)	Efficiency (%)
Transparent antenna	6.86	4.05	52.37
Optically Invisible antenna	6.23	0.98	29.83

Source: From [21]. ©2019 IEEE. Reproduced with permission.

10.3.1 Packaging Strategy

Figure 10.13 depicts the detailed topology of the mmWave 5G AoD module consisting of 28 GHz phased-array antenna elements for foreside beamforming. To devise the phased-array AoD, the optically invisible antenna layer and the antenna substrate are positioned between the backlight and the front glass panel of the display as shown in Figure 10.13. A 40 μm-thick COP is used as the

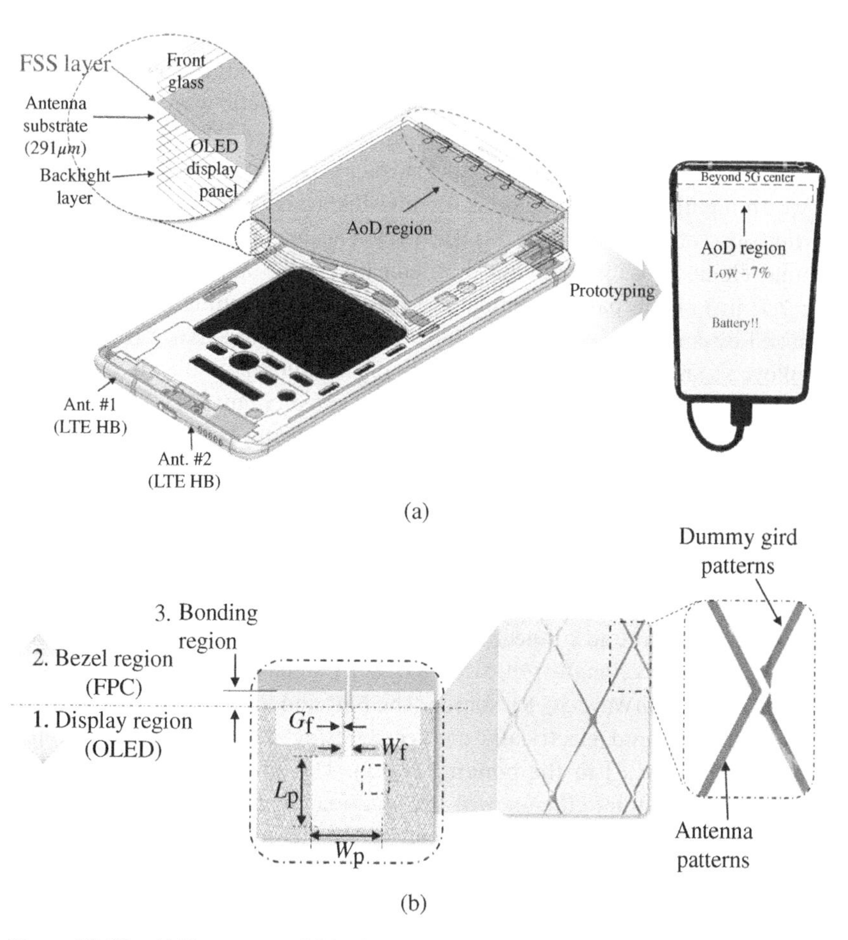

Figure 10.13 (a) The proposed 5G NSA wireless communication systems consisting of mmWave 5G and legacy antenna modules. (b) The detailed view of the mmWave 5G phased-array AoD within the display panel. *Source:* From [22]. ©2020 IEEE. Reproduced with permission.

antenna substrate during the fabrication. Gorilla glass by Corning functions as the front glass panel of the mmWave 5G NR cellular handset prototype. A linear arrangement of broadside planar patch antenna elements is designed and implemented above the light-emitting layer of the OLED display within the mmWave 5G NR cellular handset prototype. The devised AoD module spans across three areas within the cellular handset: (i) The view area of the display panel (display region), (ii) The bezel area of the cellular handset (bezel region), and (iii) The bonding region as shown in Figure 10.13b. A region denoted as the "dummy grid region," which consists of identical diamond-shaped mesh unit cells, surrounds the antenna patterns in the display region. The simulated surface current distributions (Figure 10.13b) confirm that electromagnetic wave distributions are suppressed on the surface. This creates distinct electrical boundaries within this region and mitigates the effect of the surface wave on the radiation patterns. It is worth mentioning that the optically invisible transmission line features an insertion loss of 0.4 dB/mm at 28 GHz [21]. The operating frequency of the antenna element is predominantly determined by adjusting the length (L_P = 2.7 mm) of the patch antenna and impedance matching can be further optimized by controlling the width (W_P = 2.8 mm). An impedance transformer is employed to match the input impedance of the antenna to the characteristic impedance of the signal line using a flexible printed circuit (FPC) board. This provides an electrical connection between the antenna panel and the main board of conventional cellular handsets. The width (W_f) of the antenna feedline is 0.3 mm and the edge-to-edge distance (G_f) between the ground pad and the signal line is 0.11 mm. The bonding region is realized through a conductive joint between the antenna panel (display region) and the FPC (bezel region), accurately emulating real-life configurations. The feedlines consisting of signal lines and the electrical ground connections are designed and pattered on the surface of the FPC to bi-directionally route the 28 GHz RF signals between the mmWave 5G AoD and the mmWave 5G RF source. The FPC and the AoD module are separately fabricated and electrically connected using conventional anisotropic conductive film (ACF) in the bonding region. The simulated and calculated measured antenna total efficiency of the antenna element features 27.7% and 30.1%, respectively, at 28 GHz. This demonstration includes two types of phased–array antenna configurations: (i) 1×4 arrangement, and (ii) 1×8 arrangement. In the case of the 1×4 phased-array AoD module, each antenna element is individually connected to each feed line. In the case of the 1×8 phased-array topology, four pairs of antenna elements comprise a sub-array structure using 1 : 2 power divider circuits to further enhance the antenna directivity, as shown in the AoD region of Figure 10.14a. In both antenna array configurations, the spacing between adjacent antenna elements is determined to be 7 mm to minimize electromagnetic interference (EMI).

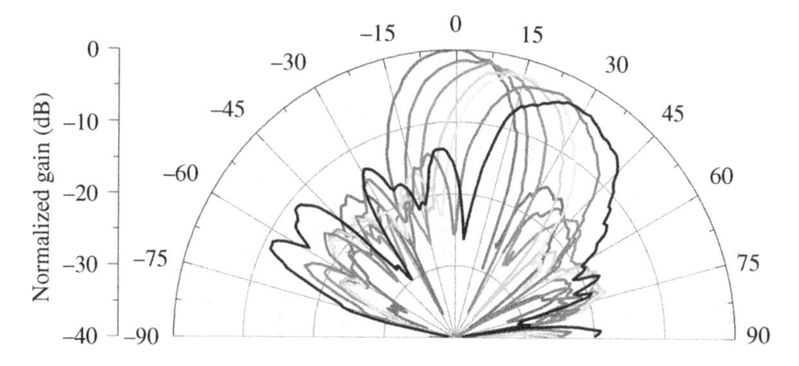

Figure 10.14 Measured beamforming far-field radiation patterns of the mmWave. 5G AoD integrated within a fully operating cellular handset prototype at 28 GHz during the RX mode (0° is defined as the foreside). *Source:* From [22]. ©2020 IEEE. Reproduced with permission.

10.3.2 Component- and System-Level Verifications

A mmWave 5G NR cellular handset prototype featuring the fabricated phased-array AoD is designed and assembled. The Android-based prototype is fully operational and resembles a conventional smartphone as presented in Figure 10.13. It includes all essential components such as the legacy wireless modules, cameras, and batteries to investigate and ascertain the validity of the AoD module under real-life scenarios. To evaluate the foreside antenna gain and antenna beamforming characteristics, each RF port of the 1×4 and 1×8 AoD prototypes is assembled using 2.92 mm connectors to the mmWave 5G beamforming test board. Although noise emitted by the display panels often results in undesired EMI during wireless modes, the measured results confirm that the display noise has a negligible effect on AoD beam formulation at 28 GHz. It is worth mentioning that the far-field radiation patterns of the display ON mode are measured one hour after the ON state of the OLED panel. Figure 10.14 illustrates the measured and normalized far-field radiation patterns of the identical AoD as a function of beam scanning angles, indicating that foreside beamforming is achieved. It is experimentally confirmed that the 1×4 phased–array AoD features an estimated maximum peak EIRP of 14.14 dBm and 3 dB beam scan ranges of ±22.5°. The measured boresight antenna gain of the 1×4 and 1×8 AoD prototypes is 4.13 and 8.53 dBi, respectively, with full inclusion of all loss factors within the fabricated prototype.

The system-level analysis of the mmWave 5G NR cellular handset prototype with the AoD is discussed. The mmWave 5G NR cellular handset prototype is positioned at the receiver (RX) side, as illustrated in Figure 10.15. A compact antenna test range (CATR)-based compact mmWave 5G NR chamber situated at

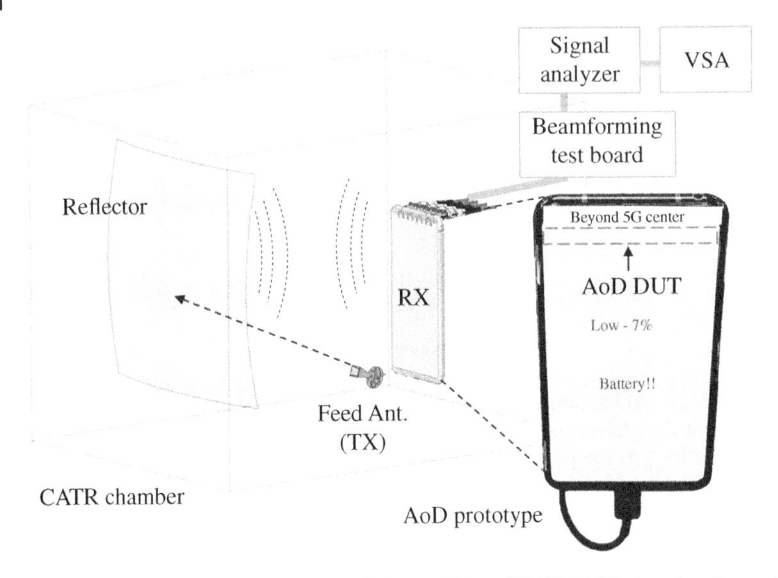

Figure 10.15 Measurement setup of the mmWave 5G NR CATR chamber for system-level analysis of the 5G NR cellular handset prototype featuring AoD for foreside scenarios. *Source:* From [22]. ©2020 IEEE. Reproduced with permission.

SK Telecom, Seongnam, Republic of Korea is employed during the entire measurement. A series of QPSK, 16-, and 64-QAM mmWave 5G NR waveform signals are generated and upconverted to 28 GHz using an analog and vector signal generator. The subcarrier spacing is fixed at 120 kHz in all scenarios. A single feed horn antenna is used for the transmitter (TX) with a total transmit power of 6 dBm. The received signal at the phased-array AoD is downconverted and demodulated using a signal analyzer running the analysis software. The 28 GHz phased-array AoD module within the 5G NR cellular handset prototype is electrically connected to the RF channels of the 5G beamforming test board, which can electronically steer the RX beam toward the TX beam (foreside). The constellation diagrams with a series of differently modulated mmWave 5G NR waveforms and bandwidth are summarized in Figure 10.16. In the case of the 64-QAM measurement setup, two component carriers (CCs) consisting of 400 MHz subcarriers are included to emulate carrier aggregation capability. It is worth mentioning that all experiments are performed without any array calibration and equalization. Therefore, the EVM is attributed to phased array error and quantization error from attenuators and phase shifters within mmWave 5G RFIC used during the experiments. Nevertheless, it can be confirmed that all measured EVM of the 5G NR cellular handset prototype satisfies the EVM requirements [25] for 5G modulation schemes defined by 3GPP.

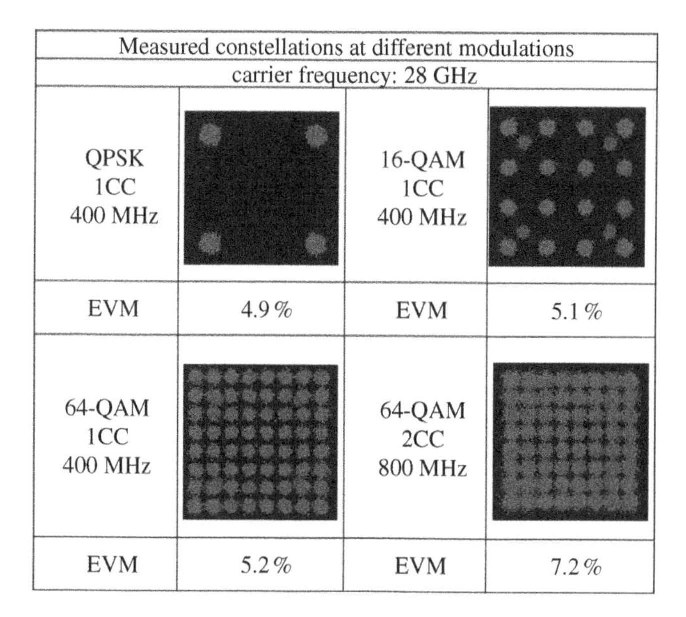

Figure 10.16 Measured constellations at different modulations. *Source:* From [22]. ©2020 IEEE. Reproduced with permission.

10.4 OLED Touch Display-Integrated Optically Invisible Phased Arrays

In Sections 10.2 and 10.3, AoD configurations have omitted the presence of the touch sensor layer due to the inherent challenges related to the coexistence of two conductive touch-sensor and antenna layers. (i) From the antenna perspective, other metallic objects (e.g. a touch sensor layer) located in close proximity to the antenna increase the capacitively stored energy of the radiation reactance, which results in the reduction of the antenna radiation efficiency. (ii) From the touch sensor perspective, a separate metallic layer (antenna layers) located near the touch sensors dramatically reduces the mutual capacitance change ratio [52], which is highly correlated to touch sensitivity, according to the presence or absence of fingers. In addition, the presence of two separate electrode layers not only reduces the entire optical transparency of display components but also causes visual quality issues such as Moiré and haze due to the spatial frequency difference between two different periodic-patterned layers.

Therefore, this section proposes a new CoD concept which attempts to integrate wireless circuits on the same layer of the touch screen sensors of display panels for the first time. I exemplify this concept using a flexible, invisible HEMSs'

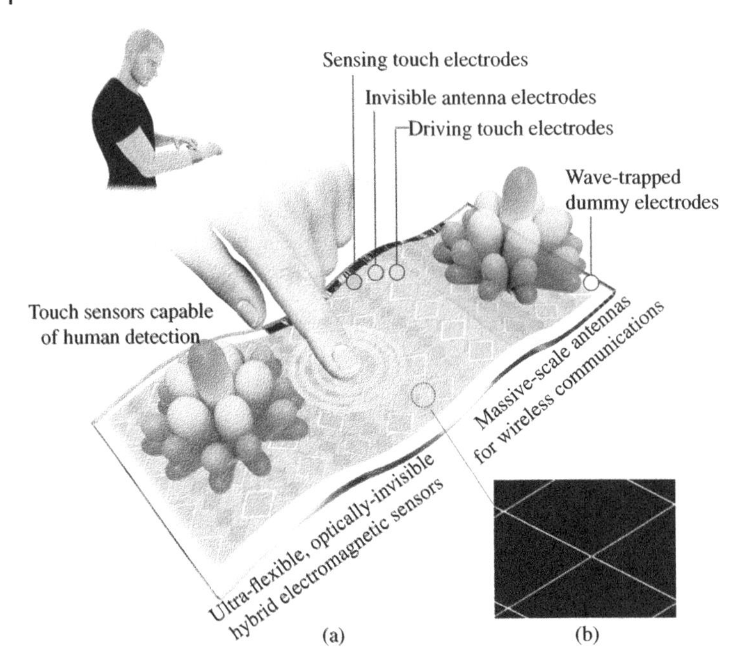

Figure 10.17 (a) Illustration of the CoD concept for future wearable devices. (b) Microscopic photograph of the CoD POC containing diamond-grid patterned electrodes. *Source:* From [23]. ©2021 IEEE. Reproduced with permission.

architecture incorporating optically invisible antennas and touch sensors as presented in Figure 10.17a.

10.4.1 Flexible, Invisible Hybrid Electromagnetic Sensor

The exemplified design of the CoD system using the flexible, invisible HEMS architecture is shown in Figure 10.17a. The HEMS architecture contains three functional blocks comprising antenna electrodes, driving/sensing touch sensor electrodes, and wave-trapped dummy electrodes, all of which are coherently arranged on a single layer. The systematic HEMS architecture is illustrated in Figure 10.18. The invisible touch sensor/antenna patterns are intertwined in a lattice form and placed on an identical layer. Independent signal paths for each circuit enable simultaneous and respective transmission of RF and touch signals. Independently adjusted gain and phase of the RF signals from the MMICs allow the adaptive beamforming operation of the active phased array antennas. I adopt an orthogonal cross-shaped touch electrode as the basic structure to reduce the design parameters and enable a robust arrangement with the integrated invisible

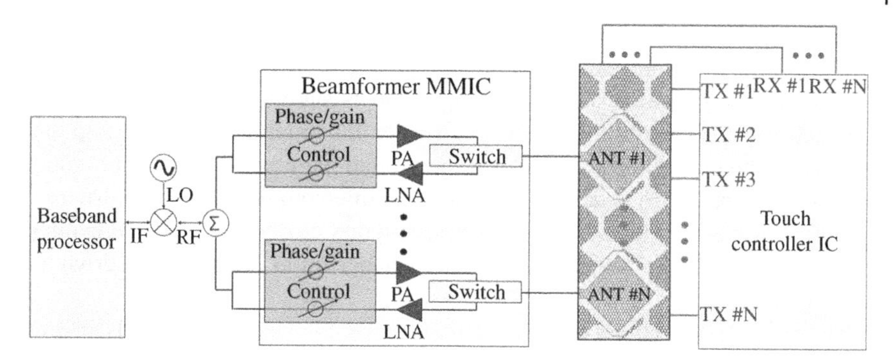

Figure 10.18 Schematics of the HEMS architecture including the touch controller IC and the beamformer MMIC. *Source:* From [23]. ©2021 IEEE. Reproduced with permission.

antenna. In this HEMS design, multiple invisible antenna elements can be formed without overlapping with driving/sensing traces in the vertical and horizontal axes of the invisible touch sensors. The entire HEMS circuit is implemented using conductive mesh alloy unit cells featuring a line width of 2.5 μm and a thickness of 2400 Å for maximal optical transparency, as illustrated in Figure 10.17b.

10.4.2 Design Process and Building Block Measurements

When devising the HEMS circuit, it becomes imperative for antenna electrodes and touch sensor electrodes to function independently amid extremely close proximity on a single layer. It is worth noting that in the case of conventional touch sensor electrodes cross talk and signal interference become inevitable due to the inherent topology. Conventional touch sensors [52] consist of driving/sensing electrodes and ITO bridges which are used to connect the sensing electrodes at the overlapping regions between the driving and sensing electrodes. When a touch contact is detected on the TSP, the electric field distribution between the driving and sensing electrodes is altered by the introduction of a new current path originating from the touch objects. This deviation in the electric field distribution changes the mutual capacitance between the driving and sensing electrodes, which validates the touch action. The mutual capacitance change ratio is defined as

$$\frac{\Delta C_{\mathrm{M}}}{C_{\mathrm{M}}} = \frac{C'_{\mathrm{M}} - C_{\mathrm{M}}}{C_{\mathrm{M}}} \tag{10.4}$$

where C_{M} denotes the equivalent mutual capacitance between the driving and sensing electrodes under consideration and C'_{M} is the corresponding mutual capacitance with the presence of the touch object. However, in the case of conventional TSPs, the addition of the antenna trace will overlap with the ITO bridges at the junctions, ultimately disabling the driving/sensing touch traces. In addition, the mutual capacitance change ratio is significantly affected by the reduced effective area of the touch sensor when the antenna is inserted into the surface of the unit cell of the touch sensor. A quad-patterned touch sensor topology (see Figure 10.19) is devised to enable the interconnection of each driving/sensing trace with the coexistence of the antenna electrodes while mitigating the reduction of the mutual capacitance change ratio between the driving/sensing traces.

The entire HEMS circuit (see Figure 10.20a–c) is composed of unit cells featuring a width (W_{unit}) of 4.35 mm and a length (L_{unit}) of 4.15 mm in consideration of scan time [53] required for touch sensor panels. It is noted that the antenna and

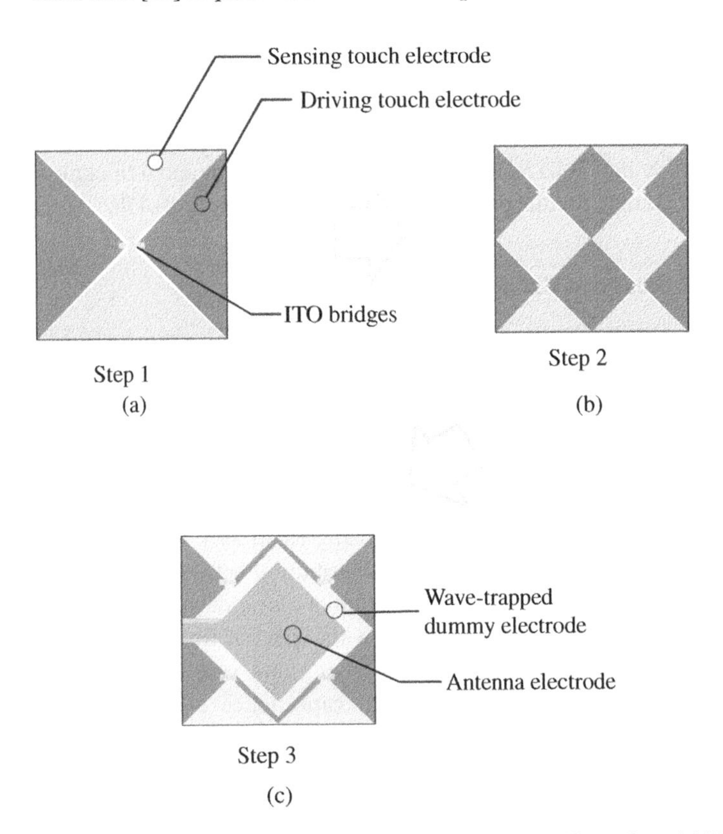

Figure 10.19 Conceptual illustration of devising a HEMS topology. (a) The unit cell of the conventional touch sensors. (b) The unit cell of the proposed quad-patterned touch sensors. (c) The unit cell of the proposed HEMS incorporating/sensing/driving touch electrodes, wave-trapped dummy electrodes, and antenna electrodes. *Source:* From [23]. ©2021 IEEE. Reproduced with permission.

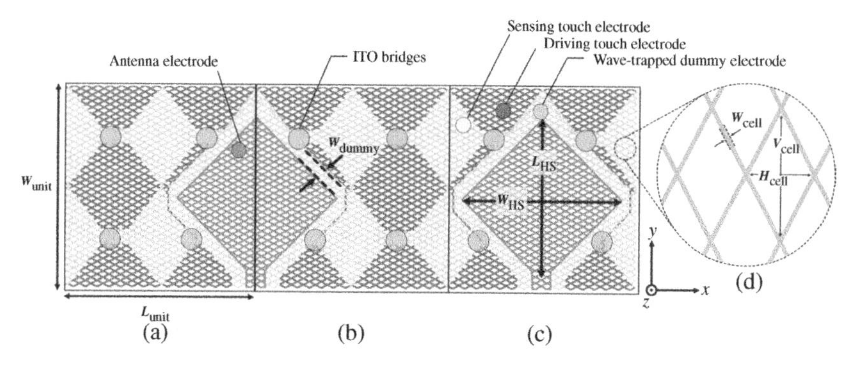

Figure 10.20 (a) The HEMS unit cell containing the left half of the antenna and dummy electrodes. (b) The HEMS unit cell containing the right half of the antenna and dummy electrodes. (c) The HEMS unit cell containing the entire antenna and dummy electrodes. (d) Magnified image of the HEMS unit cell containing diamond-grid patterned electrodes. *Source:* From [23]. ©2021 IEEE. Reproduced with permission.

dummy electrodes can either be completely integrated within a single unit cell or be divided and placed across multiple cells. Hence, the HEMS architecture is compatible with any antenna topologies and antenna array configurations due to its modular nature. The unit cells of the HEMS circuit constitute identical and periodic diamond-grid mesh structures featuring a line width (W_{cell}) of 2.5 μm, a mesh width (H_{cell}) of 131 μm, and a mesh height (V_{cell}) of 248 μm, as illustrated in Figure 10.20d.

10.4.2.1 Antenna Electrodes

A diamond-shaped patch radiator is selected for an antenna topology to be physically compatible with the quad-patterned touch sensor topology. The operating frequency of the diamond-shaped patch radiator is optimized by physically adjusting the length of the radiator (L_{HS} = 3.5 mm), and the impedance matching is performed by controlling the width of the radiator (W_{HS} = 3.47 mm). In the CoD POC, each radiator is uniformly spaced with an increment of 6.23 mm, taking into account the isolation between the elements and the beam scanning range. The stacked dielectric layer between the radiator (HEMS layer) and the ground plane (TFT layer) can be equivalently modeled as a single dielectric layer with a thickness of 288 μm, a relative permittivity of $\varepsilon_r = 2.8$, and a loss tangent of $\tan\delta = 0.01$. The feedline of the diamond-shaped patch radiator is routed and connected with the transmission lines on the FPCB consisting of a single LCP substrate layer featuring a thickness of 50 μm, a relative permittivity of $\varepsilon_r = 3$, and a loss tangent of $\tan\delta = 0.0008$. The ACF bonding process is used to create an electrical interconnection between the HEMS circuit and the RF/touch sensor packages.

To accurately characterize the antenna of the HEMS circuit, the feedline of the two antenna elements is routed to the T-junction power divider on the FPCB. RF signal feeding is established at the end of the FPCB to confirm the input impedance of the antenna. The electromagnetic simulator, Ansys HFSS, is used to extract the input impedance and the far-field properties of the antenna of the HEMS circuit. The measured input impedance bandwidth (reflection coefficient < -10 dB) of the antenna of the HEMS circuit features 1.53 GHz with a center frequency of 27.93 GHz, as illustrated in Figure 10.21b. To evaluate the radiation properties of the antenna of the HEMS circuit, the CoD POC is positioned above the jig, which is placed within the mmWave far-field anechoic chamber as illustrated in Figure 10.21a. It is confirmed from Figure 10.21c that a high correlation is achieved between the simulated and measured radiation patterns in the *yz*-plane. The measured and simulated peak realized gain (Figure 10.21d) is 2.48 and 2.31 dBi at 27.5 GHz, respectively. The calculated measured efficiency (Figure 10.21e) of the antenna of the HEMS circuit features approximate 40% at 28.5 GHz despite the inclusion of lossy dielectric materials and insertion losses of transmission lines on the FPCB. The discrepancy between simulated and measured results is mainly attributed to the manufacturing tolerances, the deviation of the substrate electrical characteristics, and the lack of detailed modeling of the actual device. In addition, it is worth noting that the radiation properties of antennas may deviate if the ground area is wider than the size of that of the designed antenna or if metallic components that have not been considered exist adjacent to the antennas.

10.4.2.2 Touch Sensor Electrodes

Considering the fact that all relevant capacitances of the touch sensor panel can be captured by electroquasistatic simulation of a small panel section due to its periodicity [54], the 5×5 HEMS unit cells are verified with consideration of the entire display material composite and stack-up. The touch sensor of the HEMS circuit is characterized using the electroquasistatic simulator, Ansys Q3D. The simulation model with the defined x–y coordinate is illustrated in Figure 10.22. Figure 10.23 shows the heat map of the mutual capacitance change ratio of the 5×5 HEMS unit cells when the touch object is positioned at (2,1) in the defined coordinate system. The mutual capacitance change ratio of the HEMS circuit at the position of the touch object features −4.43%. This value is clearly higher than a mean value of −0.25% for the remaining cells. It is noted that the absolute value of the mutual capacitance change ratio is slightly reduced in the area where the antenna and dummy electrodes coexist. However, this value can be calibrated through minor adjustments in the algorithm [55] of the touch controller IC. Nevertheless, it is clearly confirmed that touch impact is detectable for any touch position regardless of the coexistence of the antenna and dummy-grid electrodes.

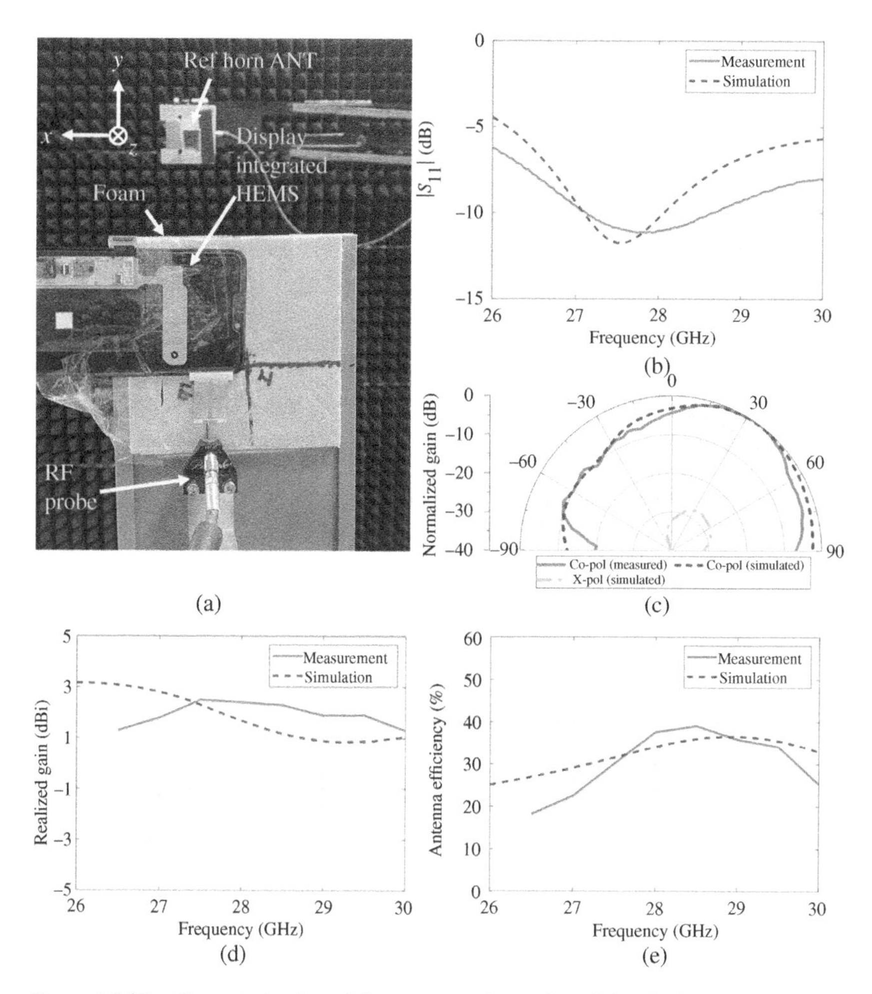

Figure 10.21 Characterization of the antenna electrodes of the display-integrated HEMS circuit. (a) mmWave far-field antenna chamber setup. (b) Measured and simulated reflection coefficients of the antenna of HEMS circuit. (c) Measured co-polarized, simulated co-polarized, and simulated *x*-polarized radiation patterns in *yz*-plane. (d) Comparison of measured and simulated peak realized gain as a function of frequencies. (e) Comparison of measured and simulated total antenna efficiency as a function of frequencies. *Source:* From [23]. ©2021 IEEE. Reproduced with permission.

10.4.2.3 Wave-Trapped Dummy Electrodes

The conductivity of the display-integrated HEMS electrodes is sufficient to guide unwanted EMI from external noise sources. In particular, it is well known that the EMI from the display components significantly deteriorates the sensitivity of RF systems [56–58]. Therefore, the wave-trapped dummy electrodes featuring a width

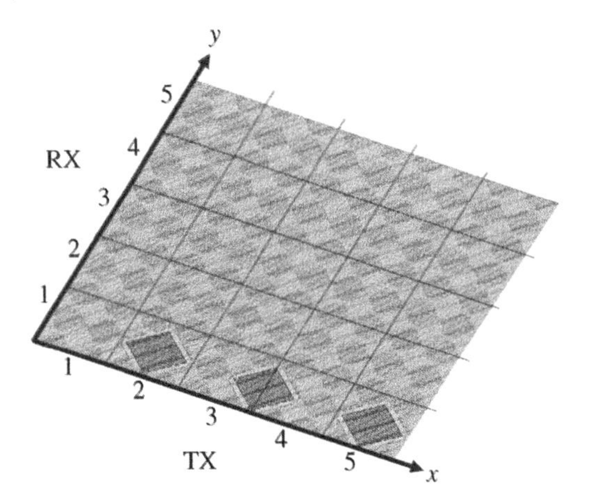

Figure 10.22 Simulation model for the 5 × 5 HEMS unit cells. *Source:* From [23]. ©2021 IEEE. Reproduced with permission.

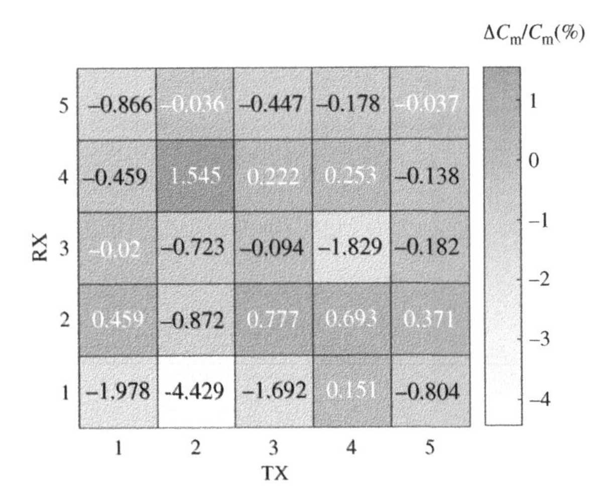

Figure 10.23 Simulated Tx–Rx mutual capacitance change ratio when the touch object is positioned at (2,1). *Source:* From [23]. ©2021 IEEE. Reproduced with permission.

(W_{dummy}) of 0.31 mm are devised to achieve high noise immunity against external EMI noises. The wave-trapped dummy electrodes are equivalent to the open-circuit boundary, which effectively suppress the near-field coupling [59]. Figure 10.24 illustrates the E-field distribution of the HEMS circuit while the touch signal is excited to the driving electrodes. E-field distributions of the HEMS circuit are calculated using the finite element method-based Ansys HFSS. It is clearly confirmed from Figure 10.24 that the lower E-field intensity is induced at

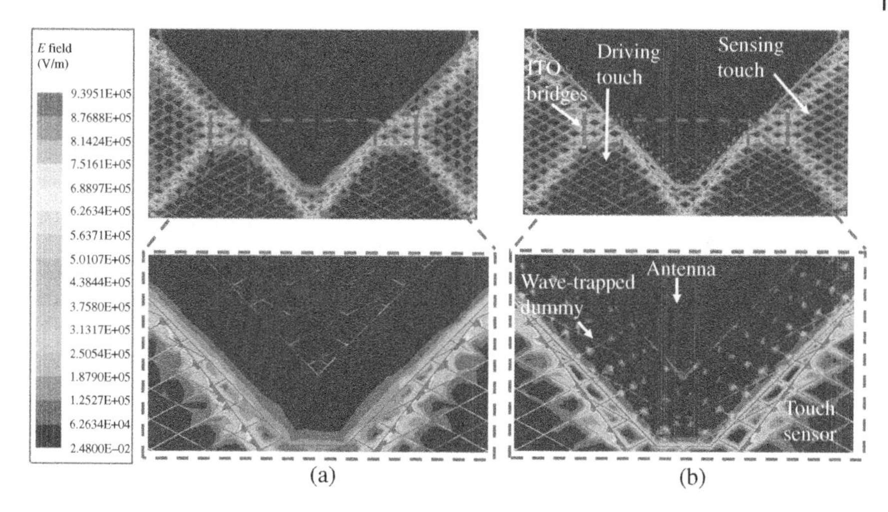

Figure 10.24 Comparison of *E*-field distribution of the HEMS circuit. (a) Without the wave-trapped dummy electrodes. (b) With the wave-trapped dummy electrodes. *Source:* From [23]. ©2021 IEEE. Reproduced with permission.

the antenna electrodes of the HEMS circuit with the wave-trapped dummy electrodes compared to that of the HEMS circuit without the wave-trapped dummy electrodes. Therefore, the electromagnetic fields are considerably shielded when the wave-trapped dummy electrodes surround the antenna electrodes, resulting in the formation of distinct electrical boundaries.

10.4.3 Demonstration within a Cellular Handset Prototype

The CoD system is demonstrated in a device testbed with an RF package and a touch sensor package. Figure 10.25 depicts the fabricated flexible, rollable, and invisible HEMS composite. The OLED display including the HEMS composite is built layer by layer. Figure 10.26 illustrates the exploded view of the CoD POC consisting of the RF package, the touch sensor package, and the display-integrated HEMS circuit. To connect with RF input/output (I/O) ports on the external rigid or flexible printed circuit board (FPCB) package, the antenna feeding section of the HEMS circuit is electrically connected with the transmission line on the FPCB. This FPCB supports a direct connection with the MMIC and can be integrated with relative ease within devices featuring stringent form factors. The touch feeding section of the HEMS circuit is also routed to the bottom edge side and connected with the semi-rigid PCB package including the touch controller IC. The RF package includes the beamformer MMIC, power distribution networks (PDNs), and low/high-speed traces. The beamformer MMIC is mounted on the top side of the RF package. The MMIC is fabricated using a 28-nm bulk CMOS

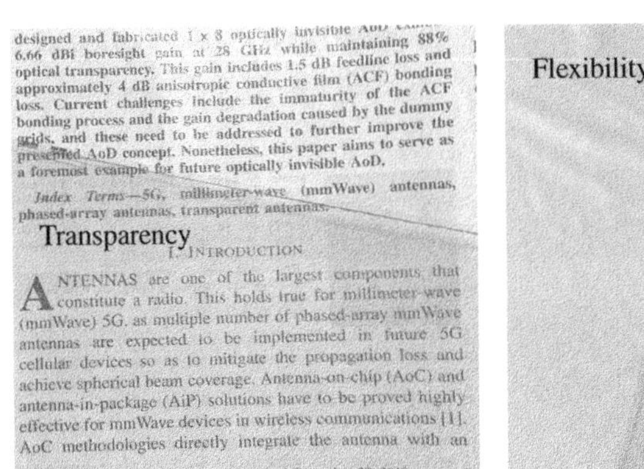

6.66 dBi boresight gain at 28 GHz while maintaining 88% optical transparency. This gain includes 1.5 dB feedline loss and approximately 4 dB anisotropic conductive film (ACF) bonding loss. Current challenges include the immaturity of the ACF bonding process and the gain degradation caused by the dummy grids, and these need to be addressed to further improve the prescribed AoD concept. Nonetheless, this paper aims to serve as a foremost example for future optically invisible AoD.

Index Terms—5G, millimeter-wave (mmWave) antennas, phased-array antennas, transparent antennas.

I. INTRODUCTION

ANTENNAS are one of the largest components that constitute a radio. This holds true for millimeter-wave (mmWave) 5G, as multiple number of phased-array mmWave antennas are expected to be implemented in future 5G cellular devices so as to mitigate the propagation loss and achieve spherical beam coverage. Antenna-on-chip (AoC) and antenna-in-package (AiP) solutions have to be proved highly effective for mmWave devices in wireless communications [1]. AoC methodologies directly integrate the antenna with an

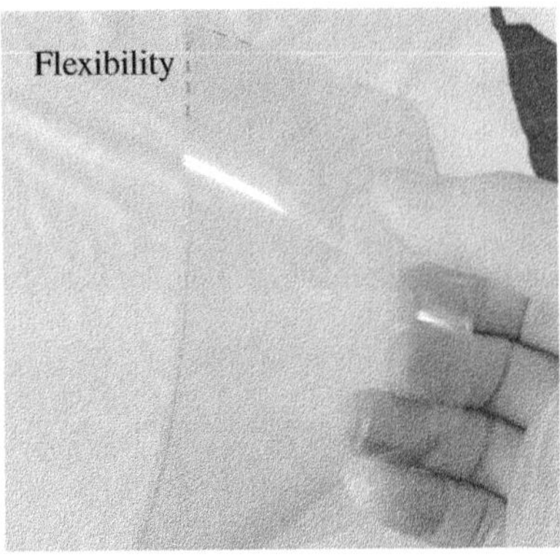

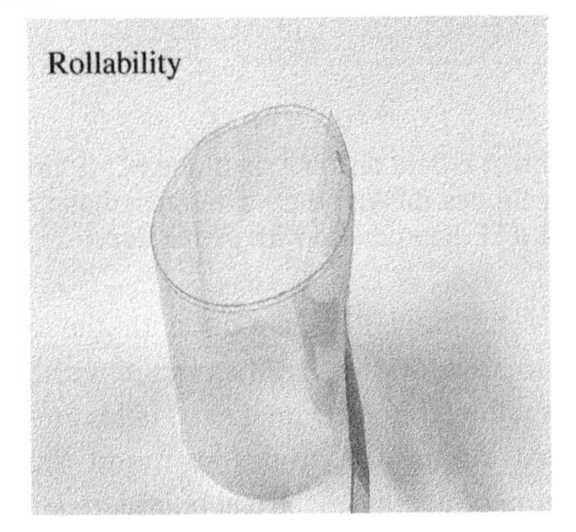

Figure 10.25 Optical photographs of the fabricated HEMS composite for freestanding (top-left), flexed curved (top-right), and rolled (bottom-middle) orientations. *Source:* From [23]. ©2021 IEEE. Reproduced with permission.

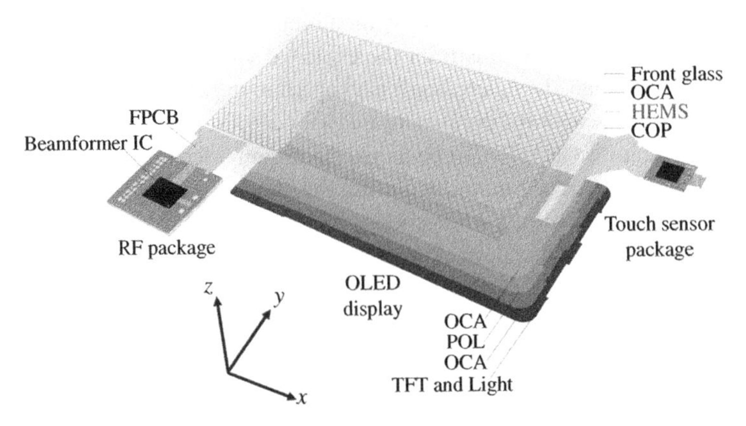

Figure 10.26 Exploded view of the CoD POC. *Source:* From [23]. ©2021 IEEE. Reproduced with permission.

technology and designed for 28 GHz 5G new radio (NR) phased-array applications. The beamformer chip directly divides/combines the RF power in RF transmitting (TX)/Receiving (RX) channels using an on-chip Wilkinson divider network. In addition, 3-bit phase shifters and variable gain amplifiers are integrated together within each TRx channel for electronically-controlled beamsteering using a high-speed serial interface. The RF power is fed by the external mixer and baseband processor.

The beamformer MMIC featuring a size of 5.47 mm × 3.75 mm is mounted on a PCB using flip-chip ball grid array technology. The RF I/O pads are routed to the edge of the PCB to connect with the radiator's feedline in the HEMS circuit, while the electrical phase of each routing transmission line is thoroughly matched to avoid any requirement (e.g. array calibration) for beamforming operations. A low-frequency 52 MHz clock signal is distributed across the beamformer chip to synchronize each part of the entire MMIC architecture. The on-PCB decoupling capacitors are utilized to satisfy the maximum dynamic current requirements by reducing the impedance of the PDN of the RF package.

System-level measurements over the air are conducted to assess the beamforming capabilities of the fabricated CoD POC within a real-life mmWave 5G NR Android-based cellular handset prototype (Figure 10.27). The calculated input weighting matrix is assigned to the beamformer MMIC by the serial peripheral interface for electronically controlled beamsteering operation. All far-field radiation patterns of the CoD POC are measured with the ON state of the OLED panel. This cellular handset prototype with the CoD POC is positioned above the antenna jig of the mmWave anechoic chamber (Figure 10.21a) for accurate measurement.

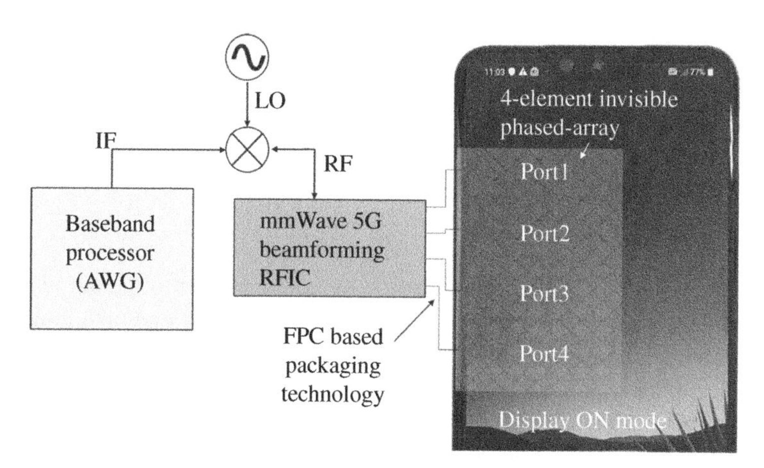

Figure 10.27 Illustration of the RF measurement setup for the CoD POC integrated within a real-life mmWave 5G NR Android-based cellular handset prototype.
Source: From [23]. ©2021 IEEE. Reproduced with permission.

RF package (Tx mode/4 RF channels)			FPCB		Estimated RF package gain
RF COM line loss*	PA Gain**	Channel line loss	Bond & mismatch loss	Line loss	
−1.45 dB	+24.9 dB	−1.97 dB	−1.5 dB	−3.30 dB	16.68 dB

NOTE: All value of each component are extracted from the measured results.
* Including 2.92 mm RF connector
** Including power-combing loss in four RF channels

(a)

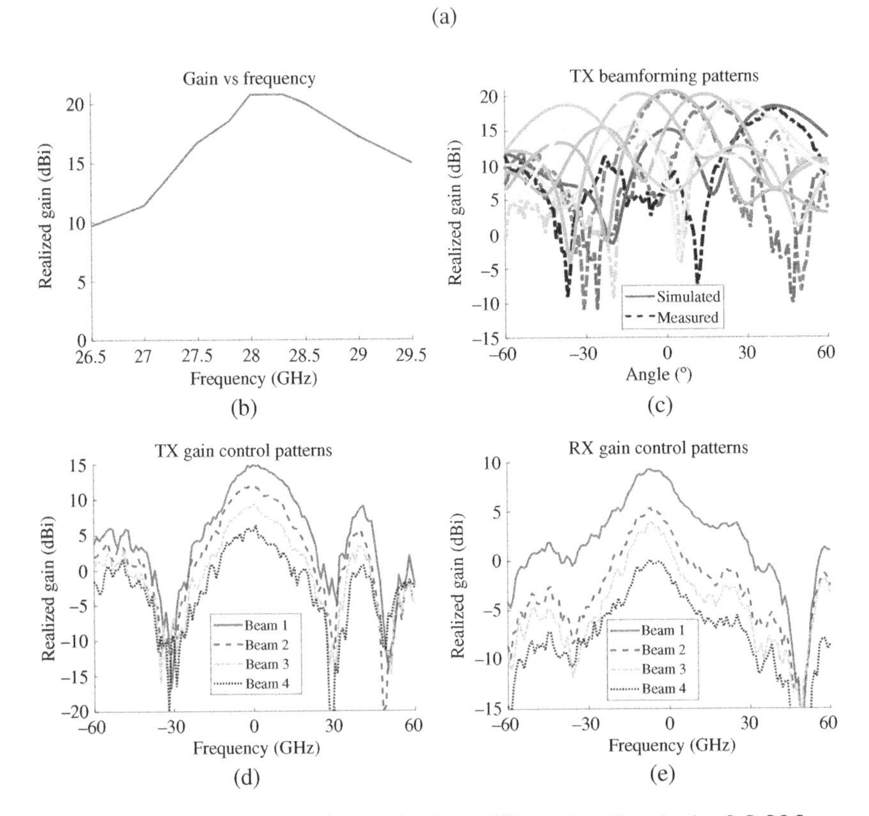

Figure 10.28 (a) Distribution loss and gain at different locations in the CoD POC component at 28 GHz. (b) Measured system gain as a function of operating frequency. (c) Measured and simulated beamsteering radiation patterns in the TX mode at 28 GHz. Measured gain control radiation patterns in the TX mode (d) and RX mode (e) at 28 GHz. *Source:* From [23]. ©2021 IEEE. Reproduced with permission.

The gain and loss of the TX channel from the RF common port to the output ports of RF channels are summarized as shown in Figure 10.28a and take into account the ohmic and power dividing loss. The well-established three-antenna method [60] is employed to perform gain calibration measurements. Therefore, the gain of the CoD POC can be measured using the known loss and gain characteristics of each component. In order to measure the CoD POC containing four RF ports, the beamformer MMIC, which can accurately distribute power to each port is configured in the measurement environment. RF characteristics of the beamformer MMIC and assembly cables/connectors are carefully extracted by conduction test using the Keysight UXA N9040B signal analyzer and Keysight E8267D signal generator. The laser alignment finder is utilized to improve the positional accuracy of the AUT in the far-field chamber setup. A piece of foam is used to support the CoD POC and minimize the coupling effect from the sample holder. The designed PAs of the beamformer MMIC feature an output P_{1dB} of 6 dBm and P_{SAT} of 7 dBm at 28 GHz. The measured peak PA gain of the RF package is 24.9 dB including a power-combing loss in four RF channels. The total ohmic loss is 8.22 dB and is composed of 6.72 dB for ohmic transmission-line loss and 1.5 dB for bond and mismatching loss between the RF package and the FPCB. The estimated RF package gain is calculated using the measured results of each component excluding the antenna array gain. Figure 10.28b illustrates the measured system gain in the TX mode over the frequency range 26.5–29.5 GHz for the CoD POC when the four elements are fully activated. A measured peak gain of 20.72 dB and a maximum estimated EIRP of 8.82 dBm are achieved at 28 GHz.

Four antenna elements with 6.23 mm center-to-center spacing between adjacent elements are activated to verify the beamforming capability of the CoD POC. The phase distribution of the RF port excitation and corresponding scanning angles are summarized in Table 10.6. The measured radiation patterns (Figure 10.28c) of the CoD POC confirm a beamsteering range of ±40° in azimuth plane without any phased array calibration. Moreover, it is meaningful that these beamsteering radiation patterns are formulated through the use of phase shifters featuring only 3-bit resolutions. It is worth noting that despite the scenario where the finger model is located upon the front glass and is centered above the first element, the antenna array of the HEMS circuit has demonstrated acceptable beam steering characteristics in different scanning angles from −30° to 48°. In addition, the MIMO array configuration can be constructed using a combination of multiple phased arrays of the HEMS circuit and phased array

Table 10.6 Phase distribution of the RF port excitation to demonstrate beamforming capability.

	Port1	Port2	Port3	Port4	Scanning angle
Beam1	0	0	0	0	0°
Beam2	0	±45°	±90°	±135°	±12°
Beam3	0	±90°	±180°	±270°	±25°
Beam4	0	±135°	±270°	±45°	±40°

*All values in degrees The power supplied to each port is set to the maximum value.
Source: From [23]. ©2021 IEEE. Reproduced with permission.

Table 10.7 Comparison with state-of-the-art mmWave phased-array AiPs.

Reference	This work	Samsung 2017 [61]	Qualcomm 2018 [62]
Type	CoD	AiP	AiP
Antenna Placement	Inside the display panel	Top and bezel region	Back cover of cellular handsets
IC integration level	RF front-end	RF front-end + RF/IF conversion	RF front-end + RF/IF conversion
Frequency (GHz)	28	60	28
Elements in array	4	4	8
Beam scanning angles (°)	±40	±40	±45
Touch operation	Yes	No	No

Source: From [23]. ©2021 IEEE. Reproduced with permission.

AiPs, which are already configured in current mobile devices. This MIMO array configuration not only increases the spectral efficiency of the entire system through spatial multiplexing but also reduces the effect of human blockage through spatial diversity technology. Gain control of the CoD POC is exemplified in both TX and RX modes, as illustrated in Figure 10.28d,e. It should be noted that these gain control radiation patterns are achieved without any calibration technology or tapering the power supplied to each port. Therefore, sidelobe level (SLL) can be further suppressed by optimizing the weighting matrix of the input port of each RF channel.

Table 10.7 summarizes the performance comparison with recently reported mmWave phased-array AiPs for cellular applications. It should be noted that the proposed CoD is the first to report touch display-integrated mmWave phased-array RF front-end, which can enhance the beamforming coverage in the foreside direction of cellular handsets.

10.5 Conclusion

In this chapter, a novel concept of an optically invisible AoD is proposed to significantly enhance the CDF, especially in the foreside direction of UE. In contrast to existing antenna component strategies, the AoD is integrated within the view area of high-resolution OLED or LCD panels of cellular handsets while remaining unnoticeable to the human eye. The fabricated 28 GHz phased-array AoD components within a real-life mmWave 5G NR Android-based cellular handset prototype satisfies the EVM requirements for 5G modulation schemes defined by 3GPP. The AoD approach is advanced to the flexible, invisible HEMS that dually functions as a radio-frequency transceiver and a high-resolution touchscreen display. The devised HEMS prototype independently features flexibility, optically invisibility (transmittance > 88%), and multi-functionality (system gain of 20.72 dB, beam scanning range of ±40°, and mutual capacitance change ratio > 4.42%). This AoD technology will pave the way for the possibility of vastly different mmWave beamforming strategies at the component, device and system level for future mobile devices. In addition, the reported results highlight the potential of engineering new advanced technologies on display platforms and provide a starting point for seamless integration of various electromagnetic sensors within display panels for future 6G wireless, radar, and sensing scenarios.

References

1 K. K. Kim, I. Ha, P. Won, D.-G. Seo, K.-J. Cho, and S. H. Ko, "Transparent wearable three-dimensional touch by self-generated multiscale structure," *Nat. Commun.*, vol. 10, no. 1, pp. 1–8, 2019.

2 P. Bhagwat, "Bluetooth: technology for short-range wireless apps," *IEEE Internet Comput.*, vol. 5, no. 3, pp. 96–103, 2001.

3 E. Ferro and F. Potorti, "Bluetooth and Wi-Fi wireless protocols: a survey and a comparison," *IEEE Wireless Commun.*, vol. 12, no. 1, pp. 12–26, 2005.

4 S. Kasera and N. Narang, *3G Mobile Networks*, NY, USA: McGraw-Hill Education, 2004.

5 E. Dahlman, S. Parkvall, and J. Skold, *4G, LTE-Advanced Pro and the Road to 5G*, 3rd Edition, Cambridge, MA, USA: Academic Press, 2016.

6 A. Gupta and R. K. Jha, "A survey of 5G network: architecture and emerging technologies," *IEEE Access*, vol. 3, pp. 1206–1232, 2015.

7 E. G. Larsson, O. Edfors, F. Tufvesson, and T. L. Marzetta, "Massive MIMO for next generation wireless systems," *IEEE Commun. Mag.*, vol. 52, no. 2, pp. 186–195, 2014.

8 X. Liu, Q. Zhang, W. Chen, H. Feng, L. Chen, F. M. Ghannouchi and Z. Feng, "Beam-oriented digital predistortion for 5G massive MIMO hybrid beamforming transmitters," *IEEE Trans. Microwave Theory Tech.*, vol. 66, no. 7, pp. 3419–3432, 2018.

9 B. Yang, Z. Yu, J. Lan, R. Zhang, J. Zhou, and W. Hong, "Digital beamforming-based massive MIMO transceiver for 5G millimeter-wave communications," *IEEE Trans. Microwave Theory Tech.*, vol. 66, no. 7, pp. 3403–3418, 2018.

10 B. M. Lee and H. Yang, "Massive MIMO with massive connectivity for industrial internet of things," *IEEE Trans. Ind. Electron.*, vol. 67, no. 6, pp. 5187–5196, 2019.

11 J. Choi, W. Hwang, C. You, B. Jung, and W. Hong, "Four-element reconfigurable coupled loop MIMO antenna featuring LTE full-band operation for metallic-rimmed smartphone," *IEEE Trans. Antennas Propag.*, vol. 67, no. 1, pp. 99–107, 2018.

12 A. Ren and Y. Liu, "A compact building block with two shared-aperture antennas for eight-antenna MIMO array in metal-rimmed smartphone," *IEEE Trans. Antennas Propag.*, vol. 67, no. 10, pp. 6430–6438, 2019.

13 P. Bahramzy, P. Olesen, P. Madsen, J. Bojer, S. Del Barrio, A. Tatomirescu, P. Bundgaard, A. S. Morris III and G. Pedersen, "A tunable RF front-end with narrowband antennas for mobile devices," *IEEE Trans. Microw. Theory Tech.*, vol. 63, no. 10, pp. 3300–3310, 2015.

14 W. Hong, K.-H. Baek, Y. Lee, Y. Kim, and S.-T. Ko, "Study and prototyping of practically large-scale mmWave antenna systems for 5G cellular devices," *IEEE Commun. Mag.*, vol. 52, no. 9, pp. 63–69, 2014.

15 S. Shakib, J. Dunworth, V. Aparin, and K. Entesari, "mmWave CMOS power amplifiers for 5G cellular communication," *IEEE Commun. Mag.*, vol. 57, no. 1, pp. 98–105, 2019.

16 M. Stanley, Y. Huang, H. Wang, H. Zhou, A. Alieldin, and S. Joseph, "A transparent dual-polarized antenna array for 5G smartphone applications," *in Proc. IEEE Int. Symp. Antennas Propag. & USNC/URSI Nat. Radio Sci. Meeting*, Boston, MA, USA, pp. 635–636, 2018.

17 L. Zhou, A. Wanga, S.-C. Wu, J. Sun, S. Park, and T. N. Jackson, "All-organic active matrix flexible display," *Appl. Phys. Lett.*, vol. 88, no. 8, p. 083502, 2006.

18 T. Sekitani, H. Nakajima, H. Maeda, T. Fukushima, T. Aida, K. Hata and T. Someya, "Stretchable active-matrix organic light-emitting diode display using printable elastic conductors," *Nat. Mater.*, vol. 8, no. 6, pp. 494–499, 2009.

19 M. Noda, N. Kobayashi, M. Katsuhara, A. Yumoto, S. Ushikura, R. Yasuda, N. Hirai, G. Yukawa, I. Yagi, K. Nomoto and T. Urabe, "An OTFT-driven rollable OLED display," *J. Soc. Inf. Disp.*, vol. 19, no. 4, pp. 316–322, 2011.

20 N. Savage, "Tomorrow's industries: from OLEDs to nanomaterials," *Nature*, vol. 576, no. 7786, p. S20, 2019.

21 J. Park, S. Y. Lee, J. Kim, D. Park, W. Choi, and W. Hong, "An optically invisible antenna-on-display concept for millimeter-wave 5G cellular devices," *IEEE Trans. Antennas Propag.*, vol. 67, no. 5, pp. 2942–2952, May 2019.

22 W. Hong, J. Choi, D. Park, M. Kim, C. You, D. Jung, and J. Park, "mmWave 5G NR cellular handset prototype featuring optically invisible beamforming antenna-on-display," *IEEE Commun. Mag.*, vol. 58, no. 8, pp. 54–60, Aug. 2020.

23 J. Park, D. Park, M. Kim, D. Jung, C. You, D. Choi, J. Lee and W. Hong, "Circuit-on-display: a flexible, invisible hybrid electromagnetic sensor concept," *IEEE J. Microwave*, vol. 1, no. 2, pp. 550–559, Spring 2021.

24 S. Hong, S. H. Kang, Y. Kim, and C. W. Jung, "Transparent and flexible antenna for wearable glasses applications," *IEEE Trans. Antennas Propag.*, vol. 64, no. 7, pp. 2797–2804, Jul. 2016.

25 User Equipment (UE) Radio Transmission and Reception; Part 2: Range 2 Standalone (Release 17), *document TS38.101-2 v17.1.0*, Mar. 2021.

26 Base Station (BS) Radio Transmission and Reception (Release 15), *document 3GPP TS 36.104 V15.1.0*, Dec. 2017.

27 User Equipment (UE) Radio Transmission and Reception; Part 1: Range 1 Standalone (Release 17), *document TS38.101-1 v17.1.0*, Mar. 2021.

28 F. Colombel, X. Castel, M. Himdi, G. Legeay, S. Vigneron, and E. M. Cruz, "Ultrathin metal layer, ITO film and ITO/Cu/ITO multilayer towards transparent antenna," *IET Sci. Meas. Technol.*, vol. 3, no. 3, pp. 229–234, May. 2009.

29 A. S. Thampy and S. K. Dhamodharan. "Performance analysis and comparison of ITO-and FTO-based optically transparent terahertz U-shaped patch antennas". *Physica E*, vol. 66, pp. 52–58, Feb. 2015.

30 X. Liu, D. R. Jackson, J. Chen, J. Liu, P. W. Fink, G. Y. Lin, and N. Neveu, "Transparent and nontransparent microstrip antennas on a CubeSat: novel low-profile antennas for CubeSats improve mission reliability," *IEEE Antennas Propag. Mag.*, vol. 59, no. 2, pp. 59–68 Apr. 2017.

31 J. S. Gomez-Diaz and J. Perruisseau-Carrier, "Microwave to THz properties of graphene and potential antenna applications," *Int. Symp. Antennas and Propag. (ISAP)*, Oct. 2012, pp. 239–242.

32 M. Tamagnone, J. S. Gomez-Diaz, J. R. Mosig, and J. Perruisseau-Carrier, "Analysis and design of terahertz antennas based on plasmonic resonant graphene sheets," *J. Appl. Phys.*, vol. 112, no. 11, Nov. 2012.

33 T. A. Elwi, H. M. AL-Rizzo, D. G. Rucker, E. Dervishi, Z. Li, and A. S. Biris, "Multi-walled carbon nanotube-based RF antennas," *Nanotechnology*, vol. 21, no. 4, Jan. 2010.

34 N. A. Vacirca, J. K. McDonough, K. Jost, Y. Gogotsi, and T. P. Kurzweg. "Onion-like carbon and carbon nanotube film antennas," *Appl. Phys. Lett.*, vol. 103, no. 7, Jul. 2013.

35 H. Rmili, J. L. Miane, T. Olinga, and H. Zangar, "Design of microstrip-fed proximity-coupled conducting-polymer patch antenna," *11th Int. Symp. Ant. Techn. Appl. Elec. (ANTEM)*, Jun. 2005, pp. 1–4.

36 N. J. Kirsch, N. A. Vacirca, E. E. Plowman, T. P. Kurzweg, A. K. Fontecchio, and K. R. Dandekar, "Optically transparent conductive polymer RFID meandering dipole antenna," *IEEE Int. Conf. RFID*, Apr. 2009, pp. 278–282.

37 A. Mehdipour, I. D. Rosca, A. R. Sebak, C. W. Trueman, and S. V. Hoa, "Carbon nanotube composites for wideband millimeter-wave antenna applications," *IEEE Trans. Antennas Propag.*, vol. 59, no. 10, pp. 3572–3578, Oct. 2011.

38 T. Rai, P. Dantes, B. Bahreyni, and W. S. Kim, "A stretchable RF antenna with silver nanowires," *IEEE Electron Device Lett.*, vol. 34, no. 4, pp. 544–546, Apr. 2013.

39 Q. H. Dao, R. Tchuigoua, B. Geck, D. Manteuffel, P. von Witzendorff, and L. Overmeyer, "Optically transparent patch antennas based on silver nanowires for mm-wave applications," *in Proc. IEEE Int. Symp. Antennas Propag. (APSURSI),* Jul. 2017, pp. 2189–2190.

40 Q. L. Li, S. W. Cheung, D. Wu, and T. I. Yuk, "Optically transparent dual-band MIMO antenna using micro-metal mesh conductive film for WLAN system," *IEEE Antennas Wireless Propag. Lett.*, vol. 16, pp. 920–923, Sep. 2017.

41 N. M. Jizat, S. K. A. Rahim, Y. C. Lo, and M. M. Mansor, "Compact size of CPW dual-band meander-line transparent antenna for WLAN applications," *IEEE Asia-Pacific Conf. Appl. Elec. (APACE)*, Dec. 2014, pp. 20–22.

42 M. A. Malek, S. Hakimi, S. K. Abdul Rahim, and A. K. Evizal, "Dual-band CPW-fed transparent antenna for active RFID tags," *IEEE Antennas Wireless Propag. Lett.*, vol. 14, pp. 919–922, Dec. 2014.

43 T. W. Turpin and R. Baktur, "Meshed patch antennas integrated on solar cells," *IEEE Antennas Wireless Propag. Lett.*, vol. 8, pp. 693–696, Jun. 2009.

44 S. Sheikh, "Circularly polarized meshed patch antenna," *IEEE Antennas Wireless Propag. Lett.*, vol. 15, pp. 352–355, Jun. 2015.

45 S. Hong, Y. Kim, and C. Won Jung, "Transparent microstrip patch antennas with multilayer and metal-mesh films," *IEEE Antennas Wireless Propag. Lett.*, vol. 16, pp. 772–775, Aug. 2017.

46 S. H. Kang and C. W. Jung, "Transparent patch antenna using metal mesh," *IEEE Trans. Antennas Propag.*, vol. 66, no. 4, pp. 2095–2100, Apr. 2018.

47 J. Hautcoeur, L. Talbi, and K. Hettak, "Feasibility study of optically transparent CPW-fed monopole antenna at 60-GHz ISM bands," *IEEE Trans. Antennas Propag.*, vol. 61, no. 4, pp. 1651–1657, Apr. 2013.

48 W. Hong, S. Lim, S. Ko, and Y. G. Kim, "Optically invisible antenna integrated within an OLED touch display panel for IoT applications," *IEEE Trans. Antennas Propag.*, vol. 65, no. 7, pp. 3750–3755, Jul. 2017.

49 W. Hong, S. Lim, S. Ko, and Y. G. Kim, "OLED-embedded antennas for 2.4 GHz Wi-Fi and bluetooth applications," *in Proc. IEEE Int. Symp. Antennas Propag. (APSURSI)*, Jul. 2017, pp. 2551–2552.

50 G. Oster, M. Wasserman, and C. Zwerling, "Theoretical interpretation of Moiré patterns," *J. Opt. Soc. Am.*, vol. 54, no. 2, pp. 169–175, Feb. 1964.

51 E. Karrer and U. Smith, "Diffusion of light from a searchlight beam," *J. Opt. Soc. Am.*, vol. 7, no. 12, pp. 1211–1234, Dec. 1923.

52 S. Kim, W. Choi, W. Rim, Y. Chun, H. Shim, H. Kwon, J. Kim, I. Kee. S. Kim, S. Lee, J. Park, "A highly sensitive capacitive touch sensor integrated on a thin-film-encapsulated active-matrix OLED for ultrathin displays," *IEEE Trans. Electron Devices*, vol. 58, no. 10, pp. 3609–3615, 2011.

53 K. Lim, K.-S. Jung, C.-S. Jang, J.-S. Baek, and I.-B. Kang, "A fast and energy efficient single-chip touch controller for tablet touch applications," *J. Disp. Technol.*, vol. 9, no. 7, pp. 520–526, 2013.

54 A. Lüttgen, S. K. Sharma, D. Zhou, D. Leigh, S. Sanders, and C. D. Sarris, "A fast simulation methodology for touch sensor panels: formulation and experimental validation," *IEEE Sens. J.*, vol. 19, no. 3, pp. 996–1007, 2018.

55 M. D. Tenuta and L. W. Bokma, "Touch sensor panel calibration," U.S. Patent 8 890 854, Nov. 2014.

56 E. Song, H.-B. Park, and H. H. Park, "An evaluation method for radiated emissions of components and modules in mobile devices," *IEEE Trans. Electromagn. Compat.*, vol. 56, no. 5, pp. 1020–1026, 2014.

57 C. Hwang, S. Kong, T. Enomoto, J. Maeshima, K. Araki, D. Pommerenke, J. Fan, "LCD baseband noise modulation estimation for radio frequency interference in mobile phones," *in Proc.* IEEE Asia-Pac. Int. Symp. Electromagn. Compat., Seoul, South Korea, 2017, pp. 226–228.

58 C.-L. Lin, P.-C. Lai, P.-C. Lai, T.-C. Chu, and C.-L. Lee, "Bidirectional gate driver circuit using recharging and time-division driving scheme for in-cell touch LCDs," *IEEE Trans. Ind. Electron.*, vol. 65, no. 4, pp. 3585–3591, 2017.

59 D. Sievenpiper, L. Zhang, R. F. J. Broas, N. G. Alexopolous, and E. Yablonovitch, "High-impedance electromagnetic surfaces with a forbidden frequency band", *IEEE Trans. Microwave Theory Tech.*, vol. 47, no. 11, pp. 2059–2074, Nov. 1999.

60 K. Selvan, "Preliminary examination of a modified three-antenna gain-measurement method to simplify uncertainty estimation," *IEEE Antennas Propag. Mag.*, vol. 45, no. 2, pp. 78–81, 2003.

61 W. Hong, K.-H. Baek, and S. Ko, "Millimeter-wave 5G antennas for smartphones: overview and experimental demonstration," *IEEE Trans. Antennas Propag.*, vol. 65, no. 12, pp. 6250–6261, Dec. 2017.

62 J. D. Dunworth, A. Homayoun, B-H. Ku , Y-C. Ou, K. Chakraborty, G. Liu, T. Segoria, J. Lerdworatawee, J. W. Park, H-C. Park, H. Hedayati, D. Lu, P. Monat, K. Douglas and V. Aparin, "A 28 GHz bulk-CMOS dual-polarization phased-array transceiver with 24 channels for 5G user and basestation equipment," *in Proc. IEEE Int. Solid-State Circuits Conf.*, Feb. 2018, pp. 70–71.

Index

Microwave and Millimeter-Wave Antenna Design for 5G Smartphone Applications, First Edition. Wonbin Hong and Chow-Yen-Desmond Sim.

Published 2023 by John Wiley & Sons, Inc.

n

o

p

r

s

Printed and bound by CPI Group (UK) Ltd, Croydon, CR0 4YY
16/04/2025
14658419-0002